Springer Geochemistry/Mineralogy

For further volumes:
http://www.springer.com/series/10171

Ali Akbar Ramezanianpour

Cement Replacement Materials

Properties, Durability, Sustainability

Ali Akbar Ramezanianpour
Concrete Technology Center
Department of Civil Engineering
Amirkabir University of Technology
Tehran
Iran

ISSN 2194-3176 ISSN 2194-3184 (electronic)
ISBN 978-3-642-36720-5 ISBN 978-3-642-36721-2 (eBook)
DOI 10.1007/978-3-642-36721-2
Springer Heidelberg New York Dordrecht London

Library of Congress Control Number: 2013945143

Printed on acid-free paper

Springer is part of Springer Science+Business Media (www.springer.com)

Preface

Sustainability is an important issue all over the world. Carbon dioxide emission has been a serious problem in the world due to the greenhouse effect. Today many countries agreed to reduce the emission of CO_2. Many phases of cement and concrete technology can affect sustainability. Cement and concrete industry is responsible for the production of 7 % carbon dioxide of the total world CO_2 emission. The use of supplementary cementing materials (SCM), design of concrete mixtures with optimum content of cement, and enhancement of concrete durability are the main issues toward sustainability in concrete industry.

The aim of this book is to present the latest findings in the properties and application of supplementary cementing materials and blended cements currently used in the world in concrete. It consists of eight chapters. The book opens with the Chap. 1 on natural pozzolans, a material which has been used for over 2500 years. This is followed by the Chap. 2 on fly ash, a by-product of the combustion of pulverized coal and the most consumed artificial pozzolans in the cement industry all over the world. The Chap. 3 deals with the granulated ground blast furnace slag, a by-product of the metallurgical industry used as a hydraulic binder in large projects located in severe environments. In Chap. 4, silica fume a by-product in the production of silicon and silicon alloys and a super pozzolan for enhancement of concrete durability is discussed. The Chap. 5 deals with metakaolin (MK), commercially available since the mid-1990s, and one of the recently developed supplementary cementing materials. Metakaolin differs from other supplementary cementitious materials (SCMs), like fly ash, silica fume, and slag, in that it is not a by-product of an industrial process. It is produced by heating kaolin, one of the most abundant natural clay minerals, to temperatures of 650–900 °C. Rice husk ash (RHA), an artificial pozzolan from the combustion of rice husk in a control process and a very reactive material is explained in the Chap. 6. Increasing the rice production and the usage of rice hull in electricity generation and also incorporation of rice husk ash in concretes with enhanced durability, are the main reasons for the better future of rice husk ash. The Chap. 7 deals with the properties of Portland limestone cements. Perhaps the main advantage of producing Portland-limestone cements is that by introducing limestone into cement, the total volume of cement would increase, or in other words, the amount of clinker required to produce a certain amount of cement would decrease. This would result in a substantial amount of energy saving in the production of cement as the consumption of natural raw materials and the fuel needed for production of clinker would be reduced. Finally, the role of supplementary cementing materials on sustainable development is included in the Chap. 8 of this book. Supplementary cementing materials have proven to be economic environmental alternatives to typical concrete mixes. More researches are needed for further applications of supplementary cementing materials to reduce energy consumption for the production of clinker cements and reduction of carbon dioxide emission in the cement and concrete industry.

Each chapter begins with a introduction followed by production, physical, chemical, and mineralogical properties, hydration reactions and pozzolanic activity, effects on properties of fresh concretes, effects on the mechanical properties of hardened concretes, effects on durability of concretes and their applications in concrete.

Author has tried to include his 35 years of experience in the field of concrete technology specially the engineering properties and durability of concretes containing cement replacement materials in this book. The author's findings on natural and artificial pozzolans in numerous research projects in Iran and all over the world during these years are discussed in different chapters.

The author wishes to thank Prof. R. D. Hooton, Dr. Amir. M. Ramezanianpour, Mr. M. Mahdikhani, and Mr. H. Bahrami, for their assistances in preparation of some parts of the book. Particular thanks are extended to the Springer Publisher for publishing this book. I would also like to thank Mrs. Hashemi for processing this publication.

May 2013 Ali A. Ramezanianpour

Contents

Chapter 1
Natural Pozzolans

Introduction

Pozzolan is defined as a siliceous or siliceous and aluminous material, which in itself possesses little or no cementitious value but will, in finely divided form and in the presence of moisture, chemically reacts with calcium hydroxide at ordinary temperatures to form compounds possessing cementitious properties.

Natural pozzolan is a raw or calcined natural material that shows pozzolanic properties. Volcanic ash or pumicite, tuffs, shales and opaline cherts and diatomaceous earths are examples of natural pozzolans.

The oldest example of hydraulic binder was a mixture of lime and a natural pozzolan, a diatomaceous earth from the Persian Gulf at about 5000 B.C. [1]. The next applications of natural pozzolans goes back to about 1600 B.C. in Mediterranian region on the Aegean Island of Thera, now called Santorin, and the other in 79 A.D. on the bay of Naples, Italy. Both pozzolans were volcanic ashes or pumicites with high volcanic glassy form.

The Romans and Greeks used natural pozzolan-lime mixture as binding materials for many structures over 2000 years ago. Such mortars consist of six parts of a natural pozzolan, two parts by volume of lime, and one part by volume of sand. These mortars were used as the first hydraulic cements in aqueducts, bridges, sewers, water tanks and other types of structures. Some of these structures are still exist in countries like Italy, Iran, Greece and Spain. Most recent structures such as the Suez Canal in Egypt built in 1880, the sea walls and marine structures in the islands of the Aegean Sea and the harbors of Alexandria in Egypt are among many structures built with pozzolanic materials. Many monuments from Roman's era are in use today in many parts of the Europe. This is attributed to the high durability of pozzolanic materials in various environmental conditions [2].

A. A. Ramezanianpour, *Cement Replacement Materials*,
Springer Geochemistry/Mineralogy, DOI: 10.1007/978-3-642-36721-2_1,

Natural Pozzolan Classification

There have been several classifications for pozzolans and natural pozzolans. Establishment of a precise classification of natural pozzolans seems very difficult because this common name includes materials which are very different in terms of chemical composition, mineralogical nature and geological origin. Their common property is their reaction with lime and hardening by time. One of the proposed classification for natural pozzolans is shown in Fig. 1.1. This classification is based on the origin of pozzolans [3]. In this classification, Natural pozzolans are those materials which do not require any further treatment apart from grinding to react with lime. In this classification, artificial pozzolans are the materials with low pozzolanic activity and need further treatments to achieve pozzolanic activity.

In another classification, natural pozzolans are classified in four categories based on the principal lime reactive constituent present. They are volcanic tuff, unaltered volcanic glass, calcined clay or shale, and raw or calcined opaline silica [4].

Production

Pyroclastic Rocks

Explosive volcanic eruptions which project minute particles of melted magma into the atmosphere result in pyroclastic rocks formation. The rapid pressure decrease during eruption causes the gas release from magma. Therefore a microporous structure is usually observed in such rocks. Due to the rapid cooling process, a glassy state is formed. The ground deposit materials are mixed with other materials from the base of the volcano. The pozzolanic activity of the melt magma is low due to the slow rate of cooling and more crystal forming. Figure 1.2 shows the microstructure of Bacoli pozzolan in Italy [5].

Incoherent natural pozzolans have been formed in Italy since 1500 B.C. after a tremendous explosive volcanic eruption. It has been widely used as Santorin earth since ancient times. The incoherent glassy rhyolites have been found in the United States [6], Indi [7] and Turkey [8]. Rhenish trass which is a natural pozzolan of volcanic origin is included among the tuffs or compact and coherent materials, but incoherent layers of glass have been found in its deposits. It has been found in the Valley of the Rhine River in Germany and has been used for many years [9, 10].

Volcanic pozzolans are usually deposited in compact layers known as tuffs which originate from weathering and cementation of loose particles by diagenetic or other natural processes. Weathering can cause zeolitisation or orgillation. It means that it can transform the glass of the pozzolans into zeolitic minerals or clay minerals. The degree of transformation reached by the original deposit depends on the intensity of the diagenetic actions as well as on their duration. Zeolitisation improve the properties of pozzolans whereas argillation usually reduces the pozzolanic properties [11].

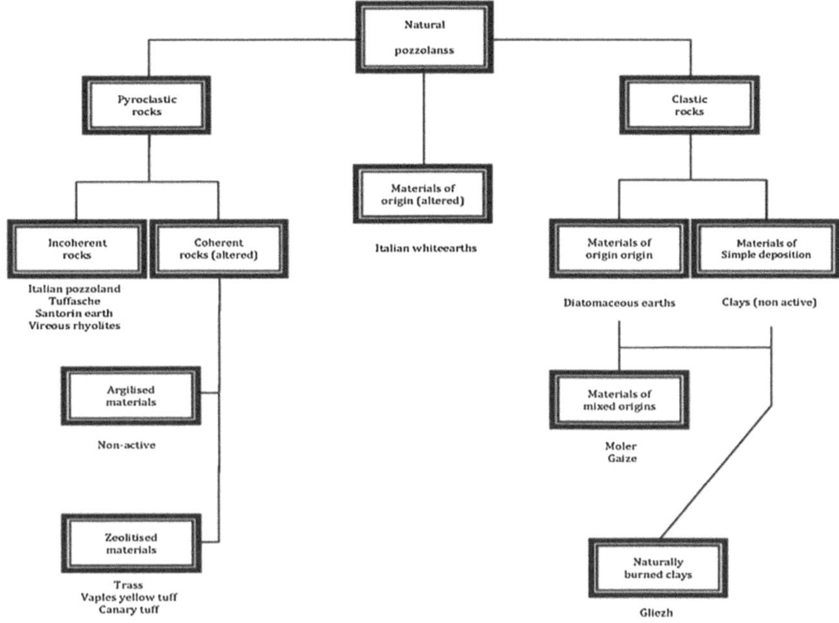

Fig. 1.1 Classification of pozzolans

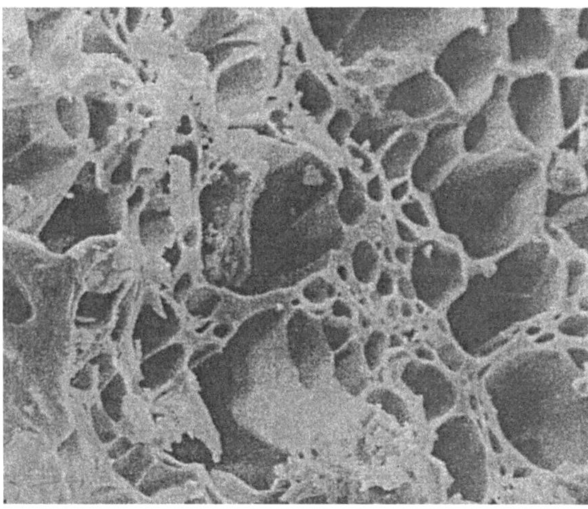

Fig. 1.2 SEM image of Bacoli pozzolan [5]

The transformation of volcanic pozzolan, incoherent and mostly glassy, into compact and zeolitised tuffs is found in some deposits. An example is the Naples natural pozzolan. Deposits of volcanic tuffs have been discovered in Italy, Greece, United States, Spain and Iran. Figure 1.3 shows the deposit of volcanic tuffs known as trass in the Albors Mountain, north of Iran.

Fig. 1.3 Deposit of volcanic tuffs known as trass in the Albors Mountain, north of Iran

Another type of an incoherent natural pozzolan is pumicite. It is a finely divided volcanic ash composed of angular and porous particles of siliceous glass and varying proportions of crystal fragments differing from pumice only in grain size. They occur as stratified or massive deposits, usually as lake beds. Pumicite was also discovered in USA, Iran, and Canada.

Diatomaceous earths and some clays are sedimentary rocks which show pozzolanic properties by reacting with lime. Clays originate from the alteration of igneous rocks. Diatomaceous earth usually forms from the siliceous skeletons of microorganisms known as diatom deposited in fresh or under sea waters. Diatoms and clay minerals are sometimes deposited together under water. Some types of clays such as montmorillonite group can react with lime to produce calcium silicate and calcium aluminate hydrates. They are not usually used as pozzolan for replacement with cement due to their higher water demand in concrete mixtures resulting in low strengths.

Deposits of diatomaceous earths can be found in Canada, United States, Germany, Algeria, Iran, Russia and Denmark [12]. A diatomaceous earth consisting of amorphous opal and montmorillonite has been found in Europe which can react with lime before and after calcination to present pozzolanic activity [13].

Physical, Chemical, and Mineralogical Properties

Physical Properties

Physical properties of natural pozzolans vary widely. The fineness, specific surface area, the shape of particles, and density of natural pozzolans influence the properties of fresh concrete and the strength development of hardened concrete.

Fineness of natural pozzolans is usually measured by wet sieving method. The amount of the sample retained when wet sieved on a 45 μ sieve is determined in accordance of ASTM Method C430. Several standards specify the maximum residue in percentage retained on a 45 μ sieve. Depending on the grinding system and grinding time, natural pozzolans are produced from a few microns up to 200 μ. Particle size distribution of natural pozzolans can be measured by laser particle size analyzer, X-ray sedigraph, and coulter counter. In some cases the agglomeration of a number of small particles may form a large particle. In most natural pozzolans productions, they contained particles greater than 1 μ.

The specific surface area of natural pozzolans, which is the area of a unit mass, is measurable by different techniques. The most common is the Blaine specific surface area technique, which measures the resistance of compacted particles to an air flow.

A laser particle size analyzer can also be used for determination of the specific surface area of natural pozzolans. The Brunauer-Emmett-Teller (BET) nitrogen adsorption technique has also been used for determining the specific surface of the particles, but the results obtained by this method are usually higher than the results obtained by the Blaine specific surface area technique or particle size analyzer.

The specific surface area of blended natural pozzolans is usually similar to that for the normal Portland cement at about 300–400 kg/m^2. Larger surface area with smaller particles of natural pozzolans can be produced by separate grinding. Higher surface area of about 500 kg/m^2 has increased the pozzolanic activity of most natural pozzolans.

The shape of particles of natural pozzolans is usually angular or irregular. Some of them have micro porous character and increase the water demand for constant consistency in concrete mixtures. Figure 1.4 shows the Scanning Electron Micrograph of the Jadjroud Trass as an example of the shape of natural pozzolans.

The specific gravity of natural pozzolans is determined according to ASTM C188 test method similar to Hydraulic cements. If natural pozzolan contains water soluble compounds, the use of a non-aqueous solvent, instead of water is recommended. The specific gravity of different natural pozzolans varies over a wide range, like the other physical properties. The specific gravity of volcanic ashes varies between 1.8 and 2.4 and reaches to 2.9 for the high density compacted tuffs.

Chemical and Mineralogical Properties

The chemical composition of natural pozzolans varies within wide ranges and depends on their sources. Silica and alumina are usually high in incoherent volcanic pozzolans. The next oxide elements are iron, calcium and magnesium oxides. The alkali content is not high but may vary between 3 and 10 %. Loss on ignition is generally low but may reach to 9 % in some pozzolans. The chemical analysis of some incoherent volcanic pozzolans is shown in Table 1.1 [14, 15].

Fig. 1.4 Scanning electron micrograph of the Jadjroud Trass

The chemical composition of pozzolanic tuffs also varies considerably depending on their origin. The silica or silica plus alumina contents are very similar to pyroclastic pozzolans. Minor elements are very variable and their loss on ignition is usually high. Table 1.2 shows the chemical composition of a number of pozzolanic tuffs at different countries.

The chemical composition of some diatomaceous earths is shown in Table 1.3. Their chemical composition varies with their clay content. Those diatomaceous earths with high clay content are rich in alumina while the others with low clay are rich in silica.

The mineralogical composition of natural pozzolans also varies at different sources. Owing to their formation process, the natural pozzolans consist of crystalline and non-crystalline particles as glassy particles. Table 1.4 shows the mineral composition of some volcanic pozzolans [6, 14].

As examples in the Italian pozzolans a pomiceous glass containing sanidine and small amount of plagioclase and pyroxene have been observed. In others augite,

Table 1.1 Chemical analyses of some incoherent volcanic pozzolanas (%)

Pozzoalna	Country	SiO_2	Al_2O_3	Fe_2O_3	CaO	MgO	Na_2O	K_2O	TiO_2	SO_3	LOI
Bacoli	Italy	53.08	17.89	4.29	9.05	1.23	3.08	7.61	0.31	0.65	3.05
Barile	Italy	44.07	19.18	9.81	12.25	6.66	1.64	1.12	0.97	Tr.[a]	4.42
Salone	Italy	46.84	18.44	10.25	8.52	4.75	1.02	6.35	0.06	Tr.	3.82
Sentorin earth	Greece	65.1	14.5	5.5	3.0	1.1					3.5
Rhine tuffash	Germany	58.91	19.53[b]	2.85	2.48	1.33	6.12	4.53			2.21
Rhyolite pumicite	USA	65.74	15.89	2.54	3.35	1.33	4.97	1.92			3.43
Higashi Matsuyama	Japan	71.77	11.46	1.14	1.10	0.54	1.53	2.55			6.50
Trass	Iran	68.3	12.7	1.0	2.5	1.0	1.92	2.02		Tr.	8.50

[a] *Tr.* trace

[b] $Al_2O_3 + TiO_2$

Table 1.2 Chemical analyses of pozzolanic tuffs (%)

Pozzoalna	Country	SiO_2	Al_2O_3	Fe_2O_3	CaO	MgO	Na_2O	K_2O	SO_3	LOI
Rhine trass	Germany	52.12	18.29	5.81	4.94	1.20	1.48	5.06		11.10
Bavaria trass	Germany	62.45	16.47	4.41	3.39	0.94	1.91	2.06		7.41
Selyp trass	Hungary	55.69	15.18	6.43	2.83	1.01			0.26	16.33
Yellow tuff	Italy	54.68	17.70	3.82	3.66	0.95	3.43			9.11
Gujarat tuff	India	40.9	12.0	14.0	14.6	1.45				12.06
Trass K	Bulgaria	71.63	10.03	4.01	1.93	1.22	2.35	3.05		
Zeolite (clinoptilolite)	Japan	71.65	11.77	0.81	0.88	0.52	1.80		0.34	9.04
Zeolite (mordenite)	Japan	41.11	11.79	2.57	2.07	0.15	1.66		0.27	9.50
Jajroud trass	Iran	68.3	12.70	1.20	2.68	1.35	1.95	2.10	0.33	8.70

Table 1.3 Chemical analyses of silica-rich pozzolans of different origin (%)

Pozzoalna	Country	SiO_2	Al_2O_3	Fe_2O_3	TiO_2	CaO	MgO	Na_2O	K_2O	SO_3	LOI
Diatomaceous earths											
Moler	Denmark	75.6	8.62	6.72		1.10	1.34	0.43	1.42	1.38	2.15
Diatomite	USA	85.97	2.30	1.84		Trace	0.61	0.21	0.21		8.29
Diatomite	USA	60.04	16.30	5.80		1.92	2.29				11.93
Mixed origin											
Secrofano	Italy	85.50	3.02	0.44	1.22	0.58		0.16	0.26	0.77	7.94
White hearth (a)	Italy	90.00	2.70	0.70		0.20					6.10
White hearth (b)	Italy	84.25	4.50	1.55		2.40					8.40
Beppu white clay	Japan	87.75	2.44	0.41	1.10	0.19	0.23	0.11	0.11		
Gaize	France	79.55	7.10	3.20		2.40	1.04			0.86	5.90

mica, olivine and zeolitic minerals, fluorite and clay minerals were traced. The rhyolites found in the USA contain from 65 to 95 % glass and variable amounts of calcite, quarts, feldspar, sanidine, hornblende and montmorillonite.

The Japanese natural pozzolan is mainly consists of glass and small quantities of quarts, feldspar and plagioclase. Small quantities of clay minerals and zeolites have also traced in some places.

The mineralogical composition of tuffs is rather complex since the glass or the original pozzolan is transformed by an outometamorphic process in zeolitic compounds such as hershelite, cabazite and phillipsite. The main minerals in Rhine and Bavarian trasses are glass, quarts and feldspar. The tuffs may also contain variable amounts clay minerals and zeolite. The zeolitazation and transformation of volcanic pozzolan, incoherent and mostly glassy, into compacted tuffs has been observed in many deposits. In Naples pozzolan, the tuff is separated from the pozzolan by an intermediate layer. The chemical composition of the layers is relatively similar with minor variation in alkali and iron contents. In the Rhine trass deposit, the chemical composition of the layers is very similar but the

Table 1.4 Minerals in some volcanic pozzolans

Pozzolan	Country	Active phases	Inert phases
Bacoli	Italy	Glass	Quartz, feldspars, augite
Barile	Italy	Partially decomposed glass	Pyroxenes, olivine, mica
Salone	Italy	Glass analcime	Analcime
Vizzini	Italy	Glass	Leucite, pyroxenes, alkali, feldspars, mica
Volvic	France	Glass	Feldspars, quartz, diopside
Santorine earth	Greece	Glass	Quartz, anorthite, labradorite
Rhine trass	Germany	Glass (55–60) Glass (62–67)	Quartz (9 %), feldspar (15 %)
Bavaria trass	Germany	Chabasite (3 %) Analcime (5 %)	Quartz (19 %), feldspar (15 %)
Rhyolite pumicite	USA	Glass (80 %)	Clay (5 %), calcite, quartz, quartz (9 %), feldspar, etc. (15 %)
Higashi Matsujama	Japan	Glass (97 %)	Quartz (1 %), anorthite (1 %)
Jajroud trass	Iran	Glass (65 %)	Quartz (22 %), rhyolite, chert, chalcedony

microscopic structures vary significantly [9]. The loose material rich in glass adjacent to the compact material was also found in the Tuffasche pozzolan. The loose tuffasche consists of mostly glass and small amount of quarts and sanidine and no trace of zeolitic materials. The compact tuffasche beneath the loose layer also seems glassy but contains a large amount of diffused zeolites. The zeolite minerals in compacted trass consist of mainly analcime and small amount of phillipsite and cabazite.

The hydrothermal treatment of some natural pozzolans has converted them to zeolitic minerals. Their chemical composition has not changed but the volcanic glass has transformed to zeolitic minerals. The rate of transformation and the type of formed minerals depend upon the thermal treatment. In some cases the glassy tuff has completely converted to the zeolite.

The mineralogical composition of diatomaceous earths has shown that opal content of this material varies between 25 and 100 % and the remainder is clay minerals and other minerals as feldspar and quarts. Its pozzolanic activity with lime is very high due to their high surface area and high amorphous silica content.

Pozzolanic Activity of Natural Pozzolans

Pozzolanic reaction is defined as the reaction between the active phases of the pozzolan with lime. The amount of amorphous material usually determines the reactivity of natural pozzolans. It is fairly difficult to determine the reactive constituents of natural pozzolans and pozzolanic reaction is usually assessed by the consumption of lime in a mixture of lime-pozzolan or measurement of the silica and alumina soluble in acid.

Evaluation of Pozzolanic Activity

Evaluation of the pozzolanic activity of fly ash and other pozzolans falls into three categories: chemical, physical, and mechanical.

There are two sorts of chemical evaluations. The first measures the amount of $SiO_2 + Al_2O_3 + Fe_2O_3$ rendered soluble as a result of the pozzolanic reaction, either in acids or in alkalis. The second measures the reduction of calcium ions in a pozzolan-saturated lime solution. With this method, the pozzolanicity is assessed by comparing the quantity of $Ca(OH)_2$ in a liquid phase in contact with the hydrated cement with the quantity of $Ca(OH)_2$ capable of saturating a inedium of the same alkalinity. This is the method of the International Organization for Standards (ISO) recommendation R863-.1968, Pozzolanicity Test of Pozzolanic Cements. This method does not directly indicate the contribution of the pozzolan to the strength of the cement structure with time. Another technique dissolves a pozzolan in a mixture of nithc and hydrofluoric acid and measures the heat generated, as an indication of pozzolanic activity. Raask and Bhaska modified this method to assess the pozzolanicity of fly ashes. They used hydrofluoric acid only, which selectively reacted with silica. They monitored the reaction by measuring the solution's electrical conductivity.

The XRD technique has been used to monitor the progress of the lime uptake in pozzolan Portland cement containing pozzolans. The results obtained by this method indicated good correlation between the lime combined in the reaction and the compressive strength of mortars at 6 months and 1 year.

The methods of assessment based on compressive strength (mechanical methods) also evaluate pozzolanic activity. These methods assess the strength properties of concretes containing pozzolans and are more useful in practice than the evaluation of pozzolan-lime or pozzolan Portland cement reactivity. ASTM C 311 describes the strength activity index with portland cement and with lime.61 For both, the compressive strength of the control mixture is compared with the strength of pozzolan-containing mixture at ages of 7 or 28 days.

The problem of the slow rate of strength development during the strength activity and pozzolanic activity tests has been overcome in a method proposed by Lea. Lea introduced an accelerated test, which measures the strength of pozzolan Portland cement mortars cured at 50 °C, and compared the results with the strength of specimens cured at 18–24 °C. The difference in strengths relates to pozzolanic activity. Lea correlated the difference in the strength of mortar specimens with the strength of pozzolanic-cement concretes after 180 and 365 days of curing at 18 °C.

Attempts have been made to compare the pozzolanicity and the pozzolanic activity indices measured by various methods. In most cases, the relationships between the results obtained in chemical and mechanical techniques were poor. Watt and Thorne 63 reviewed a number of chemical methods in connection with their studies of fly ash activity and compared the results obtained by four methods. In a CANMET study the pozzolanic activity of 11 Canadian fly ashes with portland cement was determined using both ASTM C 311 and CSA A23.5 standard test procedures. The two tests are similar, except that the ASTM test involves curing at 38 °C for 28 days, whereas the CSA A23.5 test involves accelerated curing at 65 °C for 7 days there was a high degree of correlation between the results of the CSA Pozzolanic test (curing of 65 °C for 7 days) and those of the ASTM Pozzolanic test (curing of 38 °C for 28 days): the correlation coefficient is 0.98.

Factors Affecting Pozzolanic Activity

There are several parameters influencing the pozzolanic activity of natural pozzolans. The nature of active phases and their contents in the pozzolan, surface area and fineness of the particles, the lime pozzolan mixture and the amount of mixing water, curing system and temperature are the most important factors affecting the reactivity.

In the zeolitic pozzolans which are usually more reactive than the glassy ones, clinoptilotite and herschelite are more reactive than analcime. The glasses of different pozzolans show different potential in combining lime. The lime pozzolan combination of Bavarian, Jadjroud and Rhine trasses were different. Table 1.5 shows the lime phases combination of the trass pozzolan [16].

Reactivity of Thermally Treated Pozzolans

Pozzolanic reactivity of natural pozzolans may change by thermal treatment. This is attributed to the chemical and mineralogical changes of the pozzolan. Some pozzolans lose water in glassy or zeolitic phases and the destruction of the crystal structure in clay minerals. This leads to the enhancement of pozzolanic activity of the pozzolans. The temperature and duration of heating may decrease the specific surface area and causes devitrification and recrystallisation. This results in lower pozzolanic activity of the pozzolan. In the heating process of natural pozzolans, if the temperature of calcination is increased gradually, combined lime initially increases and later decreases. In the heating process, a reduction in the surface area of the pozzolan is occurred. This means that for every pozzolan, there is an optimum thermal treatment. Researches on different natural pozzolans show that the optimum temperature is about 700–800 °C. Owing to the densification and

Table 1.5 Lime-binding capabilities of the principal trass minerals and their contributions to trass-lime binding [16]

Mineral component	Lime reaction (mg CaO/g)	Free alkali		Average amount in trass (%)	Calculated lime reaction (mg CaO/g trass)
		Na_2O (mg/g)	K_2O (mg/g)		
Rhenish trass					
Quartz	43	1.5	0.4	13	5.5
Feldspar	117	1.1	0.2	15	17.5
Leucite	90	1.3	1.8	6	5.4
Analcime	190	10.7	3.0	7	13.3
Kaolin	34	0.3	2.1	2	0.7
Glass phase	364	18.0	24.0	55	200.0
Total	–	–	–	98	242.5
Glass phase					
Bavarian	727	6.0	6.0	66	179.0
Obsidian glass	176	3.7	3.1	–	–

crystallization at higher temperatures, more stable phases are formed. The result is combination of lower lime and a decrease in the amount of silica and alumina dissolved in acid solution.

The pozzolanic activity of natural microsilica consisting of opal and clay did not improve by heating up to 700 °C On the contrary, the pozzolanic activity of diatomites containing high quantities of clay minerals improved at same temperature [6]. Modification observed on the thermal treatment of Jadjroud trass at 600 °C.

Hydration Reactions and Hydration Products

The setting and hardening of Portland cement occur as a result of the reaction between the compounds of cement and water. The major compounds of cement that react with water to produce reaction products are tricalcium silicate (C_3S), dicalcium silicate (C_2S), tricalcium aluminate (C_3A), and tetracalcium aluminoferrite (C_4AF). The hydration products from the two calcium silicates are similar and differ only in the amount of calcium hydroxide formed, as shown below:

$$\begin{aligned} &2C_3S + 6H \rightarrow C_3S_2H_3 + 3CH \\ &\text{tricalciumsilicate} + \text{water} \rightarrow \text{C-S-H} + \text{calciumhydroxide} \end{aligned} \tag{1.1}$$

$$\begin{aligned} &2C_2S + 4H \rightarrow C_3S_2H_3 + CH \\ &\text{dicalciumsilicate} + \text{water} \rightarrow \text{C-S-H} + \text{calciumhydroxide} \end{aligned} \tag{1.2}$$

The reaction of C_3A with water is very fast and involves reactions with sulphate ions supplied by the dissolution of gypsum. The reactions can be represented by the following:

$$(C_3A + 3CSH_2 + 26H \rightarrow C_3A(CS)_3H_{32} \quad \text{tricalcium silicate + gypsum + water} \rightarrow \text{ettringate} \tag{1.3}$$

$$C_3A + \ CSH_2 + 10H \rightarrow C_3ACSH_{12} \quad \text{monosulphalumiate hydrate} \tag{1.4}$$

C_4AF forms hydration products similar to those of C_3A, with iron substituting partially for alumina in the crystal structures of ettringite and monosuipho-aluminate hydrate.

In the absence of sulphate C_3A may form C_3AH_6 or C_4AH_{19} as shown by reactions (1.5) and (1.6):

$$C_3A + 6H \rightarrow \ C_3AH_6 \tag{1.5}$$

$$C_3A + CH + 18H \rightarrow \ C_4AH_{19} \tag{1.6}$$

Hydration of Natural Pozzolans with Lime

The reaction products of lime-pozzolan system are similar to the compounds of the hydration of Portland cement. Natural pozzolans contain oxides similar to Portland cement oxide elements and produce similar silicate and aluminate hydrates. Differences are minor and usually appear in the quantity of the reaction products.

The reaction products of natural pozzolans in a saturated lime solution are mainly calcium silicate hydrate and calcium aluminate [17]. Zeolitic compounds produce similar reaction products. With the addition of gypsum and more water, the reaction of pozzolan with lime is accelerated.

In the long term reaction of lime with a natural pozzolan besides the calcium silicate hydrates and calcium aluminate hydrates, the other compounds such as hydrogarnet, gehlenite, and carboaluminate can be observed. With the presence of gypsum in the lime-pozzolan system, ettringite is also formed.

The ratio between calcium oxide and silica in the calcium silicate hydrates vary in the lime-pozzolan system. It depends on the pozzolan to lime ratio in the mixtures, the type of pozzolan, temperature and the time of reaction and the measurement technique.

The main product of the natural pozzolan-lime system is amorphous calcium silicate hydrates at temperature between 50 and 90 °C which is similar to the hydration products of Portland cement [18].

Similar compounds have been observed in other natural pozzolans. Reaction of burned kaolin and lime produces calcium silicate hydrate, gehlenite hydrate ($C_4 ASH_8$) and small amount of calcium aluminate hydrate [19].

Hydration of Natural Pozzolan with Clinker

Hydration of pozzolan with the clinker phases is less complicated than that with Portland cement. This reaction modifies the properties of reaction products and affects the mechanical properties and durability of pozzolanic cements.

The kinetics of early hydration of pozzolan-clinker system is investigated by isothermal transmission calorimetry. The progress of hydration at longer ages is usually monitored by XRD, DTA, DSC, and optical and electronic microscopy. The interpretation of the pozzolan-clinker reaction system is easier when the reaction of each clinker compounds is discussed separately.

The reaction of natural pozzolan with tricalcium aluminate changes the initial rate of evolution of the heat of hydration since it reduces the intensity of the second peaks. This is attributed to the diminution in the hydration rate of the C_3A [20, 21]. The decrease in the second peak intensity and its delay is affected by the specific surface area of the pozzolans and some other factors such as the dissolution of alkalis and the surface activity.

The kinetics of reaction, formation of portlandite and composition of the hydrates of C_3S hydration are also affected by the addition of natural pozzolans.

The heat evolution rate measurement is a useful tool to investigate the reaction of calcium silicates with the pozzolans especially at early ages. The changes in the heat evolution peaks of the C_3S-pozzolan blends are shown in Fig. 1.5. In general, natural pozzolans have an accelerating effect on the hydration of calcium silicate (C_3S). It was found that the dormant period does not change but the second peak is slightly delayed and its height is significantly enhanced [22].

Hydration of Pozzolan-Cement Mixture

Hydration of cement containing clinker and gypsum with natural pozzolans is rather complicated. The reason is the different rate of reaction of clinker and gypsum and the type of pozzolans used. Pozzolans usually start to react with cement after a few days when 80 % of the C_3S of cement has hydrated [21].

Natural pozzolans generally modify the kinetics of Portland cement hydration from the beginning of the hydration process and can alter heat of hydration, combined water, calcium hydroxide content and the degree of hydration of calcium silicates of the mixture.

The products of the reaction of natural pozzolans with cement are similar to those obtained in the reaction of cement paste. The differences are the amount of

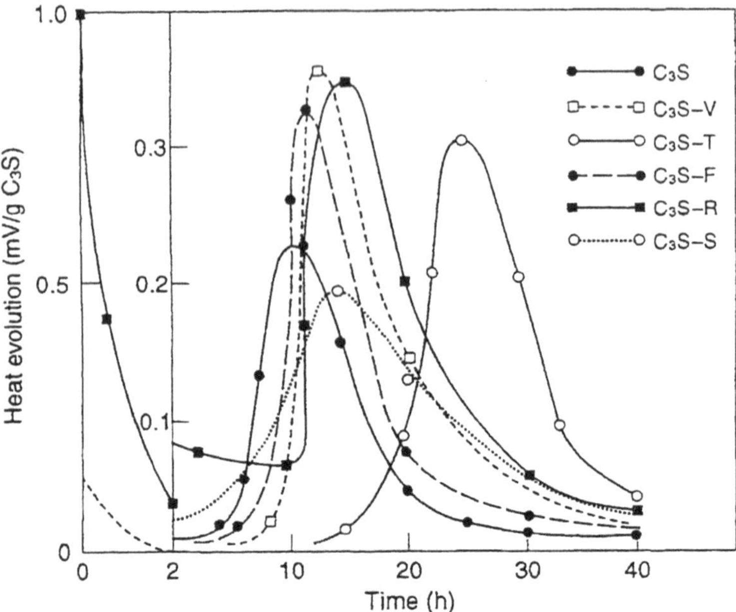

Fig. 1.5 Heat evolution curve in 6:4 C_3S-pozzolana blends (22). *V*, *F*, *R*, and *S* are natural pozzolans, *T* is fly ash. w/b = 0.4

reaction products and the morphology of the reaction products. The main reaction products are summarized below:

- Calcium silicate hydrate (CSH)
- Tetracalcium aluminate hydrate
- Ettringite
- Monosulfoaluminate
- Calcium hydroxide
- Gehlenite (C_2ASH_8)
- Calcium carbonate.

Calcium silicate hydrate (CSH) appears after one day because its hydration is accelerated when cement is mixed with pozzolans [23]. Uncertainties regarding the composition of CSH gel as observed in the products of C_3S hydration are also observed when considering the composition of the calcium silicate hydrate formed by hydration of Portland cement and Portland cement pozzolan mixture. Some researchers have indicated the difference between the inner and outer CSH products. The inner product is believed to be a pure calcium silicate hydrate having a C/S ratio between 1.5 and 2.1 [23, 24]. This ratio is appeared to be higher in the outer products and varies between 1.6 and 2.7. In the mixture of pozzolan-cement, besides the CSH deriving from the hydration of clinker silicates, secondary CSH from the reaction between pozzolan and hydrolysis lime is also observed.

Tetracalcium aluminate hydrate may be observed in the paste depending on the Al_2O_3/SO_3 ratio of the cement. It is usually carbonated due to the reaction with carbon dioxide or limestone in the form of calcite as filler added to the cement.

Similar to the hydration of Portland cement, ettringite is formed rapidly in the natural pozzolan-cement mixture. Ettringite has been observed from early hours of reaction and up to one year in such mixtures. Ettringite can be transformed to monosulfate after 3 days. The transformation of ettringite depends on the SO_3 availability in the pore solution and the carbon dioxide content of the cement paste. Carbon dioxide can react with calcium aluminate hydrate and produce carboaluminate at a slow rate. This reaction prevents or slows down the reaction of calcium aluminate with ettringite to convert it into monosulfate [25].

Calcium hydroxide or portlandite can be observed in the pozzolan-cement mixture even after 5 h of reaction. The entire calcium hydroxide released from the calcium silicates after hydration may not be consumed by the insufficient amount of pozzolan in the mixture. This has been proved by the existence of calcium hydroxide at longer ages in the pozolan-cement mixture. The reaction of pozzolan with calcium hydroxide also relates to its pozzolanic activity.

Calcium carbonate is also formed when carbon dioxide reacts with the hydrated cement compounds. It also has been observed in the mixture of pozzolan-cement. Carbonation of cementitious materials is mainly attributed to the reaction of calcium hydroxide and carbon dioxide [26].

Effects of Natural Pozzolans on the Properties of Fresh Concrete

Natural pozzolans can affect the properties of fresh concretes. Cohesiveness, consistency or workability, segregation, bleeding and setting time of pozzolanic cement concretes differ from the Portland cement concretes. These properties vary in different pozzolanic cements by variation in the particle shape, roughness and fineness of the natural pozzolans.

Most natural pozzolans improve the cohesiveness of the mixture by producing a more plastic paste which helps its consolidation and flowability. Segregation of the pozzolanic cement concretes also improves by the improvement of its cohesiveness. Consistency and workability of pozzolanic cement concretes improve when round and smooth particles of pozzolans are used in the mixture. In contrast, the irregular shape and rough texture particles would lower the workability of concretes. Higher fineness and surface area of pozzolans particles also demand more water for constant consistency and should be avoided in large amounts in the concrete mixtures without water reducing admixtures. Example of such materials is the diatomaceous earths which is rich in very fine silica particles demanding more water for certain workability. Superplasticisers are necessary for such materials to adjust workability without adding more water to the concrete mixture. Other natural pozzolans with similar fineness of cement usually do not change the workability at a significant level.

The rate of bleeding decreases by increasing the ratio of surface area of solids to the volume of water in the concrete mixtures. Addition of natural pozzolans usually increases the ratio of surface area of solids to the volume of water and hence improving the bleeding of concrete.

Segregation of the pozzolanic cement concretes is also improved when finely divided particles of pozzolans are used in the mixture. However, coarse natural pozzolans with rough texture and poor shape require more water for specific consistency. Higher water in the mixture increases the bleeding and segregation of fresh concrete and reduces the strength at hardened stage.

The setting time of cement paste and hence of concrete is affected by cement type, cement content and fineness, water content of the paste, water soluble alkalies, The type of amount of admixtures used, ambient and concrete temperature, content of pozzolan, fineness, and chemical composition of the pozzolan. Setting time of the cement paste can be adjusted by controlling these factors. The use of natural pozzolans may extend the setting time of the cement paste especially at higher level of replacement with cement.

Ramezanianpour et al. investigated the setting time of cements containing various natural pozzolans. In almost all cement pastes the setting time extended when more pozzolans were replaced with cement [27, 28].

Effects of Natural Pozzolans on the Mechanical Properties of Hardened Concrete

Strength of Mortars and Concretes

As mentioned earlier, the reaction of calcium hydroxide from Portland cement with pozzolan produces calcium silicate hydrates and hardens the mortar and concrete. The partial replacement of pozzolan with cement initially reduces the rate of hardening of cement, but at later ages the strength of Portland pozzolan cements is similar or sometimes higher than the corresponding control Portland cements. Many researchers have found similar trend for different natural pozzolans investigated at different conditions [15, 28]. Figure 1.6 shows the effect of a natural pozzolan on the compressive strength of mortars up to one year [29].

The reduction in the early strength of cement-natural pozzolans mixture is common. Exceptions are those pozzolans which are chemically very reactive. Metakaolin is a very active natural pozzolan and can produce higher early strength concrete. This is discussed in more detail in the chapter of metakaolin.

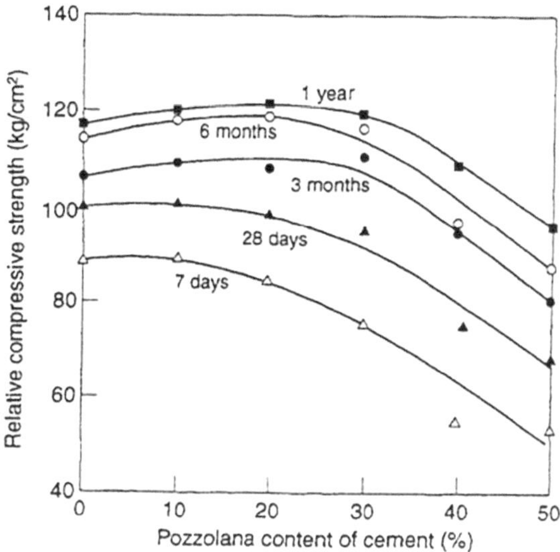

Fig. 1.6 Effects of substituting Portland cement with the pozzolan on the compressive strength of ISO mortar. Values expressed as percentage of the 28-day strength of reference cement [29]

Compressive strength of 4 different natural pozzolans with replacement level of 10–25 % at various ages is seen in Fig. 1.7 [30]. The comparison of the data shows that after 180 days of water curing, the highest compressive strength was 51.2 MPa for K20 concrete, while for control concrete (CTL), the compressive strength was 50.5. From the figure, it is found that natural pozzolans replacements reduced the compressive strengths of concrete. This is reasonable due to the reduction of cement content in the mix with the increase of natural pozzolans content. However the compressive strength increases with age.

Figure 1.8 shows the results of an investigation for the development of compressive strength of pumice and trass natural pozzolans up to one year in moist conditions. In almost all three levels of replacement (10, 22.5 and 30 %) the 7 day strength of concrete containing natural pozzolans were lower than that of control concrete. However at longer ages and beyond 180 days, higher compressive strength can be observed in most of concretes containing various percentages of natural pozzolans [31].

The strength of natural pozzolanic cement concretes depends on many factors such as the type of pozzolan, its particle size and specific surface area, fineness and pozzolanic activity of pozzolan, and the amount of pozzolans replaced with cement.

Regarding the production of pozzolanic cement concretes, the moist curing of mortar and concrete mixtures plays an important role on the strength development of pozzolanic cements.

The effect of curing on the strength gain of pozzolanic cement concretes has been investigated by many researchers. Figures 1.9 and 1.10 show the results of the curing effect on mortar specimens made with Portland pozzolan cements containing 10, 20, and 30 % Santorin earth as a natural pozzolan [15].

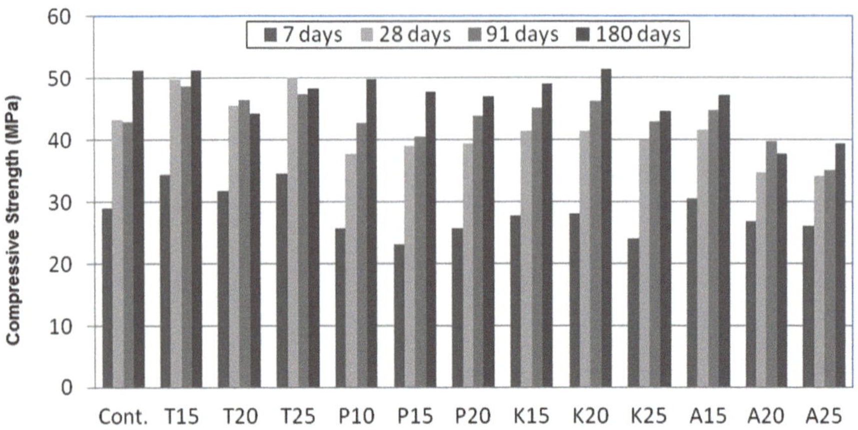

Fig. 1.7 Compressive strength of 4 different natural pozzolans at various ages [30]

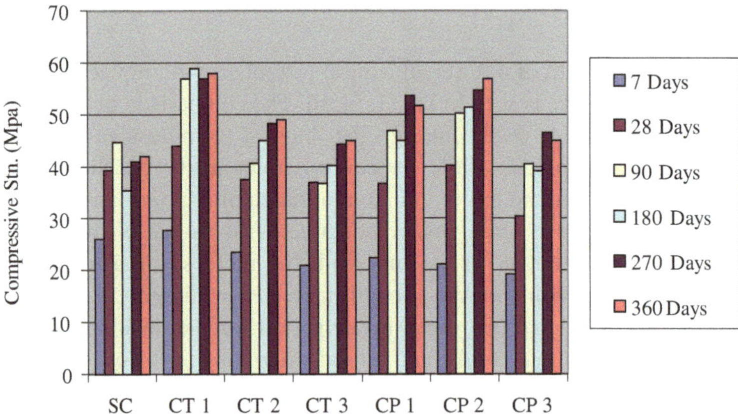

Fig. 1.8 Compressive strength of pumice and trass natural pozzolans up to 1 year [31]

It is clearly seen that this kind of natural pozzolans contributes to the strength of mortars after 7 days of curing. The compressive strength of the mortar mixtures containing 10 % pozzolan was even higher than that the control mortar mixture. The compressive strength of mortars containing 10 and 20 % pozzolan replacement was higher than the control concrete after 90 days. It is interesting to note that mortars with 30 % replacement achieved similar strength to the control concrete after one year.

The strength development of natural pozzolans has been compared with other artificial pozzolans in Fig. 1.11. It is clearly seen that the diatomaceous earth can increase the strength of concretes due to its highly reactive silica and higher fineness. The lowest strength belong to the mixture of Sacrofano natural pozzolan with larger particles and lower pozzolanic activity [32].

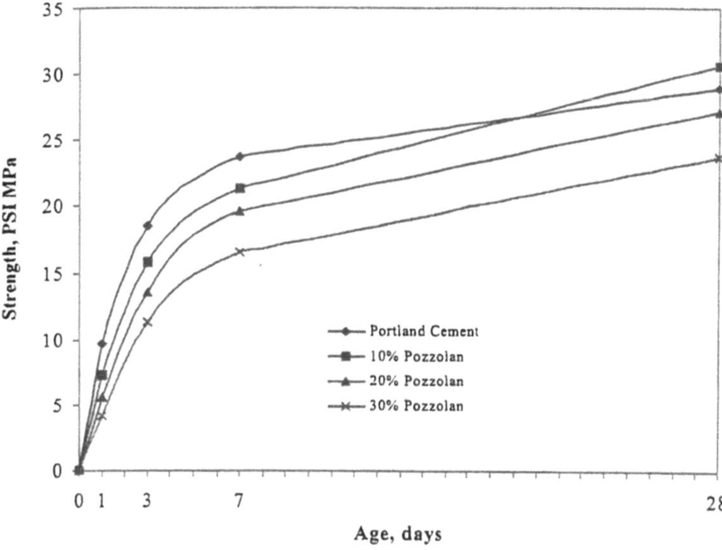

Fig. 1.9 Effect of curing time on compressive strength of mortar cubes up to 28 days made with Portland-pozzolan cements containing Santorin earth [15]

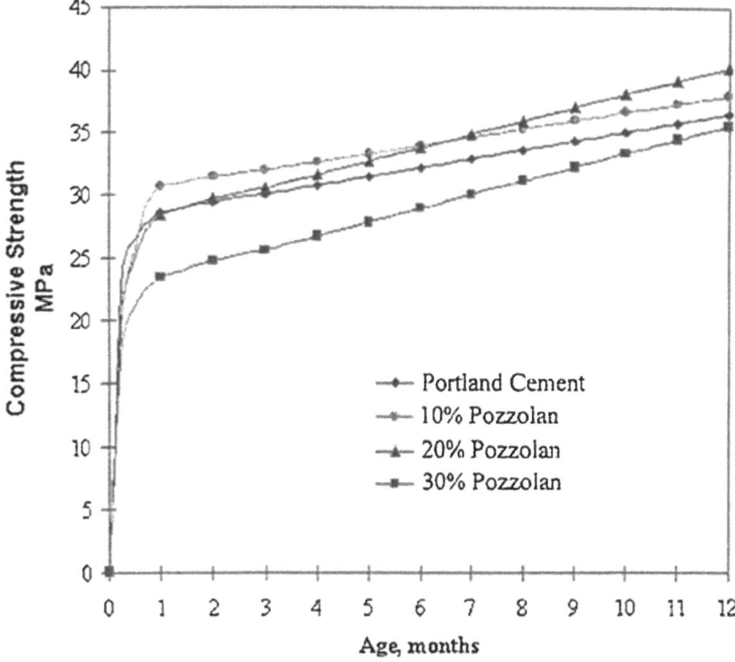

Fig. 1.10 Effect of curing time on compressive strength of mortar cubes up to 12 months made with Portland-pozzolan cements containing Santorin earth [15]

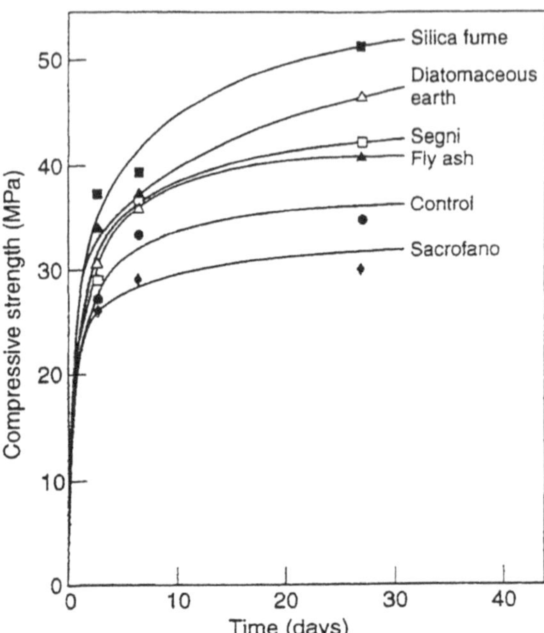

Fig. 1.11 Strength development of different 20 % pozzolana-cement mixes. Paste with w/c = 0.5; 40 × 40 × 160 mm specimens cured for 72 h in water and then in 65 % relative humidity at 20 °C

The low rate of strength gaining of some natural pozzolans is not a major problem for cement manufactures. They can easily adjust the proportion of their natural pozzolans in the blended process to achieve the required strength of their standard specifications. In the case of addition of pozzolan at site, it is hard to modify the properties of the pozzolan-cement mixture and the only possible effective measure to optimize the properties of the mixture is to increase the cement content or to add a very high reactive pozzolan.

Ramezanianpour et al. [33] also studied the mechanical properties and durability of concretes containing zeolite as a natural pozzolan. Figure 1.12 shows the effect of zeolite on compressive strength of concrete. It is clear that regardless of replacement level, zeolite decreased concrete strength at 7 days. This reduction is about 6 % for 10 % replacement and around 9 % for 15 % replacement. Compressive strength increased at 28 days due to the pozzolanic reaction. It is apparent that the optimum replacement level was about 10 % while it didn't show considerable difference at 15 % replacement.

Modulus of Elasticity

There is general agreement that modulus of elasticity of concrete is greatly influenced by strength of concrete. In Portland-natural pozzolan concretes with a

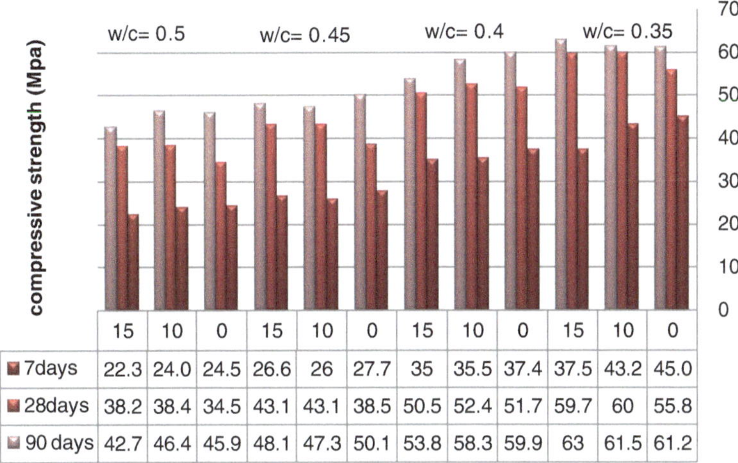

	15	10	0	15	10	0	15	10	0	15	10	0
7days	22.3	24.0	24.5	26.6	26	27.7	35	35.5	37.4	37.5	43.2	45.0
28days	38.2	38.4	34.5	43.1	43.1	38.5	50.5	52.4	51.7	59.7	60	55.8
90 days	42.7	46.4	45.9	48.1	47.3	50.1	53.8	58.3	59.9	63	61.5	61.2

Fig. 1.12 Compressive strength of control and zeolite concretes at different ages with various w/c ratios [33]

lower strength at early ages, the modulus of elasticity in general is slightly lower than that of concretes without pozzolans. However at longer ages the modulus of elasticity of concretes containing highly reactive natural pozzolans may be similar or slightly higher than that for control concrete [4].

Effect on Volume Changes of Concrete

The drying shrinkage of mortars and concretes vary with the variation of water and aggregate type and contents. The replacement of cement with normal percentages of natural pozzolan does not change the shrinkage significantly. This is clearly shown in Table 1.6 that the difference between the shrinkage of normal Portland cement and cements containing natural pozzolan is not significant. Shrinkage of mortars containing various natural pozzolans varies between 800 and 1050 microstrain in drying condition after 90 days [34].

Figure 1.13 also shows that the drying shrinkage of concretes containing 10–30 % Santorin earth is relatively similar to the control concrete [15]. As mentioned before, the modulus of elasticity of natural pozzolan cement concretes is lower than that for control concrete at early ages. Therefore, the cracking tendency due to drying shrinkage of pozzolanic cement concretes is lower than that for control concretes.

Volume change of concrete under load known as creep is related to many factors including the strength and the amount and time of loading. The strength of concretes containing natural pozzolans is usually lower than the control concrete and hence higher specific creep is expected for such concretes at early ages. This is

Table 1.6 Typical shrinkage values for normal mortar (μm/m) for specimens cured for 24 h and then kept in a climatic cabinet at 20 °C and 50 % relative humidity [34]

Cement		Curing period (days)		
		7	28	90
CEM I	32.5	427	755	890
CEM I	42.5	396	733	873
CEM I	52.5	453	842	1043
CEM I	32.5	329	653	793
CEM I	42.5	381	725	904
CEM I	52.5	434	797	991
CEM I	42.5	430	685	810
CEM I	52.5	461	770	988
CEM IV	32.5	428	784	943
CEM IV	42.5	420	743	915
CEM IV A		393	737	889
CEM IV B	32.5	440	765	900
CEM IVA	42.5	425	706	890

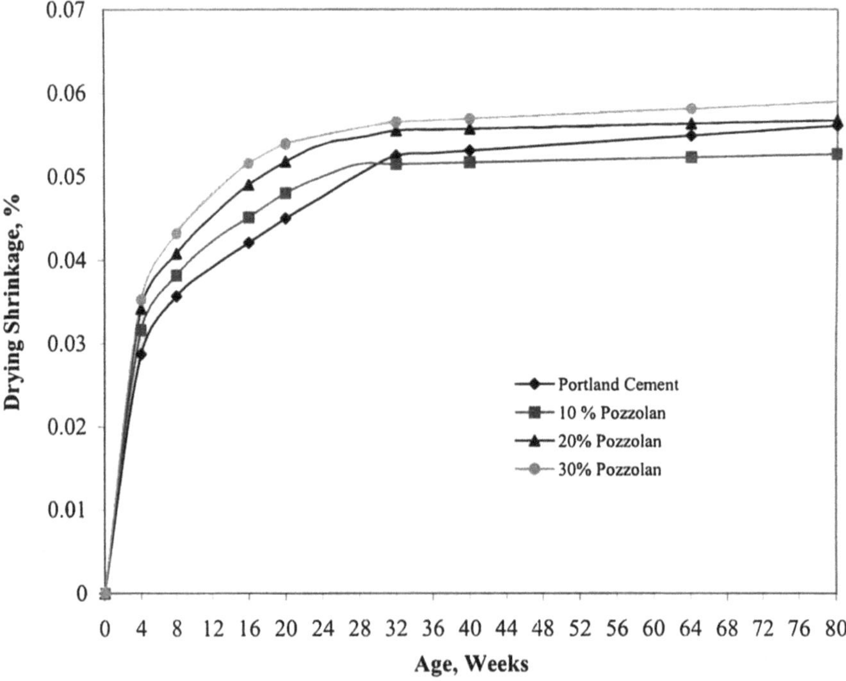

Fig. 1.13 Drying shrinkage of concrete prisms made with cements containing various amounts of Santorin earth [15]

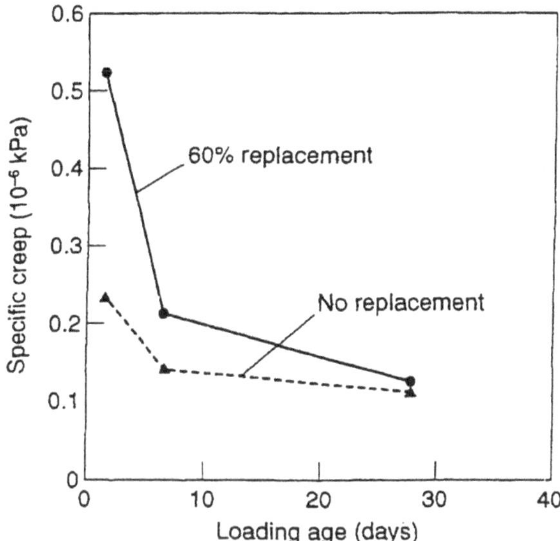

Fig. 1.14 The effects on creep of loading age and replacement of Portland cement by pozzolana [35]

clearly seen in Fig. 1.14 where high amount of natural pozzolans is replaced with cement, and concrete elements are loaded at early ages [35]. It is worth noting that after sufficient curing period, both concretes show similar specific creep if loaded at 28 days. This is related to the strength gain of pozzolanic cement at longer ages.

Microstructure, Porosity and Permeability

Microstructure of cement paste containing natural pozzolans seems to be similar to the plain Portland cement paste except for the unreacted particles of pozzolan. The hydration product at the early stages is calcium silicate hydrate which forms as a thin shell around the clinker particles. Then at following stages, thin needles of ettringite, thin plates of portlandite and monosulfate are formed. Most natural pozzolans do not show any reactivity at the first 24 h of the hydration of cement. Fibrillar calcium silicate hydrates and ettringite needles cover both clinker and pozzolan particles. After that and at further hydration process, calcium hydroxide is formed on the surface of particles.

After a few days, calcium silicate hydrates are observed as outer and inner shells. These shells can be observed on the surface of both reacted and non-reacted grains of pozzolans. These shells are formed due to the process of dissolution of clinker minerals and precipitation on the pozzolan particles.

After about 4–6 weeks the calcium silicate hydrates fill the pores and the pozzolan particles have reacted with calcium hydroxide to create new calcium silicate hydrates.

Table 1.7 Total porosity (%) by mercury intrusion of cement pastes; blending component = 30 % w/c ratio = 0.32 [36]

Sample	Portland cement	Blending component	Porosity			
			28 days drying		7 months drying	
			Rapid[a]	Slow[b]	Rapid[a]	Slow[b]
CEM 1	100		17.0	14.70	13.10	10.60
Filler	70	30	17.80	21.40	15.30	13.80
Vizzini pozzolana	70	30	18.70	21.60	14.30	12.90
Qualiano pozzolana	70	30	20.00	18.80	11.30	10.70
Casteggio pozzolana	70	30	17.80	16.30	13.60	11.60
Barile pozzolana	70	30	17.70	17.90	13.50	12.70
Segni pozzolana	70	30	17.50	19.30	13.40	34.20
Bacoli pozzolana	70	30	17.80	18.70	13.30	11.40

[a] Dried at 70 °C for 16 h under vacuum at the residual pressure of 5 mbar
[b] Dried at 20 °C in 4 successive stages (relative humidity of 55, 53, 10, 0.01 %)

The porosity of the cement paste with and without pozzolans is related to the water cement ratio of the mixture and its moist curing duration. High temperature also increases the volume of large pores. The reactivity of natural pozzolan and the amount of replacement influence the porosity of the cement paste.

The porosity of cement pozzolan pastes have been determined by several methods such as mercury intrusion porosimetry, nitrogen adsorption, helium pycnometry and methanol displacement. The total porosity of pozzolanic cement paste measured by mercury intrusion method was found higher than that for the parent plain cement paste [36]. This is clearly shown in Table 1.7. The total porosity of the cement pozzolan paste decreases by time similar to the Portland cement paste but remains higher at longer ages.

Permeability, as the intrusion of fluids into cement paste by applied pressure, is high at early ages for cement pastes containing pozzolan when compared with the parent cement paste. Moist curing at longer ages of pozzolanic cement paste reduces the permeability even lower than the plain cement paste. However, the porosity of pozzolanic cement paste remains higher than the plain cement paste even at longer ages. Results of the porosity and permeability of cement pastes containing natural pozzolans at different water to cement ratio and different times are shown in Table 1.8 [37]. The water penetration of cement paste containing Santorin earth as a natural pozzolan is also shown in Table 1.9 at different ages. The effect of the amount of pozzolan and the duration of hydration are clearly observed in reducing the porosity and water penetration of cement paste [15].

The higher porosity and lower permeability of cement paste containing natural pozzolans is attributed to the slow pozzolanic reaction of the pozzolan and disconnection of the channels between large pores. The secondary reaction products of pozzolan with cement paste may not fill the large pores created at initial stages of reaction and reduces the permeability only by disconnecting the channels

Table 1.8 Permeability and Porosity of cement pastes hardened for up to 90 days

Curing (days)	Permeability ($m^2 \times 10^{-17}$)					Porosity (%)				
	1	3	7	28	90	1	3	7	28	90
w/c ratio = 0.32										
OPC paste	5.60	0.30	0.12	0.00	0.00	20.8	19.7	14.4	9.8	5.9
BPC paste	1.94	0.70	0.12	0.06	0.00	29.5	26.3	21.3	14.7	7.1
w/c ratio = 0.40										
OPC paste	18.70	0.59	0.07	0.07	0.00	33.3	28.6	20.9	16.8	11.1
BPC paste	14.30	1.80	0.42	0.02	0.00	39.3	29.8	24.7	20.6	10.9
w/c ratio = 0.50										
OPC paste	214.00	14.70	2.35	0.19	0.00	43.6	37.8	32.2	20.8	14.5
BPC paste	218.00	22.30	3.74	0.06	0.00	44.4	42.4	40.0	24.4	22.4

Table 1.9 Depth of penetration of water into hydrated cement pastes

Age	Depth of penetration, mm			
	Portland cement	10 % Santorin earth	20 %	30 %
28 days	25	23	23	22
90 days	25	23	23	22
1 year	25	23	18	15

between the large pores. Therefore, the porosity of pozzolanic cement remains still higher but its permeability decreases and is lower than the parent cement paste.

Diffusion of different ions and in particular chloride ions in cement paste is also an important parameter controlling the corrosion of steel bars in reinforced concrete structures. Apart from many factors affecting the chloride diffusion of cement pastes, cement type also influences this parameter. Chloride diffusion coefficient in Sulfate resistant Portland cement paste is higher than that for the parent plain cement. This is attributed to the low content of C_3A in sulfate resistant Portland cement and hence its lower chloride binding capacity [38]. On the contrary, the use of pozzolanic cement has decreased the chloride diffusion coefficient of blended cements when compared with the plain parent cement paste. The effect of pozzolanic materials and temperature on the chloride diffusion coefficient of cement pastes is depicted in Table 1.10 [39].

Although the C_3A content of pozzolanic cement paste is lower than that of plain cement and hence it has lower chloride binding capacity and should give higher diffusion coefficient, but its chloride diffusion coefficient is lower. This lower diffusion coefficient in the cement paste containing pozzolans is attributed to its lower permeability than the parent cement paste. The Chloride diffusion coefficient and the permeability coefficient of Portland pozzolanic cement pastes were 3–10 times lower than that of the plain Portland cement [40, 41].

The microstructure and porosity of mortars incorporating natural pozzolan do not differ significantly with that for the pozzolan cement paste. The reaction of pozzolan with calcium hydroxide reduces the amount and size of portlandite in the

Table 1.10 Coefficients for diffusion of chloride ion into cement pastes

Sample	D ($cm^2s \times 10^8$)	Temperature (°C)
Portland cement paste	1.23	10
Portland cement paste	2.51	25
Portland cement paste	4.85	40
Pozzolanic cement paste	0.83	10
Pozzolanic cement paste	0.90	25
Pozzolanic cement paste	0.97	40

Table 1.11 The effect of natural pozzolans on the sorptivity of concrete ($m/s^{0.5}$) at various ages [30]

		Cont	T15	T20	T25	P10	P15	P20	K15	K20	K25
S (10^{-6}) ($m/s^{0.5}$)	91 days	0.74	0.47	0.52	0.41	0.82	0.63	0.62	0.55	0.4	0.52
	180 days	0.54	0.47	0.47	0.46	0.52	0.47	0.45	0.46	0.37	0.43
Water height (mm)	91 days	45	17	6	13	40	25	20	15	18	8
	180 days	30	25	15	22	23	18	20	14	24	10

cement paste and also at the transition zone. This improves the homogeneity of the paste containing pozzolan. The c-axes of the calcium hydroxide crystals for the pozzolan cement paste and at the aggregate interface are relatively parallel to the surface of aggregate.

The porosity of mortars containing natural pozzolans is usually higher than the plain Portland cement mortars especially at early ages. The differences in porosities depend upon the pozzolanic activity and the length of moist curing of mortars.

Permeability of mortars containing natural pozzolan also depends upon several factors such as the type of pozzolan, the level of replacement, mix design and curing duration in humid environments. The permeability of pozzolanic mortars was found higher than that of Portland cement mortars at early ages. However under similar mixture designs the permeability of mortars containing natural pozzolans was lower under moist conditions and at longer ages. Results of the water sorptivity test on concretes containing different natural pozzolans are shown in Table 1.11. It is clearly seen that the use of trass natural pozzolan (T15, T20, T25) has decreased the capillary adsorption of well cured specimens at longer ages [30].

The water penetration of concrete containing natural pozzolans were found lower than the control concretes at longer ages when properly cured in moist conditions [42]. This is also clearly seen in Fig. 1.15 where water penetration depth of concrete specimens containing trass was lower than that of normal concrete at 180 and 360 days [43].

The average test results for the water penetration depth of the concretes containing four different natural pozzolans are illustrated in Fig. 1.16 [30]. At 90 days, except for A15 concretes, the other specimens provided lower water penetration depth than the normal concretes. For example, P20 and T20 provided a

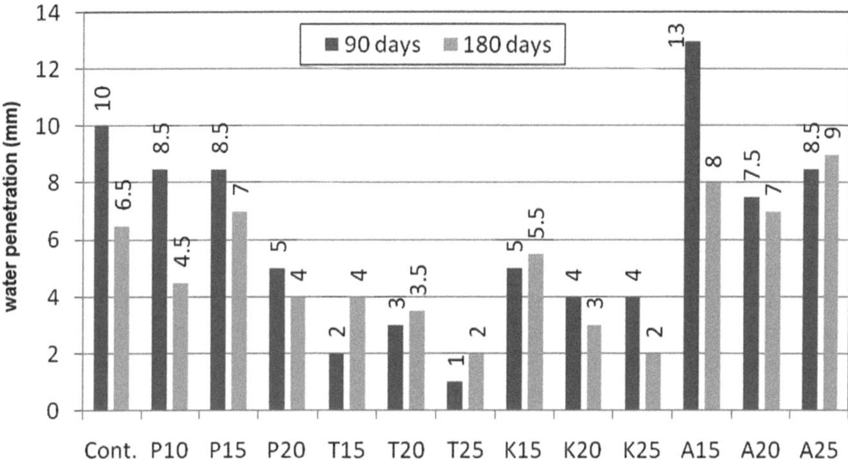

Fig. 1.15 The effect of natural pozzolans on the water penetration (mm) at various ages [30]

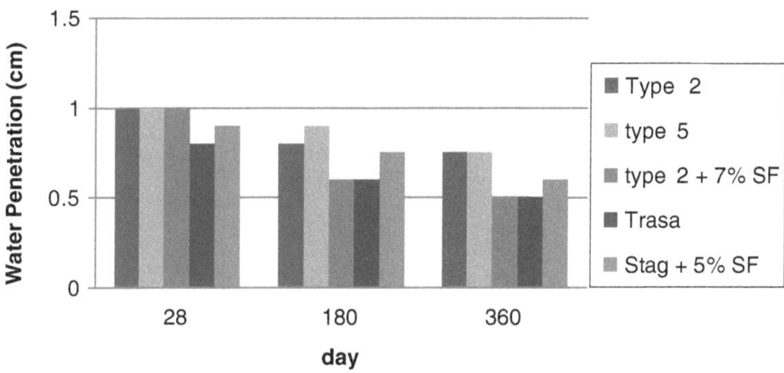

Fig. 1.16 Water penetration in concretes versus age (w/cm = 0.35)

water penetration depth close to 5 and 3 mm, while normal concrete provided 10 mm water penetration depth. As expected, the lower depth was obtained at 180 days for all concretes. Nevertheless, at this age, the water penetration depth varied from 1 to 9 mm, which is relatively low. Upon hydration of cement, natural pozzolans have the capability of partially obstructing voids and pores.

Effect of Natural Pozzolans on Durability of Concrete

Failure of concrete after a period less than the lifetime for which it was designed may be caused by the environment or by a variety of internal causes. External causes may be physical or chemical: weathering, extremes of temperature, Abrasion, or chemical action in the cement, aggregate, or reinforcement components. Internal causes may lie in the choice of materials or in inappropriate combinations of materials and may be seen as alkali-aggregate expansion or other forms of failure. Of all the causes of lack of durability in concrete the most widespread is excessive permeability. Permeable concrete is vulnerable to attack by almost all classes of aggressive agents. To be durable, Portland cement concrete must be relatively impervious.

Increasingly, concrete is being selected for use as a construction material in aggressive or potentially aggressive environments. Concrete structures have always been exposed to the action of sea water. In modern times, the demands placed on concrete in marine environments have increased greatly, as concrete structures are used in arctic, temperate, and tropical waters to contain and support the equipment, people, and products of oil and gas exploration and production. Concrete structures are used to contain nuclear reactors and must be capable of containing gases and vapors at elevated temperatures and pressures under emergency conditions. Concrete is increasingly being placed in contact with sulphate and acidic waters. In all of these instances, the use of fly ash as concrete material has a role, and an understanding of its effect on concrete durability is essential to its correct and economical application.

The following sections of this chapter seek to provide a general view of the present knowledge regarding the durability of natural pozzolan concrete. The subject matter is vast, complex, and as yet incompletely understood.

Effect of Natural Pozzolan on Carbonation of Mortars and Concretes

Carbonation is the reaction of carbon dioxide with all of the hydrates of cement paste. The dominating reaction is with calcium hydroxide to produce calcium carbonate.

$$CO_2 + Ca(OH)_2 \rightarrow CaCO_3 + H_2O.$$

Carbonation of dense and compact concrete decreases the total porosity of concrete and hence improves the microstructure of cement paste. This enhances the durability and strength of concrete by lowering the permeability (see Fig. 1.17).

However in reinforced concrete, carbonation reduces the durability. Reaction of carbon dioxide with calcium hydroxide reduces the alkalinity of mortar and concrete and increases the risk of corrosion of bars and spalling of concrete cover.

The progress of carbonation in concrete depends on environmental conditions and concrete quality. Relative humidity, carbon dioxide concentration in the air, an temperature are the most important parameters affecting carbonation. Water

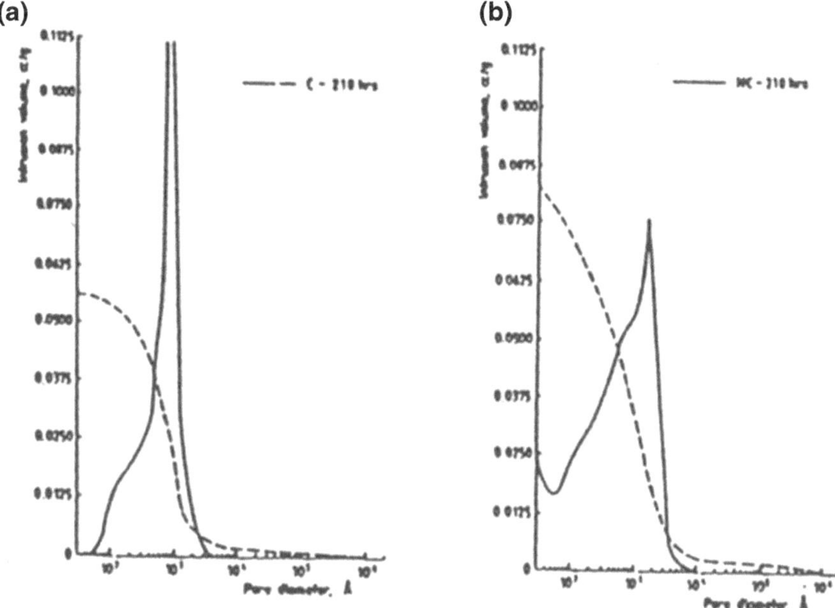

Fig. 1.17 Pore size distribution of a trass mortar mix. **a** Non-carbonated part. **b** Carbonated part

cement ratio, permeability and moist curing duration are the main factors affecting the concrete quality and hence its carbonation depth.

It is generally believed that carbonation of concrete containing natural pozzolan is higher than that for the plain concrete. It is attributed to the lower portlandite content of the pozzolanic cement paste. However, some researchers have reported similar carbonation for both pozzolan and non pozzolan concretes having the same compressive strength at 28 days.

The slow pozzolanic reaction of most natural pozzolans results in higher carbonation when compared with the normal Portland cement concretes at early ages. The length of moist curing of pozzolanic concretes also alters their carbonation depths. The lower the length of moist curing is the higher the carbonation depth in pozzolanic cement concretes. Figure 1.18 depicts the effect of curing on the carbonation of concretes containing a natural pozzolan.

In an accelerated carbonation measurement of concretes containing natural pozzolan, the depth of carbonation was found higher when compared with the concretes maintained in normal atmospheric condition. This can be attributed to the higher concentration of carbon dioxide, lower relative humidity and above all shorter curing in an accelerated test methods. Figure 1.19 shows the accelerated carbonation of concretes containing a natural pozzolan in comparison with other pozzolans [44].

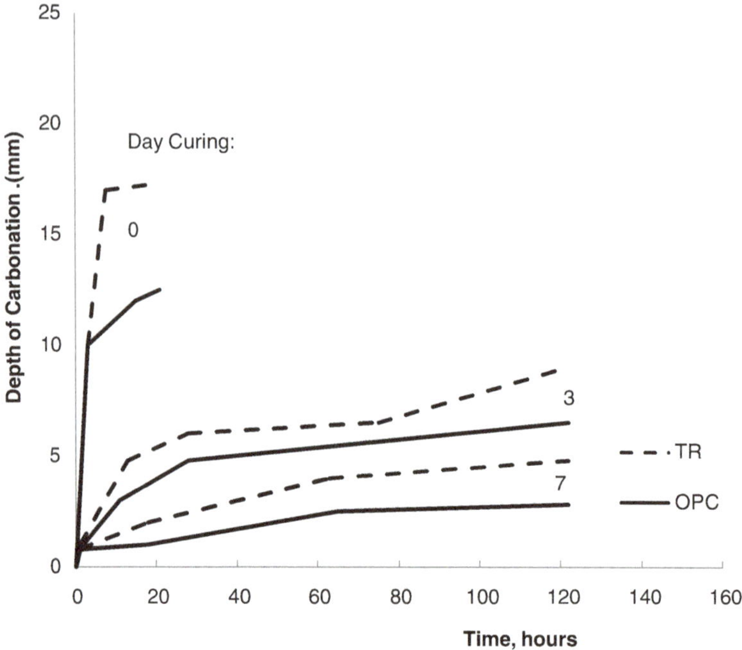

Fig. 1.18 Effect of curing time on the carbonation of opc and trass mixes carbonated at 90 % R.H. and 90 % CO_2 concentration

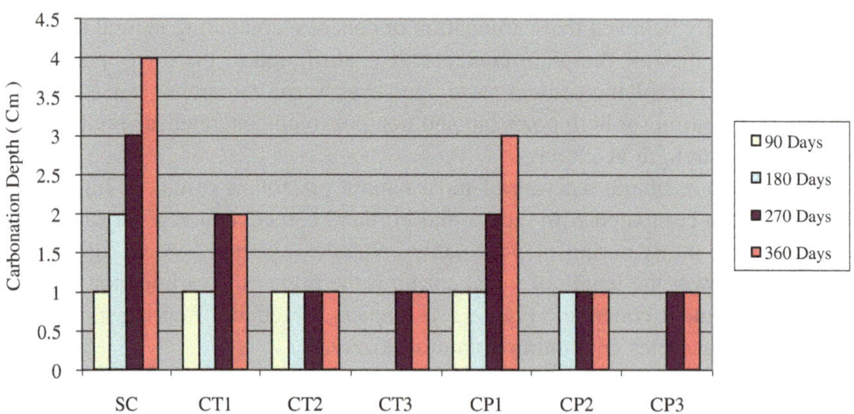

Fig. 1.19 Depth of carbonation-time relationships for mixes carbonated at 50 % R.H., 90 % CO_2 concentration and not cured

Effect of Chloride Ions on Durability of Pozzolanic Cement Mortars and Concretes

Chloride ions affect the durability of both reinforced and non reinforced concretes. Contaminated materials can increase the internal chloride content of concretes while the brackish water from sea water or any other salty water are usually the external sources of chloride ions. Penetration of chloride ions into concrete leads to the higher leaching of calcium hydroxide and hence higher porosity and permeability of concretes. Deposition of chloride crystals and further swelling and expansion decreases the strength properties of concrete. The effect of calcium chloride is more deleterious than that of the sodium chloride due to the formation of higher expansive crystals of calcium chloride compounds.

As stated earlier the replacement of cement with natural pozzolan reduces the diffusion of chloride ions at later ages. Results of the rapid chloride penetration test (RCPT) of concretes containing natural pozzolans are shown in Table 1.12. It is clearly seen that with a sufficient moist curing and at longer ages, pozzolanic mortars and concretes perform better than the normal cement concretes [45]. All concrete mixtures containing 10 %(CT1, CP1), 22.5 %(CT2, CP2), and 30 %(CT3, CP3) natural pozzolans have shown a considerable reduction in Chloride diffusion when compared with that of control concrete (SC).

Chloride profile of concrete mixtures containing trass as a natural pozzolan after 5 years in the tidal zone of the Persian Gulf region is shown in Fig. 1.20. This clearly shows lower chloride penetration with the incorporation of pozzolans.

In the investigation of long term performance of concrete mixtures containing trass, the electrical resistance and micro-cell current density were measured at 1, 4 and 9 years [46]. Result at 4 years in tidal region is shown in Table 1.13. Concrete containing trass cement depicts lower current density and higher electrical resistivity than that of the control concrete. Therefore less corrosion of reinforcements is expected in trass concrete at longer ages.

Ramezanianpour et al. [33] also studied the durability of concretes containing zeolite as a natural pozzolan. As shown in Fig. 1.21, electrical resistivity of specimens considerably increases versus time. It is clear that mixtures containing zeolite have higher resistivity than control mixtures at various w/c ratios. The

Table 1.12 RCPT test results (coulomb)

Code	90 days	180 days	270 days	360 days
SC	3311	3008	2754	2405
CT1	2401	2159	2002	1894
CT2	2311	2042	1896	1693
CT3	2154	1853	1745	1596
CP1	2673	2211	1915	1887
CP2	2400	2010	1705	1552
CP3	2443	2018	1805	1608

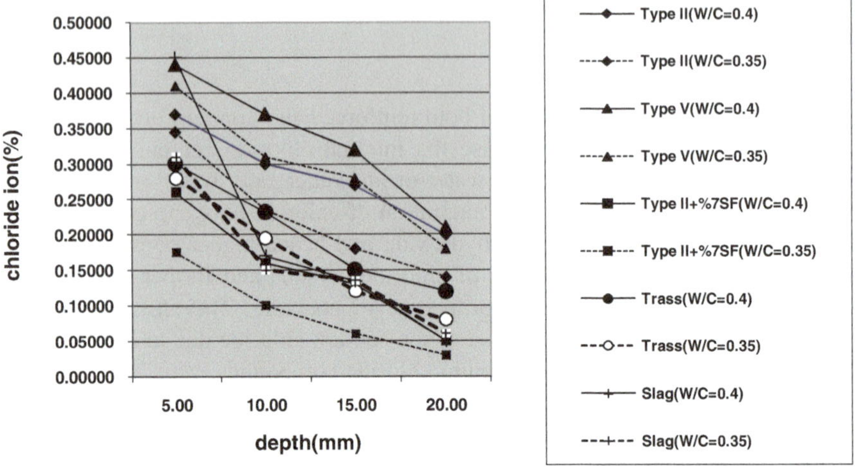

Fig. 1.20 Profile of acid soluble chloride ion concentrations (w/cm) = 0.35, in the tidal region at 5 years

Table 1.13 Current density and concrete resistance at 4 years in the tidal region

Mixture	w/cm	Current (μA/cm^2)			Resistance (kohm)		
		3.5 cm	5 cm	7 cm	3.5 cm	5 cm	7 cm
Control	0.40	2.84	2.38	2.79	0.8	1.2	1.2
Control	0.35	2.03	1.56	1.22	0.8	1.5	1.7
Trass	0.40	1.90	2.37	1.68	1.1	1.3	1.5
Trass	0.35	1.99	1.74	1.55	1.3	1.4	1.8

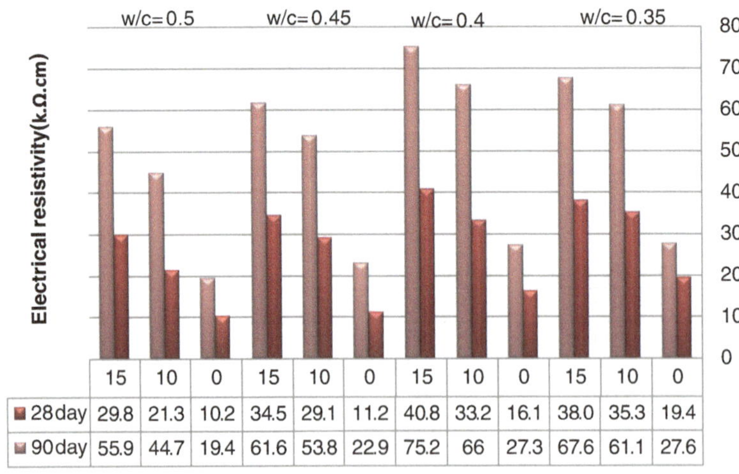

Fig. 1.21 Electrical resistivity of control and zeolite concretes at different ages with various w/c ratios [33]

resistivity of concrete specimens increases when zeolite content up to 15 % increases. According to the report of CEB committee it can be seen that the corrosion rate of reinforcing steel is low when the resistivity of concrete exceeds 20 KΩ-cm. All the zeolite mixtures have resistivity higher than 20 KΩ-cm.

Sulfate Resistance of Mortars and Concretes Containing Natural Pozzolan

Sulfate ions such as ammonium, magnesium, sodium and calcium react with cement compounds and cement hydrates. They may cause expansion, loss of strength and produce musky mass with no resistance. Ammonium sulfate [$(NH_4)_2SO_4$] is the most harmful sulfate to cement based materials. Magnesium sulfate ($MgSO_4$) reacts with all cement compounds including calcium silicate hydrates to form brucite [$Mg(OH)_2$)] and gypsum ($CaSO_4 \cdot 2H_2O$) giving ettringite at the later ages. Calcium sulfate ($CaSO_4$) reacts with calcium aluminate hydrates to produce expansive ettringite ($3CaO \cdot Al_2O_3 \cdot 3CaSO_4 \cdot 32H_2O$). Sodium sulfate ($Na_2SO_4$) reacts on calcium hydroxide to produce expansive gypsum which is in the presence of aluminates may also produce ettringite.

Natural pozzolanic cements generally increase the resistance of concrete in sulfate bearing soils, ground water, natural acid water, and sea water. Combination of sulfate resistant cement and a natural pozzolan may not increase sulfate resistance and may also reduce the resistance if active aluminum compounds are present in the pozzolan.

Replacement of natural pozzolans with Portland cement reduces both the calcium aluminate in cement and calcium hydroxide in the hardened cement paste which are major compounds responsible for the sulfate attack.

Sulfate resistance of various natural pozzolans has been investigated at the USBR since 1974 [47]. The results of accelerated tests in 2.1 % sodium sulfate solution to predict the service life of various concretes are shown in Fig. 1.22. From the test results up to 24 years, it was shown that 8 years of continuous service exposure is comparable to one year of accelerated testing using a criteria of 0.5 % expansion or 40 % reduction in the modulus of elasticity. By extrapolation of the results, it is possible to estimate of the concretes in sulfate solutions.

Figure 1.23 shows the effect of various percentages of an Italian natural pozzolan on the expansion of 1:3 mortar samples stored for more than 5 years in 1 % magnesium sulfate solution. Significant reduction in the expansion of mortars containing natural pozzolan is attributed to the reduction of calcium hydroxide and also the lower permeability of such mixtures when compared with the plain mortars [29].

In an investigation the replacement of 30 % calcined volcanic glass with Portland cement reduced the expansion of Portland cement having a C_3A content ranging between 9.4 and 14.6 % in sulfate solution. The expansion of mortars was

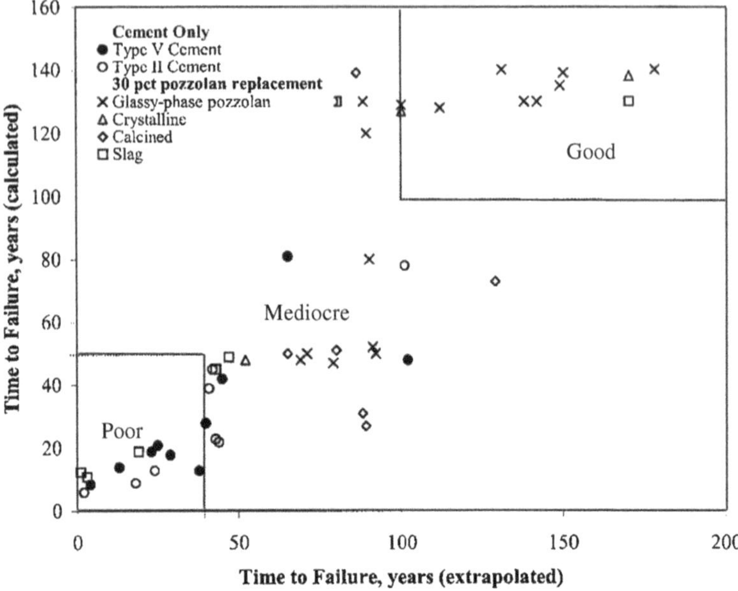

Fig. 1.22 Accelerated sulfate resistance tests to predict service life of concrete [47]

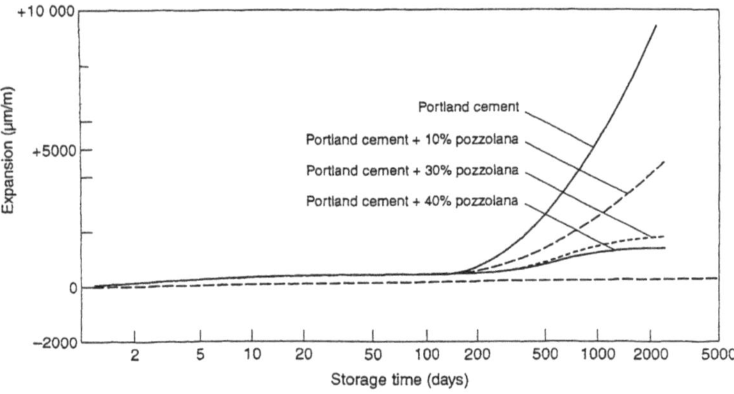

Fig. 1.23 Effect of substituting Portland cement for pozzolan on the expansion of 1:3 Mortar samples 2 × 4 × 25 cm stored in 1 % $MgSO_4$ solution

less than 1 % after one year exposure to sodium sulfate solution [48]. The strength of the mortar samples containing natural pozzolans increases usually increases at early ages in sodium sulfate solution but reduces at longer ages [49].

The sulfate attack of magnesium sulfate is considered to be more severe than that of sodium sulfate. Natural pozzolans usually increase the resistance of concretes and mortars in sodium sulfate exposure, but they may worsen the performance of sulfate resisting Portland cements in magnesium sulfate solutions. This opposite action depends on the type and properties of natural pozzolans.

Combination of Sulfate and Chloride Attacks by Sea Water

Sea water contains several salts including sulfate and chloride ones. It is agreed that sulfate salts of the sea water affect the durability of concrete while the cause of corrosion of reinforced concretes is due to chloride ion penetration of sea water. The attack on concrete is observed by expansion, followed by cracking, spalling and finally mass loss. Deterioration of concrete by sea water is less than that from sea composition and from the sulfate and magnesium contents. The reason is the reactions of bicarbonate and magnesium ions of the sea water with the cement hydrates and formation of calcium carbonate (aragonite) and brucite. These reaction products which are insoluble fill the pores and reduce the penetration of other ions including chloride and sulfate ions into the mortars and concretes [50]. Besides, gypsum and ettringite with their expansive formation are more soluble in chloride bearing solutions and less stable in an environment having less pH as observed in the brucite formation.

Replacement of natural pozzolans with Portland cement at sufficient level usually reduces the expansion of concretes in sea water. The replacement of up to 20 % of a natural pozzolan with cement slightly improved the expansion of mortars in artificial sea water [51].

Long term service life and better performance of concretes containing natural pozzolans have been investigated in many projects [52]. In an investigation of the durability of concretes containing trass as a natural pozzolan in the Persian Gulf Region, chloride penetration, corrosion of reinforcement, sulfate attack and carbonation of reinforced concrete specimens were assessed [53]. Results after 4 years in the exposure site show a very good performance for the trass concrete when compared with the control concrete. Sulfate attack and carbonation depth were very low and penetration of chloride ions was only a few millimeter. There was no sign of corrosion of reinforcements even after 9 years while corrosion occurred on the reinforcement bars of control concrete made with type 5 Portland cement.

Effect of Natural Pozzolans on Suppressing the Alkali Aggregate Reaction

Deterioration of concrete due to alkali silica reaction was first discovered by Stanton in 1940 [54]. The most common reaction is the reaction between certain type of silica in siliceous or siliceous limestone aggregates and cement alkalies.

This reaction forms a gel which should tend to absorb water from the environment and to expand. As a result, map cracking, gel exuding from cracks, pop-out and spalling occur on concrete elements. The high alkali cement, reactive aggregate and high humidity and moist environment are the three major conditions for the reaction to take place. Expansion also increases by increasing the cement content and the water to cement ratio in concrete mixtures [55, 56].

The term alkalies refers to the sodium and potassium phases of cement and expressed as sodium oxide equivalent which is sum of the percentage of sodium oxide and 0.658 times the percentage of potassium oxide. The use of low alkali cement (less than 0.6 % of sodium oxide alkali equivalent) can control the expansion of reactive aggregates. However lower alkali content and less than 0.6 % sodium oxide equivalent alkali is necessary for certain very reactive aggregates. Alkalies can also come from external sources such as sea water and de-icing salts. In such applications, even the low alkali cement may not prevent the deterioration of concrete due to delayed disruptive alkali aggregate reaction [57].

The expansion of concretes due to alkali silica reaction may also be mitigated by the addition or substitution of finely reactive materials such as natural pozzolans with Portland cement [58].

The effect of natural pozzolans on both alkali silica and alkali carbonate reactions has not been fully investigated. It is said that insufficient proportion of natural pozzolan may increase the detrimental effects of the alkali silica reaction [59]. However, expansion reduction has been observed in many concrete mixtures when natural pozzolans and volcanic tuffs both in the raw state and after thermal treatment have been used [60–62].

It has been found that the percentage by solid volume of the pozzolan needed to replace Portland cement for adequate reduction of expansion varied from 20 % with diatomite and between 20 and 30 % with calcined shale [60]. For volcanic glasses, the level of replacement between 30 and 35 % was necessary for such reduction in expansion. In an investigation shown in Fig. 1.24 for the control of alkali aggregate reaction, Portland cement with 1 % equivalent sodium oxide blended with 20–30 % Santorin earth as a natural pozzolan was the appropriate level of replacement [15].

The effect of other types of natural pozzolans on reducing the alkali aggregate reaction is also shown in Figs. 1.25 and 1.26 [47, 61].

In an investigation it was found that the use of 20–40 % zeolite or fly ash was similarly efficient in controlling deleterious expansion of mortar containing highly reactive aggregate [63].

In an investigation the results of concrete prism test indicates that the expansion of specimens can be controlled by addition of trass as a natural pozzolans at 20 and 30 % replacement [63]. This is clearly seen in Fig. 1.27. In general, the effect of natural pozzolans to suppress alkali silica reaction depends on the reactivity of aggregate, the level of natural pozzolan replacement, the equivalent alkali content of cementitious composition and the activity index of natural pozzolan. The maximum reduction of concrete prism expansion compared with control sample increases with an increase in activity index of natural pozzolan. It was also

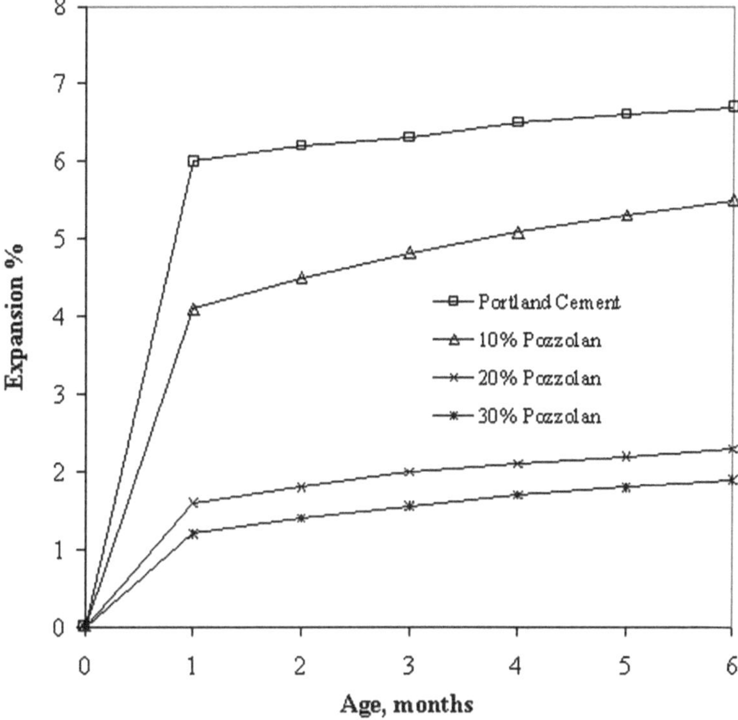

Fig. 1.24 Control of alkali-silica expansion by Santorin earth [15]

concluded that the expansion of specimens is declined by increase of natural pozzolan replacement level, while in some cases a pessimum amount in natural pozzolan replacement level exists.

The reduction of expansion due to alkali silica reaction of blended cements with pozzolans is attributed to the following properties of mortars and concretes containing such materials:

- Lower alkalinity and pH of the pore solution
- Higher alkali content of hydration products
- Lower calcium hydroxide content
- Lower calcium to silicate (C/S) ratio of calcium silicate hydrate (CSH)
- Lower permeability and hence lower ion mobility.

Replacement of Portland cement by pozzolanic materials generally reduces the alkali and OH concentration in the pore solution. There might be some pozzolans which increase the alkalinity of the solution and may not be able to reduce the expansion of reactive aggregate concretes. Investigation shows that if the alkali concentration falls below a certain safe level, deleterious expansion does not occur. The threshold of alkali or hydroxide concentration which is suitable for preventing

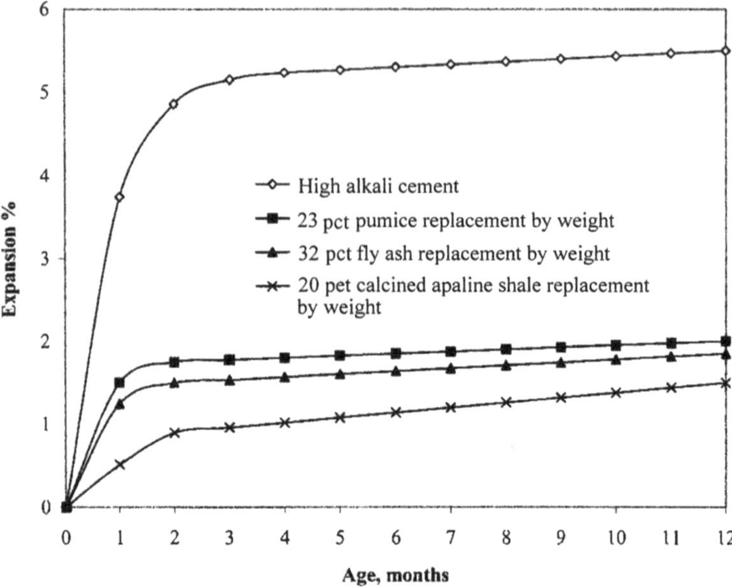

Fig. 1.25 Effect of pozzolan on reactive expansion of mortar made with cement and crushed Pyrex glass sand [47]

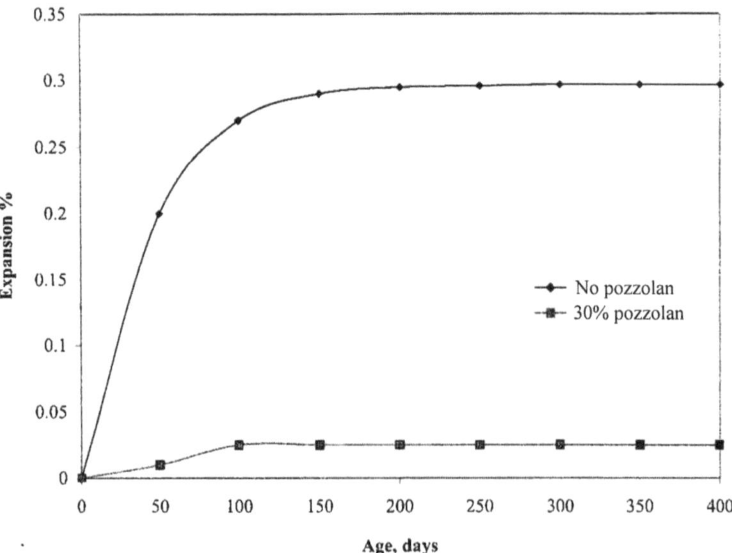

Fig. 1.26 Effectiveness of pozzolan in reducing expansion due to alkali-silica reaction [61]

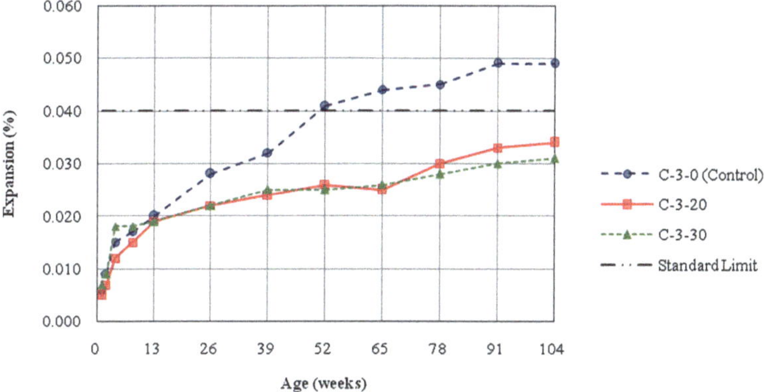

Fig. 1.27 Effect of a natural pozzolan on expansion reduction of concrete prisms up to 2 years [72]

deleterious expansion due to AAR is not yet fully established. It varies between 0.2 and 0.65 N for different aggregates. The decrease in the alkali content of the pore solution is parallel with the increase in the alkali content of the solid phase. Consumption of calcium hydroxide by pozzolans reduces the C/S ratio of CSH and this allows more alkalis to be incorporated in calcium silicate hydrates. The combination of both calcium hydroxide and alkalis reduces the pH of the solution. This could account for the expansion reduction of the mortars and concretes. This effect has also been observed in some pozzolans having high alkali content.

Hossain investigated the alkali-silica reaction, autoclave expansion, and strength activity index of Portland cement were volcanic ash (VA) and pumice powder (VP) were used as additives. Cement was replaced in varying percentages by VA and VP (0–40 %).

Alkali-silica reaction and strength activity index tests were conducted as per ASTM C 311, and autoclave expansion tests as per ASTM C 151.

Based on the test results, he concluded that (i) for the effectiveness in the alkali-silica reactions as per ASTM C 618, the expansion (the ratio between the length change of the test mix and that of the control mix) at day 14 should have a maximum value of 100 %. Mix with 10 % VP or VA content showed the maximum length change and the highest reduction in length value. Mixes with 30 % and 40 % VP or VA content, on the other hand, satisfied the ASTM C 618 requirement; (ii) The autoclave expansion of the paste mixes decreased with the increase in VP and VA content. All mixes containing up to 40 % VP or VA, satisfied the ASTM C 618 requirement of 0.8 % maximum; and (iii) the strength activity index (SAI), which is the ratio between the strength of the tested samples and the strength of the control samples, is calculated at 7 and 28 days. The SAI value ranged between 78 and 100 % at 7 days and between 67 and 100 % at 28 days. SAI values for mixes with 10, 20 and 30 % VA or VP content are more than 75 % as required by ASTM C 618 [64].

Freezing and Thawing of Concretes Containing Natural Pozzolans

Frost resistance of concrete mainly depends on the concrete properties and the freezing temperature of the environment. In a permeable concrete, water penetrates through the capillary pores and freezes at temperature below freezing point. The result of the freezing is the expansion of concrete. The opposite action takes place when temperature rises above the freezing point. Concrete deteriorates in the freezing and thawing cycles. Permeability of concrete plays an important role on the frost resistance of concrete. Higher water cement ratio concrete mixtures contain higher capillary pores and hence lower resistance to the cycles of freeze–thaw action. Curing of concrete before exposure to the cold environment is also necessary to reduce the capillary pores and to increase the frost resistance. The best solution for increasing the durability of concretes against freeze–thaw damage is the addition of air entraining agents. Distribution of air bubbles in hardened concrete control the expansion of ice formed from freezing water and enhances the freeze-thaw durability.

Researches investigating the effect of natural pozzolans on freezing-thawing action show similar performance as for plain cements with similar strengths and air bubble contents. The only difference is a longer moist curing needed for the concrete containing natural pozzolans than that for the plain concrete before exposure to the cold environment. Long term performance of concrete dams subjected to several cycles of freezing and thawing was satisfactory. Most of these dams were built with cements containing natural pozzolan [65]. In some cases when sufficient moist curing was applied to the pozzolanic cement concretes before exposure to frost attack, higher resistance was observed at equal compressive strength to the plain concrete mixtures [66].

An investigation carried out to compare the scaling test resistance of type 2 Portland cement concrete with concrete containing a natural pozzolan and limestone filler.

Results show that the performance of type 2 Portland cement is more appropriate than the composite cement for mixture without air bubble in salt scaling test. However, the mixture containing natural pozzolan cement with entrained air bubbles shows the best performance in salt scaling test [33].

Application of Natural Pozzolans in Mortars and Concretes

Natural pozzolans in combination with lime or Portland cement have been used in several applications for many years. As a grout system it has been used for contact and curtain grouting of tunnels and dams, soil stabilization and slope protection and for preplaced aggregate concrete. Lower heat of hydration, lower permeability, higher resistance to sulfate attack and alkali aggregate reaction and better cohesiveness are the technical improvements in the application of grouts containing natural pozzolans. It has been used as a masonry cement for masonry and other structures. Manufactures of concrete masonry units have also used natural pozzolans as cement replacement materials in their production. In such applications, steam curing has been used for achieving high early strength to solve the slow reaction of natural pozzolans. In cold regions and in order to increase the resistance of masonry units to freezing and thawing effect, the use of air entraining admixture is recommended. Natural pozzolanic cements have been widely used in sulfate environment and in the severe conditions of sewage systems. Low permeable concrete pipes manufactured with natural pozzolanic cements have been used in sewage systems in some countries and performed very well in such acidic environments.

Several reports have been published on natural pozzolan use in mass concrete [33, 66]. The main advantage of the application of natural pozzolans in mass concrete is the lower heat of hydration of natural pozzolanic cements and hence controlling the thermal cracking. In addition, increased sulfate resistance and reduction of alkali aggregate reaction of concretes containing proper proportions of natural pozzolans are other reasons for widely usage of them in construction of dams as massive concrete structures. In the USA natural pozzolans have been used in the construction of Friant, Bonneville, Davis, Glen Canyon, Flaming George, and John Dams [67]. In the construction of Jagin roller compacted dam in the south of Iran, a natural pozzolan up to 30 % replacement was used. Figure 1.28 shows a picture of Jagin dam constructed with roller compacted concrete.

Recent Researches on Natural Pozzolans

Walker and Pavı conclude that amorphousness determines pozzolan reactivity to a much greater extent than any other pozzolan property. It also concludes that the specific surface area of the pozzolan governs the water demand of the paste, while amorphousness determines the strength of the paste. In contrast, the chemical composition of the pozzolan is not instrumental as a variable affecting neither pozzolan reactivity nor strength.

The results evidenced that the replacement of lime by pozzolan lowers the water demand of the paste. The flow test clearly subdivided the pozzolans into two groups of high and low water demand; and evidenced that the pozzolan's specific surface has a much greater influence on water demand than its particle size, amorphousness or the amount of pozzolan in the mix.

From the chemical index, the paper also concludes that the dominant factor that determines the initial activity of a pozzolan in suspension is amorphousness rather than SiO_2 content or specific surface. In addition, it appears from the results that the amount of lime combined by reactive crystalline phases in the pozzolans is insignificant when compared to that bound by their amorphous fraction. The behavior and properties lime/pozzolan pastes are determined by complex

Fig. 1.28 RCC Jagin Dam in Iran

relationships of interdependent variables including pozzolan surface area, particle size, chemical composition and amorphousness [68].

In Askarinejad's work natural pozzolan nanostructures were fabricated by a direct top-down sonochemical route from micrometer natural pozzolan, for the first time. Thermal gravimetry method was applied to evaluate the pozzolanic reactivity of natural pozzolans before and after the ultrasound process. TG/DTA results showed superior reactivity of natural pozzolan nanostructures to initial natural pozzolans. Compressive strength test results also showed superior strength of cement containing nano natural pozzolan to the specimen containing initial natural pozzolan. Concerning the above results, using ultrasound irradiation can be suggested as an effective method for nanomodification of natural pozzolans and also similar materials in concrete technology [69].

Bondar etc. show that the percentage of reacted pozzolans increases due to adding mineral additives with additional silicon and calcium sources while adding mineral additives with additional aluminium source does not increase the reactivity. In addition, the deficiency of oxides such as SiO_2, Al_2O_3 and CaO in a natural pozzolan such as Taftan can be compensated for by adding mineral additives to increase the level of oxides in the final product before activation [70].

In other research, Bondar etc. show that pozzolans in their natural state and after calcination can be activated and condensed with sodium silicate in an alkaline environment to synthesize a high performance cementitious construction material with a low environmental impact. They show that by correlating the parameters which give rise to the relationship of pozzolanic activity with compressive strength, a pozzolan containing the sodium zeolite clinoptilolite, can be used to prepare a moderate to high compressive strength binder by both elevated temperature curing and calcination [71].

References

1. R. Malinowski, K. Frifelt, in cooperation with Bonits, Hjerthem, and Floding, Prehistoric Hydraulic Mortar: The Upaid Period 5-4000 years BC: Technical Properties. Document D12, Swedish Council for Building Research, Stockholm, Sweden, 16 pp (1993)
2. F.M. Lea, *The Chemistry of Cement and Concrete* (Chemical Publishing Inc., New York, 1971), pp. 414–453
3. F. Massazza, Chemistry of pozzolanic additions and mixed cements. Il Cemento **1**, 3–38 (1976)
4. P.K. Mehta, Natural pozzolans: supplementary cementing materials for concrete, CANMET-SP-86-8E, Canadian Government Publishing Center, Supply and Ser-Vices, Ottawa, Canada, KIA A0S9 (1987)
5. N. Parravano, V. Caglioti, Research on pozzolanas. La Ricerca Scientifica 271–292 (1937)
6. R.C. Mielenz, L.P. Witte, O.J. Glantz, Effect of calcination on natural pozzolanas-symposium on use of pozzolanic materials in mortars and concretes. American Society of Testing and Materials Special Technical Publication no. 99, pp. 43–92 (1950)
7. Central Board of Irringation and Power Research, Scheme applied to River Valley Projects: Survey of work done on pozzolana in India. Status Report, no. 2, 177 pp (1971)
8. M.S. Akman, F. Mazlum, F. Esenli, A comparative study of natural pozzolans used in blended cement production. in Fourth International Conference on Fly Ash, Silica Fume, Slag and Natural Pozzolans in Concrete, Istanbul, May 1992. American Concrete Institute Special Publication, 132, vol. I, pp. 471–494 (1994)
9. R. Sersale, A.R. Richerche Sulla genesi, Sulla constituzione e sulla reattivita'del "trass" Renano. Silicates Industriels **30**(I), 13–23 (1965)
10. C.E. Lovewell, Pozzolans in highway concrete in admixtures in concrete, Special Report No. 119, High-way Research Board, Washington, D.C., pp. 21–32 (1971)
11. G. Malquori, Portland-Pozzolana cement. in *Proceedings of the 4th International Symposium on the Chemistry of Cement*, Washington. National Bureau of Standards, 1962; Monograph 43(2), pp. 983–1006 (1960)
12. R. Mielenz, L. Witte, O. Glantz, Effect of calcination on natural pozzolans. in *Symposium on Use of Pozzolanic Materials in Mortars and Concretes*, STP No. 99, American Society for Testing and Materials, West Con-shohocken, PA, pp. 43–92 (1950)
13. S. Johnsson, P.J. Andesen, Pozzolanic activity of calcined Moler clay. Cem. Concr. Res. **20**, 447–452 (1990)
14. U. Costa, F. Massaza, Factors affecting the reaction with lime of Italian Pozzolanas. in *Proceedings of the 6th International Congress on the Chemistry of Cement*, Moscow, September 1974, Supplementary Paper, Section III, pp. 2–18 (1974)
15. P.K. Metha, Studies on blended Portland cements containing Santorin earth. Cem. Concr. Res. **11**, 507–518 (1981)
16. U. Ludwig, H.E. Schwiete, Researches on the hydration of trass cements, in *Proceedings of the 4th International Congress on the Chemistry of Cement*, Washington 1960. US Monograph 43(2), pp. 1093–1100 (1962)
17. U. Ludwig, H.E. Schwiete, Lime combination and new formations in the trass-lime reactions. Zement-Kalk-Gips **10**, 421–431 (1963)
18. M. Collepardi, A. Marcialis, L. Massidda, U. Sanna, Low pressure steam curing of compacted lime-pozzolana mixtures. Cem. Concr. Res. **6**, 497–506 (1976)
19. J. Amborise, M. Murat, J. Para, Hydration reaction and hardening on calcined clays and related minerals: V- extension of the research and general conclusions. Cem. Concr. Res. **15**, 261–268 (1985)
20. M. Collepardi, G. Baldini, M. Pauri, M. Corradi, The effect of pozzolanas on the tricalcium aluminate hydration. Cem. Concr. Res. **8**(6), 741–752 (1978)

21. H. Uchikawa, S. Uchida, Influence of pozzolana on the hydration of C_3A, in *Proceedings of the 7th International Congress on the Chemistry of Cement*, Paris, vol. III, pp. IV-24–29 (1980)
22. K. Ogawa, H. Uchikawa, K. Takemoto, I. Yasui, The mechanism of the hydration in the system C_3S pozzolanas. Cem. Concr. Res. **10**(5), 683–696 (1980)
23. H. Uchikawa, Effect of blending components on hydration and structure formation, in *Proceedings of the 8th International Congress on the Chemistry of Cement*, Rio de Janeiro, vol I, pp. 249–280 (1986)
24. P.L. Rayment, The effect of pulverized-fuel ash and the C/S molar ratio and alkali content of calcium silica hydrates in cement. Cem. Concr. Res. **12**, 133–140 (1982)
25. D.L. Rayment, A.J. Majumdar, The composition of the C-S-H phase in Portland cement pastes. Cem. Concr. Res. **12**, 753–764 (1982)
26. F. Massazza, M. Daimon, Conference report: chemistry of hydration of cements and cementitious systems, in *Proceedings of the 9th International Congress on the Chemistry of Cement*, New Delhi, pp. 383–446 (1992)
27. A.A. Ramezanianpour, M. Paydaiesh, Concrete durability and pozzolanic cement role, Building and Housing Research Center, No. 274, Winter 1998
28. A.A. Ramezanianpour, H. Rahmani, Durability of concretes containing two natural pozzolans as supplementary cementing materials, in *7th International Congress on Concrete*, 8–10 July 2008, Dundee, UK
29. F. Massaza, U. Costa, Aspects of the pozzolanic activity and properties of Pozzolanic cement. II Cemento **76**(1), 3–18 (1979)
30. A.A. Ramezanianpour, S.S. Mirvalad, E. Aramun, M. Peidayesh, Effect of four Iranian natural pozzolans on concrete durability against chloride penetration and sulfate attack, in *Second International Conference on Sustainable Construction Materials and Technologies*, 28–30 June 2010, Ancona, Italy
31. A.A. Ramezanianpour, H. Rahmani, Durability of concretes containing two natural pozzolans as supplementary cementing materials, in *7th International Congress on Concrete*, 8–10 July 2008, Dundee, UK
32. J. Costa, F. Massazza, Natural pozzolanas and fly ashes: analogies and differences, in *Proceedings of Symposium N on Effects of Fly Ash Incorporation in Cement and Concrete*, Boston, 16–18 November 1981, Materials Research Society, pp. 134–144
33. A.A. Ramezanianpour, M. Jafari nadushan, M. Peydayesh, Technical, economical and environmental comparisons of Ashfalt and concrete pavements in Iran, in *2nd National Iranian Conference of Concrete*, Tehran, Iran, 7 October 2010
34. N. Pandurovic, Z. Miladinovic, Mortar and concrete shrinkage depending on the kind of cement, in *Proceedings of the International Colloquium on the Shrinkage of Hydraulic Concretes*, Madrid, vol. I, p. II-E (1968)
35. ACI Committee 225. Guide to selection and use of hydraulic cements. J. Am. Ceram. Soc. **11–12**, 901–929 (1985)
36. F. Massazza, Microstructure of hydrated pozzolanic cements, in *1st International Workshop on Hydration and Setting*, Dijon, 1991. E. & F.N. Spon, London, 1992
37. U. Costa, F. Massazza, Permeability and pore structure of cement pastes. Mater. Eng. **1**(2), 456–466 (1990)
38. C.M. Hansson, H. Strunge, J.B. Markussen, T. Frolund, The effect of cement type on the diffusion of chloride. Nordic Concrete Research Publication no. 4, p. 70 (1985)
39. M. Collepardi, A. Marcialis, R. Turriziani, Penetration of chloride ions into cement pastes and concrete. J. Am. Ceram. Soc. **55**, 534–535 (1972)
40. C.L. Page, N.R. Short, El Tarras, A diffusion of chloride ions in hardened cement pastes. Cem. Concr. Res. **11**, 395–406 (1981)
41. S. Chaterji, M. Kawamura, A critical reappraisal of ion diffusion through cement based materials: I- Sample preparation, measurement technique and interpretation of results. Cem. Concr. Res. **22**, 525–530 (1992)

42. A.A. Ramezanianpour, A.R. Pourkhorshidi, A.M. Ramezanianpour, Long term durability of pozzolanic cement concretes in sever environments, in *12th International Congress on the Chemistry of Cement*, Montreal, Canada, July 2007
43. A.A. Ramezanianpour, A.R. Pourkhorshidi, Durability of concretes containing supplementary cementing materials under hot and aggressive environments, in *8th Canmet/ACI International Conference on Fly Ash, Silica Fume, Slag and Natural Pozzolans in Concrete*, Las Vegas, USA, May 2004
44. A.A. Ramezanianpour, J.G. Cabrera, Durability of OPC Trass, OPC-PFA and OPC-Silica fume mortars and concretes, in *13th Conference on Our World in Concrete and Structures*, Singapore, August 1988
45. A.A. Ramezanianpour et al., Diffusion Coefficient of Chloride Ions Under Simulated Conditions, in *6th International Congress on Global Construction*, Dundee, UK, July 2005
46. A.A. Ramezanianpour, A.R. Pourkhorshidi, T. Parhizkar, The role of supplementary cementing materials on durability and sustainability of concrete structures, in *8th International Congress on Civil Engineering*, Shiraz, Iran, 11–13 May 2009
47. R.J. Elfert, Bureau of reclamation experiences with fly ash and other Pozzolans in concrete, Information Circular No. 8640, U.S. Bureau of Mines, Washington, D.C., pp. 80–93 (1974)
48. K. Mather, Current research in sulfate resistance at the waterways experiment station, in *George Verbeck Symposium on Sulfate Resistance of Concrete*. American Concrete Institute Special Publication 77, pp. 63–74
49. V. Ducic, S. Miletic, Sulphate Corrosion resistance of blended cement mortars, ed. by V.M. Malhotra, *Proceedings of the 2nd International Conference on Fly Ash, Silica Fume, Slag, and Natural Pozzolans in Concrete*, Madrid, 1986. American Concrete Institute Special Publication 91: Supplementary Paper no 12, 26
50. F.W. Locher, Influence of chloride and hydrocarbonate on the sulphate attack, in *Proceedings of the 5th International Symposium on the Chemistry of Cement*, Tokyo, vol. III, pp. 328–35 (1968)
51. M. Regourd, Physico-chemical studies of cement pastes, mortars, and concretes exposed to sea water, ed. by V.M. Malhotra, in *Proceedings of the International Conference on Performance of Concrete in Matine Environment*, St. Andrews by the sea. American Concrete Institute Special Publication 65, pp. 63–82 (1980)
52. R. Guyot, R. Ranc, A. Varizat, Comparison of the resistance to sulfate solutions and to sea water of different Portland cement with or without secondary constituents, ed. by V.M. Malhotra, in *Proceedings of the 1st International Conference on the use of Fly Ash, Silica Fume, Slag, and Other Mineral By-Products in Concrete*, Montebello, Canada. American Concrete Institute Special Publication 79, vol. I, pp. 453–469 (1983)
53. A.A. Ramezanianpour, Durability and sustainability of concretes containing supplementary cementing materials, in *Tenth ACI International Conference on Recent Advances in Concrete Technology and Sustainability Issues*, Seville, Spain, October 2009
54. T.E. Stanton, Expansion of concrete through reaction between cement and aggregate. Proc. ASCE **66**, 1781–1811 (1940)
55. S. Sprung, Influences on the alkali-aggregate reaction in concrete, in *Proceeding of the Symposium on Alkali-Aggregate reaction- Preventive Measures*, Reykjavik, pp. 231–244 (1975)
56. J. Krell, Influence of mix design on alkali-silica reaction in concrete, in *Proceedings of the 7th International Conference on Alkali-Aggregate Reaction*, Ottawa, pp. 452–455 (1986)
57. M. Geiker, N. Thaulow, The mitigating effect of pozzolans on alkali silica reactions, ed. by V.M. Malhotra, in *Proceedings of the 4th International Conference on Fly Ash, Silica Fume, Slag, and Natural Pozzolans in Concrete*, Istanbul, May 1992. American Concrete Institute Special Publication 132, vol. I, pp. 533–548
58. T.E. Stanton, Studies of the use of Pozzolanas for counteracting excessive concrete expansion resulting from reaction between aggregates and the alkali in cement, in *Symposium on Use of Pozzolanic Materials in Mortars and Concretes*. American Concrete Institute Special

Publication San Francisco, American Society for Testing and Materials Special Technical Publication (1949)
59. B. Mather, Use of admixture to prevent excessive expansion of concrete due to alkali-silica reaction transportation research record number 1362, Developments in Concrete Technology (Part 2), pp. 99–103, Transportation Research Board; NAS-NRC, Washington D.C. (1993)
60. L. Pepper, B. Mather, Effectiveness of mineral admixtures in preventing excessive expansion of concrete due to alkali-aggregate reaction. Proc. ASTM **59**, 1178–1203 (1959)
61. M.N.A. Saad, W.P. de Andrade, V.A. Paulon, Properties of mass concrete containing an active pozzolan made from clay. Concr. Int. **4**(7), 59–65 (1982)
62. K. Kordina, W. Schiwick, Investigation on additives to concrete for the prevention of alkali-aggregate reaction. Betonwerk + fertigteil – Technik **66**, 328–331 (1981)
63. B. Ahmadi, M. Shekarchi, Use of natural zeolite as a supplementary cementitious material. Cem. Concr. Compos. **32**, 134–141 (2010)
64. K.M.A. Hossain, Volcanic ash and pumice as cement additives: pozzolanic, alkali-silica reaction and autoclave expansion characteristics. Cem. Concr. Res. **35**, 1141–1144 (2005)
65. R.E. Davis, A review of pozzolanic materials and their use in concretes, in *Symposium on Use of Pozzolanic Materials in Mortars and Concrete*, STP 99, American Society for Testing and Materials, West Conshohocken, PA (1950)
66. R.C. Mielenz, Mineral admixtures- history and background. Concr. Int. **5**(8), 34–42 (1989)
67. B. Mather, *Admixtures in Rapid Construction of Concrete Dams* (ASCE, New York, 1971), pp. 75–102
68. R. Walker, S. Pavía, Physical properties and reactivity of pozzolans, and their influence on the properties of lime-pozzolan pastes. Mater. Struct. **44**, 1139–1150 (2011)
69. A. Askarinejad, A.R. Pourkhorshidi, T. Parhizkar, Evaluation the pozzolanic reactivity of sonochemically fabricated nano natural pozzolan. Ultrason. Sonochem. **19**, 119–124 (2012)
70. D. Bondar, C.J. Lynsdale, N.B. Milestone, N. Hassani, A.A. Ramezanianpour, Effect of adding mineral additives to alkali-activated natural pozzolan paste. Constr. Build. Mater. **25**, 2906–2910 (2011)
71. D. Bondar, C.J. Lynsdale, N.B. Milestone, N. Hassani, A.A. Ramezanianpour, Effect of heat treatment on reactivity-strength of alkali-activated natural pozzolans. Constr. Build. Mater. **25**, 4065–4071 (2011)
72. A.A. Ramezanianpour, M. Shafikhani, M. Nili, The effect of natural pozzolans on controlling expansion of concrete due to alkali-silica reaction, in *Proceedings of International Conference on Concrete and Reinforced Concrete*, Moscow (2005)

Chapter 2
Fly Ash

Introduction

Fly ash is a by-product of the combustion of pulverized coal in thermal power plants. The dust-collection system removes the fly ash, as a fine particulate residue, from the combustion gases before they are discharged into the atmosphere.

Fly ash particles are typically spherical, ranging in diameter from <1 μm up to 150 μm. The type of dust collection equipment used largely determines the range of particle sizes in any given fly ash. The fly ash from boilers at some older plants using mechanical collectors alone is coarser than from plants using electrostatic precipitators.

The types and relative amounts of incombustible matter in the coal used determine the chemical composition of fly ash. More than 85 % of most fly ashes comprise chemical compounds and glasses formed from the elements silicon, aluminum, iron, calcium, and magnesium. Generally, fly ash from the combustion of subbituminous coals contains more calcium and less iron than fly ash from bituminous coal. Unburned coal collects with the fly ash carbon particles, the amount of which is determined by such factors as the rate of combustion, the air/fuel, and the degree of pulverization of the coal. In general, fly ash from subbituminous coals contains very little unburned carbon. Plants that operate only intermittently (peak-load stations), burning bituminous coals, produce the largest percentage of unburned carbon.

The term fly ash was first used in the electrical power industry ca. 1930, the first comprehensive data on its use in concrete in North America were reported in 1937 by Davis et al. [1]. The first major practical application was reported in 1948 with the publication by the United States Bureau of Reclamation if data on the use of fly ash in the construction of the Hungry Horse Dam. Worldwide acceptance of fly ash slowly followed these early efforts, but interest has been particularly noticeable in the wake of the rapid increase in energy costs (and hence cement costs) that occurred during the 1970s.

In 1980 [2], Manz reported the 1977 estimated world production of coal ash was 278.443 Mt, of which ~14 % was used. In a recent report dealing with the

A. A. Ramezanianpour, *Cement Replacement Materials*,
Springer Geochemistry/Mineralogy, DOI: 10.1007/978-3-642-36721-2_2,

worldwide production and use of coal ash, Manz [3] indicated that ~562 Mt of coal ash was produced in 1989, of which ~90 Mt or 16.1 %, was used. The total amount used in concrete was about 27.9 Mt consisting of 2.8 % as cement raw material, 7.6 Mt in blend cement, and 17.5 Mt for cement replacement. Compared with 1977 [2], an approximately threefold increase has occurred in the amount used as cement replacement.

In recent years, it has become evident that fly ashes differ in significant and definable ways, reflecting their combustion and, to some extent, their origin. The Canadian Standards Association (CSA) [4] and ASTM [5] recognize two general classes of fly ash:

- Class C, normally produced from lignite of subbituminous coals; and
- Class F, normally produced from bituminous coals.

Physical, Chemical, and Mineralogical Properties of Fly Ash

Physical Properties

Fly ash is a fine-grained material consisting mostly of spherical particles. Some ashes also contain irregular of angular particles. The size of particles varies depending on the sources. Some ashes may be finer or coarser than Portland cement particles. Figures 2.1 and 2.2 show the scanning electron microscope (SEM) micrographs of polished sections of subbituminous and lignite fly ashes [6]. Figure 2.3 shows a secondary electron SEM image of bituminous fly ash particles. Some of these particles appear to be solid, whereas some larger particles appear to be portions of thin, hollow spheres containing many smaller particles.

Fineness

Dry- and wet-sieving methods are commonly used in the measurement of fineness of fly ashes. ASTM designation C311-77 recommends determining the amount of the sample retained when wet sieved on a 45 μm sieve, in accordance with ASTM method C 430, except that a representative sample of the fly ash of natural pozzolan is substituted for hydraulic cement in the determination. Dry sieving on a 45 μm sieve can be performed according to a method established at CANMET [7]. Several countries specify standards for maximum residue (in percentage) retained on a 45 μm sieve as follows [7].

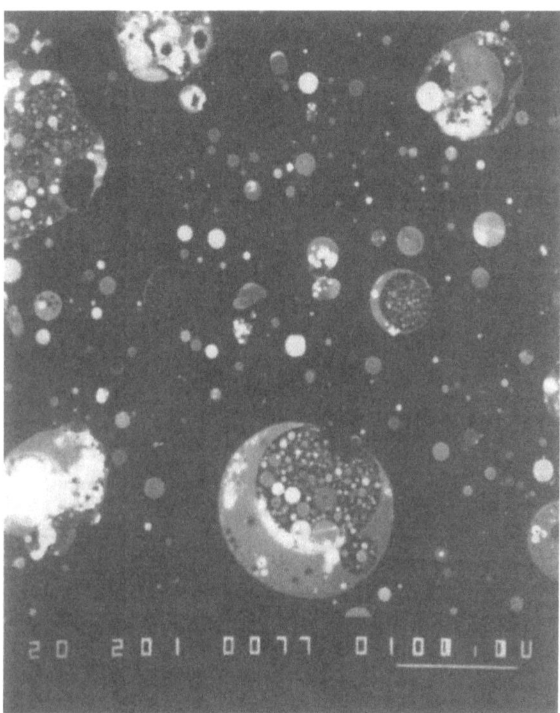

Fig. 2.1 SEM micrograph of a subbituminous ash [6] (Backscattered electron image of a polished section of the dispersed sample)

Germany	50
Australia	50
United States	34
Canada	34
Japan	25
Spain	14
United Kingdom	12.5

Results of SEM and particle-size analysis have shown that spherical and rounded fly ashes vary in size from 1.0 to 150 μm fly ashes of irregular and angular shape are usually larger.

Particle- size distribution of fly ash can be determined by various means, such as X-ray micrograph, laser particle-size analyzer, and coulter counter. In some cases, the agglomeration of a number of small particles may form a large particle. In most cases, fly ashes contain particles of >1 μm diameter. Mehta [9], using an X-ray sedimentation technique, reported particle-size distribution data for several U.S. fly ashes. Mehta found that high-calcium fly ashes were finer than the low-calcium fly ashes, and he related this difference to the presence of larger amounts of alkali sulphates in high-calcium fly ashes.

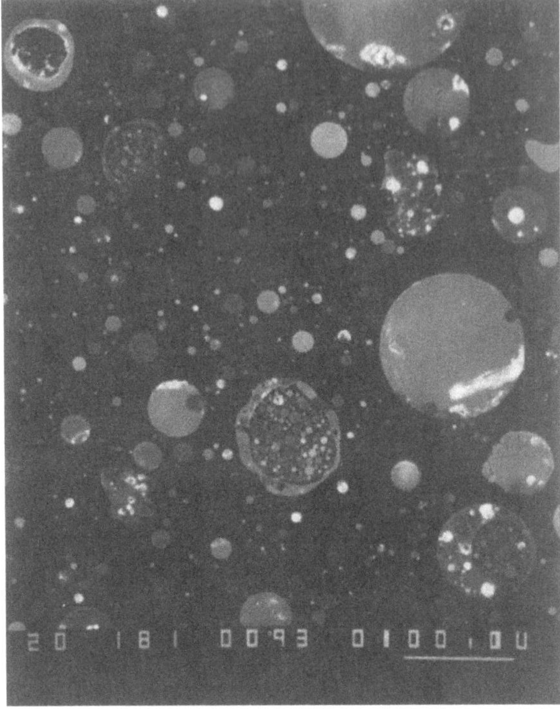

Fig. 2.2 SEM micrograph of a lignite fly ash [6] (Backscattered electron image of a polished section of the dispersed sample)

Specific Surface Area

The specific surface area of fly ash, which is the area of a unit of mass, is measurable by different technique, which measures the resistance of compacted particles to an air flow. ASTM C 204 describes this method for the measurement of the surface area of Portland cement.

Particle-size analysis can also be used for the determination of the specific surface area of fly ash; a laser particle-size analyzer is usually used for the measurement. The Brunauer–Emmett–Teller (BET) nitrogen absorption technique has also been used for determining the specific surface of the particles, but the results obtained by this method are usually higher than the results obtained by the Blaine specific surface-area technique or particle-size analysis.

Cabrera et al. [10] using the Blain technique, particle-size analysis, and the BEST technique, measured and calculated the specific surface area of various fly ashes. The results of their investigation are shown in Table 2.1.

The specific surface values measured by the BEST technique (Table 2.1) are higher than the values obtained with the Blaine technique and particle-size

Fig. 2.3 SEM micrograph of a bituminous ash [6] (Secondary electron image of the sample)

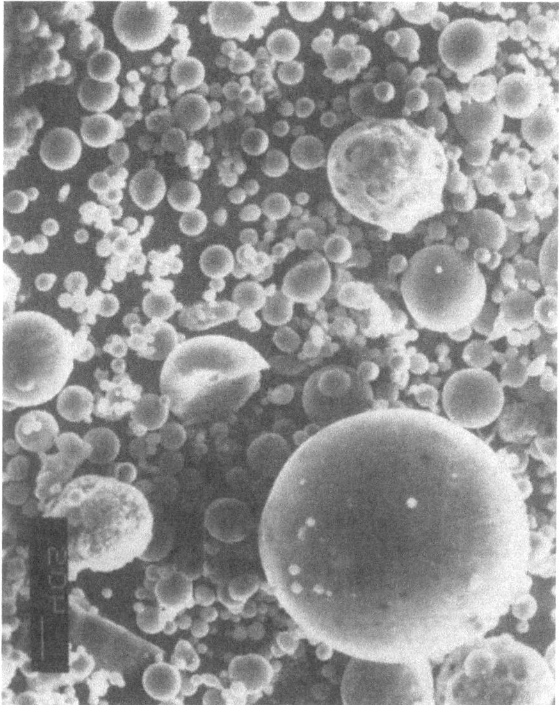

analysis. This large difference is due to the face that BET technique measures the totality of voids in the surface of particles.

In experiments conducted at CANMET [6] the specific surface area measured by the Blaine technique was found to vary from a low value of 130 m^2/kg for a bituminous ash to high value of 581 m^2/kg for a lignite ash (see Table 2.2). Fly ashes collected in electrostatic precipitators have shown surface areas of 400–700 m^2/kg [8]. The specific surface area of cyclone-collected (mechanically collected) ashes varies between 150 and 200 m^2/kg. Some modern electrostatic precipitators have collected fly ashes with a surface area of $\leq$1200 m^2/kg.

Specific Gravity

The specific gravity of hydraulic cement is determined according to ASTM C 188. This test method can also be used to determine the specific gravity of fly ashes. If fly ashes contain water-soluble compounds, the use of a non-aqueous solvent, instead of water, is recommended.

The specific gravity of different fly ashes varies over a wide range, like the other physical properties. In the CANMET investigation of 11 fly ashes [6], the specific gravity ranged from a low value of 1.90 for a subbituminous ash to a high

Table 2.1 Specific surface area of nine fly ashes, measured by three different methods Enrollment in local colleges [10]

Fly ash	Blaine (m^2/kg)	Particle-size analysis (m^2/kg)	BET (m^2/kg)
A	305	81	4070
B	413	97	3820
C	335	115	1020
D	209	92	480
E	193	NA	4700
F	671	102	8900
G	311	81	6500
H	288	NA	1240
I	254	80	970

Table 2.2 Physical properties of fly ashes [6]

Fly ash source	Type of coal[a]	Physical properties			
		Specific gravity (Le Chatelier method)	Fineness (% retained on 45 μm sieve)		Blaine specific surface area (m^2/kg)
			Wet sieving[b]	Dry sieving (Alpine jet)	
1	B	2.35	17.3 (14.9)	12.3	289
2	B	2.58	14.7 (12.7)	10.2	312
3	B	2.88	25.2 (21.7)	18.0	127
4	B	2.96	19.2 (16.6)	14.0	198
5	B	2.38	21.2 (18.3)	16.1	448
6	B	2.22	40.7 (35.1)	30.3	303
7	SB	1.90	33.2 (28.7)	26.4	215
8	SB	2.05	19.4 (16.7)	14.3	326
9	SB	2.11	46.0 (39.7)	33.0	240
10	L	2.38	24.9 (21.5)	18.8	286
11	L	2.53	2.7 (2.4)	2.5	581

[a] *B*, Bituminous; *SB*, Subbituminous; *L*, Lignite
[b] Values in parentheses do not include sieve correction factor

value of 2.96 for an iron-rich bituminous ash. Three subbituminous ashes had a comparatively low specific gravity of ~2.0, and this suggested that hollow particles, such as cenospheres or plerosferes (Fig. 2.1), were present in significant proportions in the three ashes (see Table 2.2).

Chemical Composition

The chemical composition of fly ashes depends on the characteristics and composition of the coal burned in power stations. The chemical analysis of fly ashes by means of X-ray fluorescence (XRF) and spectrometry techniques shows that SiO_2,

Al_2O_3, Fe_2O_3, and CaO are the major constituents of most fly ashes. Other elements are MgO, Na_2O, K_2O, SO_3, MnO, TiO_2, and C.

The chemical analysis of various fly ashes has indicated a wide range of compositions, reflecting wide variations in the coal used in power plants over the world. A CANMET [11] review of the chemical, physical, and pozzolanic properties of fly ash emphasized the wide variations in fly ash compositions.

In the United States, a typical chemical analysis for low-calcium fly ashes (<10 % CaO), usually formed by the combustion of bituminous coal, shows 45–65 wt% SiO_2, 20–30 wt% Al_2O_3, 4–20 wt% Fe_2O_3, 1–2 wt% MgO, $\leq$3 wt% alkalis, and $\leq$5 wt% loss on ignition (LOI) [9, 12, 13]. The high-calcium fly ashes ($\geq$10 % CaO) formed by the combustion of subbitummious and lignite coal typically contain 20–50 wt% SiO_2, 15–20 wt% Al_2O_3, 15–30 wt% CaO, 5–10 wt% Fe_2O_3, 3–5 wt% MgO, $\leq$8 wt% alkalis, and <1 wt% LOI.

ASTM C 311 describes the standard method for sampling fly ash for use as a mineral admixture in Portland cement concrete. According to this standard, silicon dioxide (SiO_2), aluminum oxide (Al_2O_3), iron oxide (Fe_2O_3), calcium oxide (CaO), free CaO, Magnesium oxide (MgO), sulphur trioxide (SO_3), available alkalis, Na_2O and K_2O LOI at 1000 °C, and moisture content at 105 °C must be determined.

Loss on ignition, the weight loss of fly ashes burned at temperatures $\leq$1000 °C, is related to the presence of carbonates, combined water in residual clay minerals, and combustion of free carbon. Carbon is the most important component of LOI. The water required for workability of mortars and concretes depends on the carbon content of fly ashes: the higher the carbon content of a fly ash, the more water is needed to produce a paste of normal consistency.

A comparison of low-calcium and high-calcium fly ashes shows that high-calcium fly ashes usually contain a smaller amount of unburned carbon (<1 %). In the case of low-calcium fly ashes, complete removal of carbon is rare. Indeed, the carbon may be encapsulated in glass, but a major portion appears to occur as cellular particles that have a very large specific surface and are, therefore, able to adsorb significant quantities not only of water, but of chemical admixtures in concrete, such as air-entraining admixtures (AEA), water-reducing admixtures, and retarders.

Several authors have reported the chemical composition of various fly ashes produced in North America. Diamond [14] discussed the range of chemical and mineralogical constituents. This study included both ASTM Type C (high-calcium) and ASTM Type F (low-calcium) ashes. Carette and Malhotra [6], in their study of 11 Canadian fly ashes, indicated a wide range of chemical composition. Table 2.3 gives the chemical composition of each fly ash as determined by inductively coupled argon plasma (ICAP) spectrometry. Manz et al. [15], examined 19 North American lignite fly ashes and characterized these for bulk and intergrain chemical composition. Table 2.4 shows the bulk chemical composition of each ash. The results of CANMET investigations [6] and the data reported by Manz et al. [15], on bituminous, subbituminous, and lignite ashes obtained from various coal-fired power plants in North America show significant differences among the chemical compositions of the fly ashes.

Table 2.3 Chemical composition of fly ashes [6]

Fly ash source	Type of coal[b]	Chemical composition (wt%)[a]												
		SiO_2	Al_2O_3	Fe_2O_3	CaO	MgO	Na_2O	K_2O	TiO_2	P_2O_5	MnO	BaO	SO_3	LOI[c]
1	B	47.1	23.0	20.4	1.21	1.17	0.54	3.16	0.85	0.16	0.78	0.07	0.67	2.88
2	B	44.1	21.4	26.8	1.95	0.99	0.56	2.32	0.80	0.25	0.12	0.07	0.96	0.70
3	B	35.5	12.5	44.7	1.89	0.63	0.10	1.75	0.56	0.59	0.12	0.04	0.75	0.75
4	B	38.3	12.8	39.7	4.49	0.43	0.14	1.54	0.59	1.54	0.20	0.04	1.34	0.88
5	B	45.1	22.2	15.7	3.77	0.91	0.58	1.52	0.98	0.32	0.32	0.12	1.40	9.72
6	B	48.0	21.5	10.6	6.72	0.96	0.56	0.86	0.91	0.26	0.36	0.21	0.52	6.89
7	SB	55.7	20.4	4.61	10.7	1.53	4.65	1.00	0.43	0.41	0.50	0.75	0.38	0.44
8	SB	55.6	23.1	3.48	12.3	1.21	1.67	0.50	0.64	0.13	0.56	0.47	0.30	0.29
9	SB	62.1	21.4	2.99	11.0	1.76	0.30	0.72	0.65	0.10	0.69	0.33	0.16	0.70
10	L	46.3	22.1	3.10	13.3	3.11	7.30	0.78	0.78	0.44	0.13	1.18	0.80	0.65
11	L	44.5	21.1	3.38	12.9	3.10	6.25	0.80	0.94	0.66	0.17	1.22	7.81	0.82

[a] By inductively coupled argon (ICAP) technique, except for Na_2O, K_2O, SO_3, and LOI

[b] *B*, Bituminous; *SB*, Subbituminous; *L*, Lignite

[c] 105–750 °C

Table 2.4 Chemical analyses for North American lignite fly ashes

Fly ash no.	Bulk chemical analysis (wt%)										
	SiO_2	Al_2O_3	Fe_2O_3	Sum	CaO	MgO	SO_3	Na_2O	K_2O	Avail. alkalis	LOI
North Dakota and Montana lignite											
81-271	25.7	15.0	9.2	49.9	26.8	7.2	8.8			1.2	0.3
81-560	30.2	12.5	4.6	47.3	23.6	7.9	9.6	7.3	0.6	5.2	1.8
82-179	42.1	12.0	8.1	62.2	18.5	5.0	4.1	8.0	1.2	3.8	0.4
83-275	45.6	15.5	7.3	68.4	20.3	5.0	1.9	1.0	1.7	0.8	0.1
85-352	39.6	14.0	10.7	64.3	15.9	5.7	2.6	4.4	1.4	2.2	0.1
87-139	27.9	10.7	9.9	48.5	21.6	5.5	12.3	5.4	1.5	3.7	1.4
86-305	35.2	20.3	6.3	61.8	25.0	6.8	1.1	0.2	0.5	0.2	0.3
Saskatchewan lignite											
85-147	50.4	21.4	3.5	75.3	11.6	3.0	0.5	0.9	3.1	3.1	0.5
86-805	46.4	24.5	4.9	75.8	13.7	4.0	0.6	1.6	0.7	0.7	0.1
87-144	47.9	21.9	4.9	74.7	13.3	2.9	1.1	1.0	2.9	2.9	0.1
Texas and Louisiana lignite											
87-146	50.3	20.2	5.5	76.0	14.4	4.0	0.7	0.9	1.2		0.2
87-147	57.9	26.3	3.9	88.1	9.6	2.1	0.4	0.0	0.4	0.3	0.3
87-154	62.3	20.9	2.2	85.3	6.1	0.7	0.5	4.1	2.1	5.5	0.2
87-155	52.2	18.0	10.5	80.7	11.9	2.5	1.3	0.2	1.4	0.5	0.1
87-156	55.5	18.6	4.3	78.4	7.0	0.8	0.3	0.6	1.9	0.3	0.1
87-159	57.5	20.6	7.0	85.1	9.1	2.6	0.2	0.4	1.4	0.3	0.1
87-219	62.0	20.1	2.0	84.1	6.9	1.2	0.6	0.9	0.9	1.5	0.4
87-239	48.9	18.5	21.8	89.1	7.3	2.6	0.5	0.4	0.9	0.8	0.1
87-157	52.8	23.6	8.9	85.3	9.5	2.7	0.4	1.1	0.8	1.6	0.0

Note ASTM C 618 specification limits: Class F fly ash: $SiO_2 + Al_2O + Fe_2O_3 = 70$ %; $SO_3 = 5$ % max; Class C fly ash: $SiO_2 + Al_2O_3 + Fe_2O_3 = 50$ %; $SO_3 = 5$ % max; LOI = 5 % max

Mineralogical Composition

In general, both the type and source of fly ash influence its mineralogical composition. Owing to the rapid cooling of burned coal in the power plant, fly ashes consist of non-crystalline particle ($\leq$90 %), or glass, and a small amount of crystalline material. Depending on the system of burning, some unburned coal may be collected with ash particles.

The X-ray diffraction (XRD) and infrared spectroscopy techniques are usually used for the determination of crystalline phases in fly ashes. The glass phases are determined by the low-angle XRD technique.

In addition to substantial amount of glassy material, each fly ash may contain one or more of the four major crystalline phase: quartz, mullite, magnetite, and hematite. In subbituminous fly ashes, the crystalline phases may include C_3A, C_4A, $\overline{S}$, calcium sulphate, and alkali sulphates [16].

As was observed with chemical composition, the mineralogical composition of fly ashes by Carrette and Malhotra [6] varied over a wide range. Figure 2.4 shows an X-ray diffractogram for one of the bituminous fly ashes. A quantitative determination of the major crystalline phase contained in the fly ashes was also made by a quantitative XRD technique. The results are given in Table 2.5.

Watt and Thorne [17] examined 14 fly ashes by microscopic and XRD techniques and found that most of them, after extraction with water, contained only four crystalline phases in significant amounts: quartz, mullite, magnetite, and hematite.

The reactivity of fly ashes is related to the non-crystalline phase glass. The reasons for the high reactivity of high-calcium fly ashes may partially lie in the chemical composition of the glass, which is different from that of the glass in low-calcium fly ashes. Diamond and Lopez-Flores [18] and Mehta [13] pointed out that

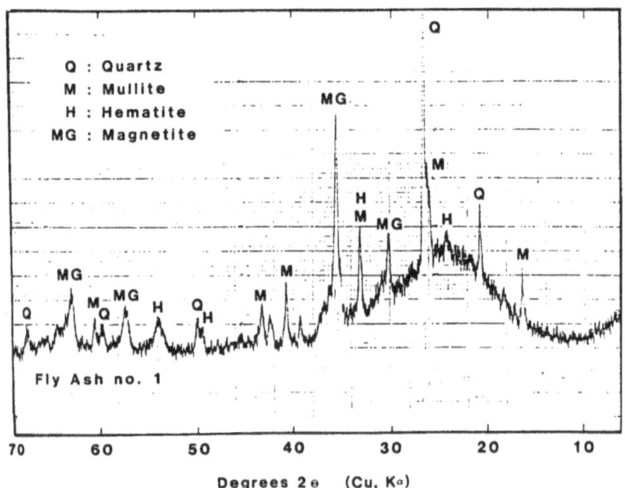

Fig. 2.4 X-ray diffractogram for a bituminous fly ash [6]

Table 2.5 Mineralogical composition of some selected fly ashes [6]

Fly ash source	Type of coal[a]	Phase composition (%)					
		Glass	Quartz	Mullite	Magnetite	Hematite	LOI (%)
1	B	72.1	4.0	12.6	6.2	1.6	3.5
4	B	70.1	3.2	3.3	17.2	4.7	1.5
5	B	55.6	6.2	19.8	5.6	3.1	9.7
6	B	54.2	8.3	23.5	4.4	2.1	7.5
7	SB	90.2	2.9	6.1	–	–	0.8
8	SB	83.9	4.1	10.2	–	1.4	0.4
9	SB	79.8	8.7	11.5	–	–	0.8
10	L	94.5	4.6	–	–	–	0.9

[a] *B*, Bituminous; *SB*, Subbituminous; *L*, Lignite

the composition of glass in low-calcium fly ashes show diffused halo maxima at 21–25° 2θ (CuK_α radiation); high-calcium fly ashes, at 30–34° 2θ.

The Fly Ash Hydration Reactions

High-calcium fly ash, which is mainly composed of glass phase and some crystalline phases (including C_2S, C_3A, $CaSO_4$, MgO, free CaO, and $C_4A_3\overline{S}$) has self-hardening properties. Ettringite, mono sulphoaluminate hydrate, and C–S–H cause hardening of the fly ash when mixed with water. Ghosh and Pratt [20] reported that the hydration behavior of C_3A and C_2S in fly ash is the same as that in cement, but the rate of formation of C–S–H from the glass phase is comparatively slow.

Low-calcium fly ash, which has very little of no self-cementing properties, hydrates when alkalis and $Ca(OH)_2$ are added. The hydration products such as C–S–H, C_2ASH_8 and C_4AH_{13} are formed, and hydrogarnet is produced at a later stage [21]. As more $Ca(OH)_2$ is supplied, more of it is fixed by silica and alumina in fly ash. The degree of hydration of fly ash increased in the presence of gypsum because the surface is activated by the destruction of the structure of the glass and crystalline phases caused by the dissociation of Al_2O_3 reacting with SO_4^{2-}.

The Effect of Fly Ash on the Hydration of Cement Compounds

Figure 2.5 shows the SEM micrograph of the fracture surface of hardened pozzolanic Cement Paste.

The hydration and reaction mechanisms of fly ash and cement compounds, such as C_3S and C_3A are more complicated than fly ash reactions with lime. Takemoto and Uchikawa [19] proposed schematic explanations of the C_3S-pozzolan reaction

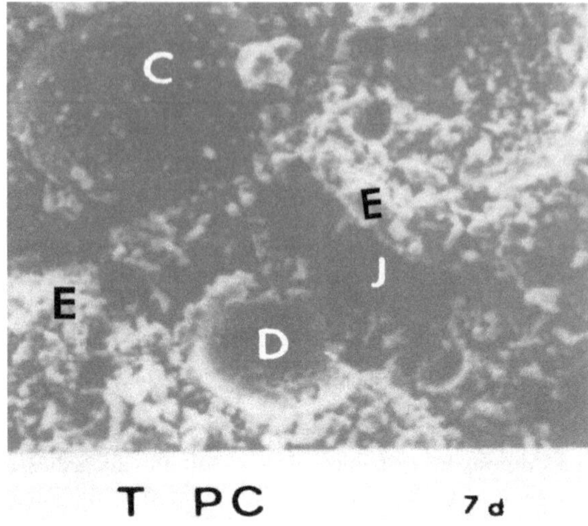

Fig. 2.5 SEM micrograph of the fracture surface of hardened pozzolanic cement paste [21]. *A*, ettringite; *B*, type I C–S–H; *C*, fly ash grain; *D*, cast-off shell of fly ash; *E*, type III C–S–H; *F*, calcium hydroxide; *G*, mono sulphoaluminate; *H*, pozzolan grain; *I*, pozzolan grain; and *J*, capillary space

and C_3A-pozzolan reaction, which are shown in Figs. 2.6 and 2.7, respectively. According to the authors, in C_3S-pozzolan system, calcium ions dissolved from C_3S run about freely in liquid and are adsorbed on the surfaces of pozzolan particles. C–S–H formed by the hydration of C_3S precipitate as the hydrates of high Ca/Si ratio on the surface of C_3S grains and as the porous hydrates of low Ca/Si ratio on the surfaces of pozzolan particles. Attack of the pozzolan surface in water brings about gradual dissolution of Na^+ and K^+, resulting in Si and Al rich amorphous layer on the surfaces. Dissolved Na^+ and K^+ increase the OH^- concentration and accelerate the dissolution of SiO_4^{4-} and layer. Due to the osmotic pressure, the layer swells gradually and the void between layer and pozzolan particle is formed. When the pressure in the void ruptures the film, $SiO_4^{4-}$and AlO_2^- diffuse into the Ca^{2+}- rich solution. Additional C–S–H and Ca–Al hydrate precipitate on the surface of outer hydrates of C_3S particles and to a slight extent on the ruptured film. Vacant space remains inside the film as the hydrates do not precipitate there because of high concentration of alkalies. For pozzolans with low alkalies, destruction of amorphous Si, Al rich film enables Ca^{2+} to move into the inside of the film and precipitate calcium silicate and calcium aluminate hydrates on the surface of pozzolan grain. Therefore no space is observed between pozzolan grains and hydrates.

Figure 2.8 shows the SEM micrograph of the fracture surface of C_3S-pozzolan paste.

The hydration of the C_3A-pozzolan system was first observed by Uchikawa and Uchida [22]. Figure 2.7 illustrates the schematic development of the C_3A-pozzolan system in the presence of calcium hydroxide and gypsum. According to Uchikawa and Uchida, the presence of the pozzoaln accelerates the hydration of C_3A by adsorbing Ca^{2+} from the liquid phase and by providing precipitation sites for ettringite and other hydrates. The C_3A-pozzolan reaction system is similar to

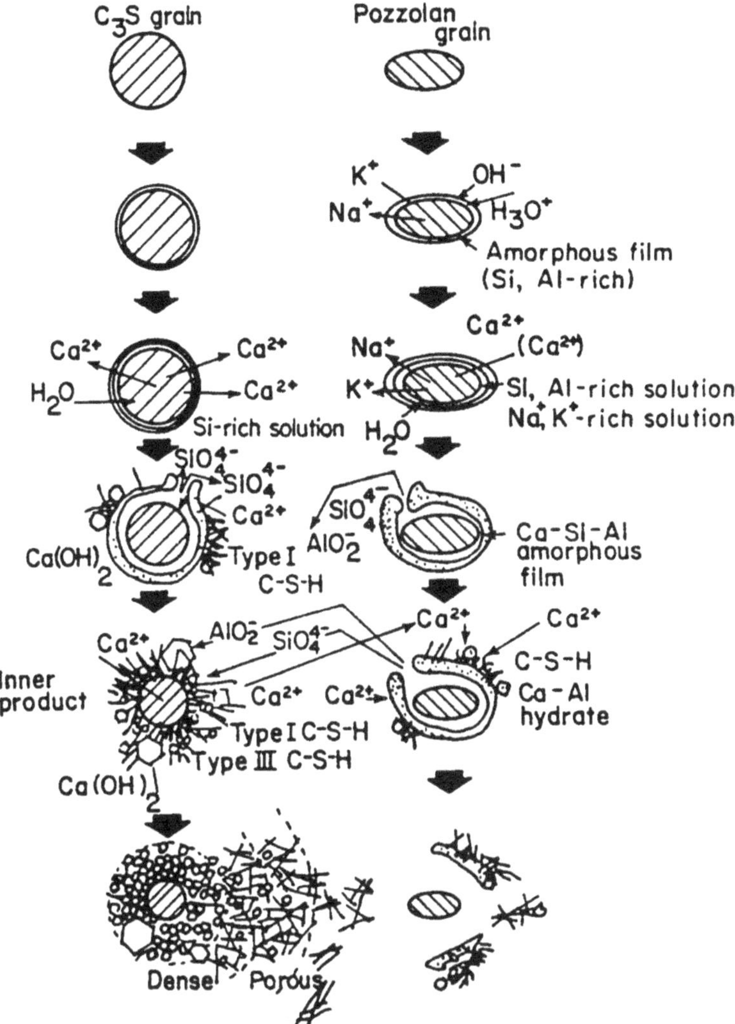

Fig. 2.6 Schematic representation of the mechanism of hydration in the C_3S-pozzolan system [19]

aluminate hydrate, and calcium silicate hydrate are formed on the surface film outside the pozzolan particles of on the surface hydrate layer of the C_3A particles, depending on the concentration of Ca^{2+} and SO_4^{2-} in solution.

As in the case of slag and some natural pozzolans, fly ash retards the hydration of C_3S at stages I and II (Fig. 2.6) and accelerates that at stage III and later (Fig. 2.6). In an investigation [23], C_3S with and without 30 % fly ash, composed of small quantities of quartz and mullite as well as the glass phase, was hydrated at 298 K with a water/cement of 0.5. The degree of hydration of C_3S was determined

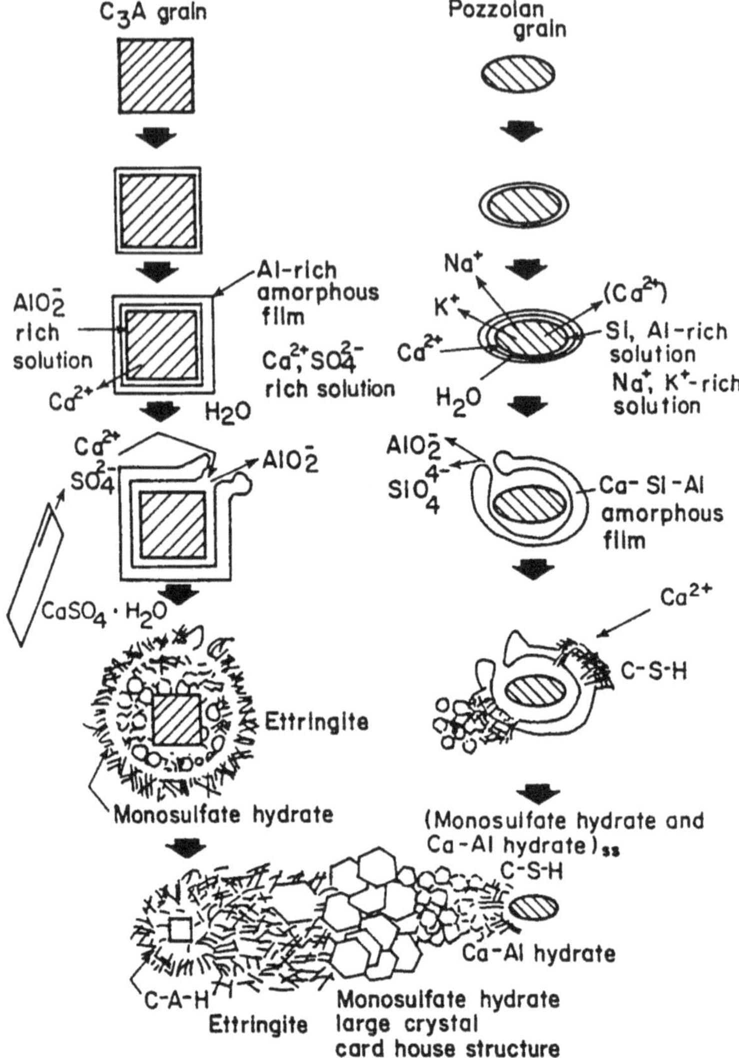

Fig. 2.7 Schematic explanation of the mechanism of hydration in the C_3A-pozzolan system in the presence of $Ca(OH)_2$ and $CaSO_4$ $2H_2O$ [19]

by XRD. The degree of hydration of C_3S alone after 1 day was 35 %, while that of the mixture with fly ash was 45 %. This result obviously shows that the hydration of C_3S at stage III (Fig. 2.6) is accelerated by adding fly ash.

Many researchers [24, 25] have reported that fly ash retards the hydration of C_3A. The degree of this retardation mainly depends on the sulphate content of fly ash, the amount of dissolved alkalis, and the calcium adsorption capacity. Ca^{2+} and $SO_4{}^{2-}$ are dissolved from some high-calcium fly ashes. Dissolved Ca^{2+} is adsorbed

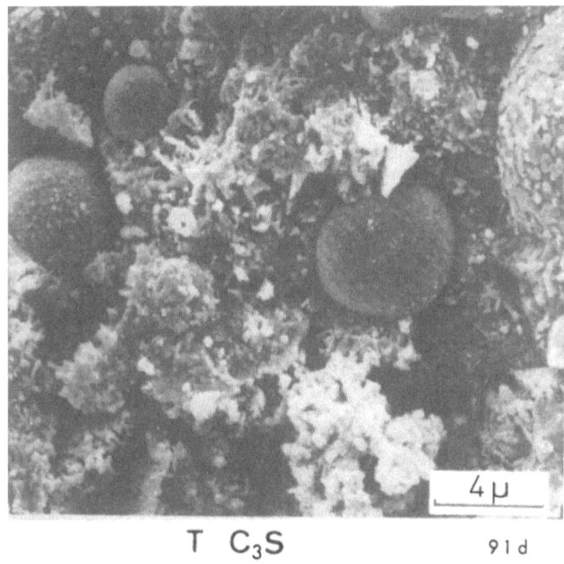

Fig. 2.8 SEM micrograph of the fracture surface of the C_3S-pozzolan paste, showing the clearance between the hydrates [19]

onto the Al-rich surface (produced by non-stoichiometric dissolution), lowering the hydration activity. Furthermore, the adsorption of SO_4^{2-} also retards the hydration of C_3A sulphate in fly ash retards the hydration more than the equivalent amount of added gypsum. This may be due to ions other than sulphate dissolved from the fly ash.

Diamond et al. [26] described the morphological aspect of the early reactions of a fly ash–cement system. Their research showed the occurrence of so-called duplex films, which rapidly develop in hydrating cement systems around exposed surfaces, such as sand grains and coarse aggregates. Such films consist of thin (0.50 μm), uniform, continuous layer of calcium hydroxide deposited quickly on the exposed surface and a thin, single layer of parallel, more or less widely spaced C–S–H gel particles. The gel is usually of type I, with the elongated particles oriented roughly perpendicular to the calcium hydroxide layer and the underlying surface. The total thickness of two layers usually >1 μm. As hydration proceeds, these duplex films may become bonded by other hydration products to other particles. The duplex film was formed on both fly ash and cement particles in a fly ash–cement system after 1 day at room temperature.

Factors Affecting Pozzolanic Reactivity of Fly Ashes

The reactivity of fly ash and other pozzolans with lime or cement is affected by inherent characteristics of the fly ash of pozzolan: chemical and mineralogical composition, morphology, fineness, and the amount of glass phase. External factors, such as thermal treatments and the addition of admixtures, also affect pozzolanic reactivity.

The sum of the silica + alumina + iron of fly ash has been stipulated by ASTM and some other standards associations as a major requirement. The silica + alumina content of fly ashes shows a good correction with long-term pozzolanic activity [27], although silica and alumina in an amorphous form only contribute to the pozzolanic activity, whereas mullite and quartz, which form by partial crystallization of the glassy phases in the fly ash, are nonreactive. Also, in most fly ashes, most of the iron oxide (Fe_2O_3) is present as nonreactive hematite and magnetite. A small amount of iron, which is present in glass, is reported to have a deleterious effect on the pozzolanic activity of fly ashes. Therefore, it has to be separated from silica and alumina when chemical requirements and pozzolanic activity of fly ashes are considered. For the above- mentioned reasons, other researchers [28, 29] have reported poor correlations between the compressive-strengths ratio (according to the pozzolanic activity index) and the sum of silicon + aluminum + iron oxides. These investigators reported that the carbon content did not significantly influence pozzolanic activity index in terms of compressive-strengths ratio.

Fineness of fly ashes is one of the most important physical properties affecting pozzolanic reactivity. The different techniques used for the measurement of the fineness of fly ash have shown different results. It is, therefore, difficult to explain the effect of fineness on pozzolanic activity.

Watt and Thorne [27, 30, 31] examined 14 well-characterized fly ashes to identify which physical and chemical characteristics determine the pozzolanic properties of fly ashes, based on the strength test. These authors developed a theoretical model that shows the relationship between the rate or degree of reaction of fly ash with lime and the fineness of the fly ash. They then compared the calculated values of the reacted material with those obtained experimentally. Their experimental results, however, suggested that the fly ashes with larger median particle size were more reactive than expected from the model. They related this to the spongy form of larger particles; microscopic observation confirmed that the larger particles became perforated during the reaction with lime.

Watt and Thorne [30] also reported that under normal and accelerated curing conditions, there were poor correlations between compressive strength and carbon content, glass content, silica + alumina content, density, and the specific surface area of the fly ash as single variables. However, the best correlation was with the specific surface area as determined by particle-size analysis. In accelerated tests performed on glasses of composition and fineness similar to those of fly ashes, the same conclusion was reached by other researchers [32].

Effects of Fly Ash on the Properties of Fresh Concrete

Fresh concrete is a concentrated suspension of particulate materials of widely differing densities, particle sizes, and chemical compositions in a solution of lime and other components. The system is not static. As soon as the cement and water mix, reactions commence that ultimately produce the binder that consolidates the concrete mass. New particles are formed, and the Original particles dissolve or are coated with cementitious products. The forces of dispersion, flocculation, and gravity compete to determine the spatial distribution of the materials in the changing mass. Heat is released during the chemical reactions, and the temperature rises. In all of these events, fly ash plays some role. Low-calcium fly ash will act largely as a fine aggregate of spherical form; high-calcium fly ash, on the other band, may participate in the early cementing reactions, in addition 10 being part of the particulate suspension.

Because concrete must be mixed and placed, frequently in a heavily reinforced formwork, it is necessary that in most cases a level of fluidity, generally termed workability, be maintained. This is determined by the rheological properties of the system, which are, in turn, influenced by all the components. Control of workability is one of the objectives of mixture proportioning.

Therefore, it is essential to understand the role of fly ash in the rheology of fresh concrete to fully exploit the potential of fly ash for improving concrete.

In practical terms, the fluidity of concrete is expressed by such properties as workability, pumpability, water needed 10 mix, and bleeding. As the use of fly ash has increased, a gradual understanding has started to form of its role in determining these properties. This chapter is largely concerned with these issues and with the other important properties of fresh concrete: temperature rise and air entrainment. Work carried out at CANMET [6], illustrates some of the effects of fly ash on the general properties of fresh concrete. The study dealt with the examination of II fly ashes from widely different sources in Canada. Table 2.3 gives the chemical properties of these ashes. A number of air-entrained concretes were prepared using the simple replacement method, and me mixtures were proportioned to obtain equal water/(cement + fly ash) at a fixed total cementitious-materials content.

In general, it is clear that at a fixed water/(cement + fly ash), slump does not always increase with the incorporation of fly ash. Another important factor revealed by the CANMET work is that the amount of air-entraining admixture (AEA) required to provide 6 % air varied greatly from one ash to another and was not always greater than that required by the control. Both of these issues are discussed in detail in this chapter.

Table 2.6 Results for time-of-set tests

Mixture no.	Nominal 28 days compressive strength (MPa)	Nominal percentage of fly ash	Time of setting (h:min)	
			Initial	Final
1	21	35	6:55	8:30
2	21	45	7:45	9:55
3	21	55	8:45	11:20
4	28	35	7:35	9:25
5	28	45	7:30	9:50
6	28	55	7:55	10:25
7	34	35	6:30	8:15
8	34	45	7:15	9:25
9	34	55	7:00	9:15

Influence of Fly Ash on the Setting Time of Portland Cement Concrete

The rate at which concrete sets during the first few hours after mixing is expressed as initial and final setting time and is determined by some form of penetrometer test. Fly ash may be expected to influence the rate of hardening of cement for a number of reasons:

- The ash itself may be cementitious (high calcium).
- Fly ash may contain sulphates that react with cement in the same way as the gypsum added to Portland cement does.
- The fly ash–cement mortar may contain less water as a consequence of the presence of fly ash, and this will influence the rate of stiffening.
- The ash may absorb surface-active agents added to modify the rheology (water reducers) of concrete, and again this influences the stiffness of me mortar.
- Fly ash particles may act as nuclei for crystallization of cement hydration products [34, 35].

There seems to be general agreement in the literature that low-calcium fly ashes retard the setting of cement. In experiments conducted at CANMET [6], the data obtained (see Table 2.6) show that all except 2 of the II ashes significantly increased both the initial and final setting times. Fly ashes with CaO content of 1.2–13.3 wt % were included in this study.

Effect of Fly Ash on Workability, Water Requirement, and Bleeding of Fresh Concrete

The small size and the essentially spherical of low-calcium fly ash particles influence the rheological properties of cement pastes, causing a reduction in the water required or an increase in workability compared with that of an equivalent

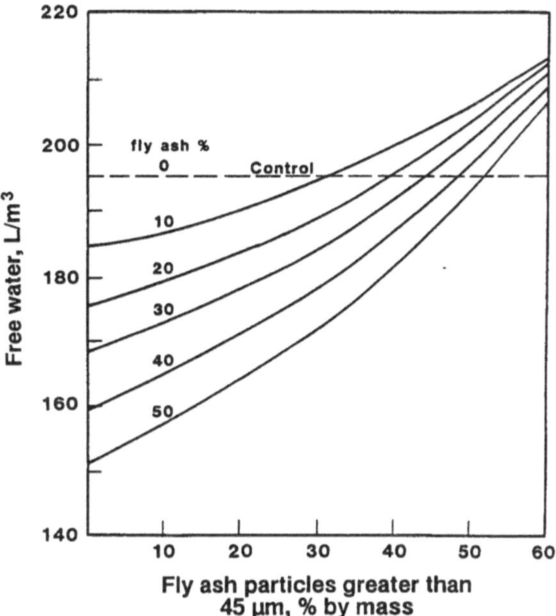

Fig. 2.9 Influence of coarse-particulate content of fly ash on the water required for equal workability in concrete [38]

paste without fly ash. As Davis et al. [1] noted, fly ash differs from other pozzolans, which usually increase the water requirement of concrete mixtures. The improved workability allows a reduction in the amount of water used in concrete. According to Owens [38], the major factor influencing the effects of ash on the workability of concrete is the proportion of coarse material (>45 μm) in the ash. Owens has shown that, for example, substitution of 50 % of the material >45 μm has no effect on the water requirement. The general effect of coarse fly ash particles on the water requirement is illustrated in Fig. 2.9.

In a comprehensive analysis of the data from several major experimental studies Minnick et al. [39] showed statistically significant correlations between the water requirement of mortars and certain characteristics of the fly ashes. The most consistent correlations with the water requirement were obtained for LOI and of fly ashes on the water requirement of mortars is related to the absorption of water by porous carbon particles; the correlation with No. 325 (45 μm) sieve residue should be expected because of particle interference effects.

Stuart et al. [40] studied the effect of high-range water-reducing admixtures and fly ashes on the water requirements of various mortars. The results showed that the fly ashes themselves were effective in reducing water content but differed considerably in their effectiveness as water-reducing admixtures.

Following up on Stuart's results, Helmuth [41] suggested that the water reduction caused by the fly ash was the result of an absorption and dispersion process an much like that of organic water-reducing admixtures. Helmuth suggested an alternative hypothesis-that the water reduction in mixtures without the

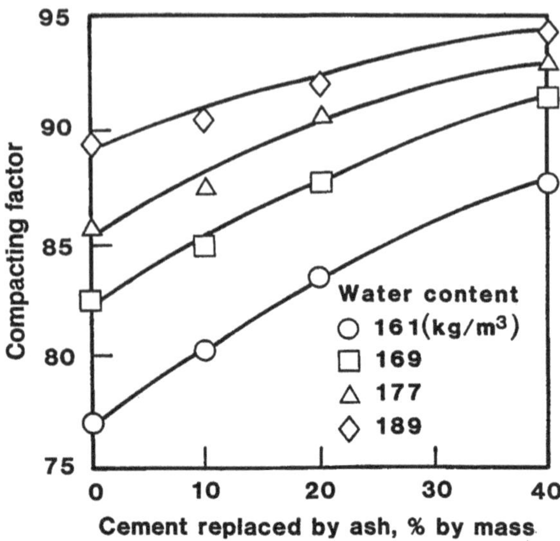

Fig. 2.10 Influence of replacement of cement by fly ash on the workability of concrete [42]

organic admixtures was a result of the adherence of a thin layer of very fine fly ash particles on portions of the surfaces of the cement particles.

Brown [42] examined the workability of four concretes with water/cement in which ash was substituted for cement. Slump, V–B time, and compacting factor were measured for each mixture. It was found that both slump and V–B workability improved with increased ash substitution. The changes were found to depend on the level of ash substitution (with small additions sometimes ineffectual) and on the water content. The data relating compacting factor and ash replacement are shown in Fig. 2.10. An empirical estimate indicates that for each 10 % of ash substituted for cement, the compacting factor changed to the same degree (as it would by increasing the water content of the mixture by 3–4 %).

In another series of experiments, Brown [42] determined the effects of ash substitution for equal volumes of aggregate of send in one concrete, keeping all other mixture proportions (and the aggregate grading) constant. The test concrete was modified by replacing either 10, 20, or 40 % of the volume of sand by ash of 10, 20, or 40 % of the volume of the total aggregate by ash. The replacement of 40 % of the total aggregate gave a mixture that was unworkable. The changes in slump and compacting factor are shown in Fig. 2.11.

Brown [42] concluded that when ash was substituted for sand or total aggregate, (workability increased to a maximum value of ~8 % ash by volume of aggregate). Further substitution caused rapid decreases in workability.

To the present, data obtained by two-point test for fly ash concrete are limited. Ellis [43] reported measurements of g and h for ash concretes proportioned for mass and volume replacement of cement. The different densities of ash and cement result in an increased volume of total cementitious material when replacement is based on mass. Figure 2.12 shows the effect of replacement on medium-

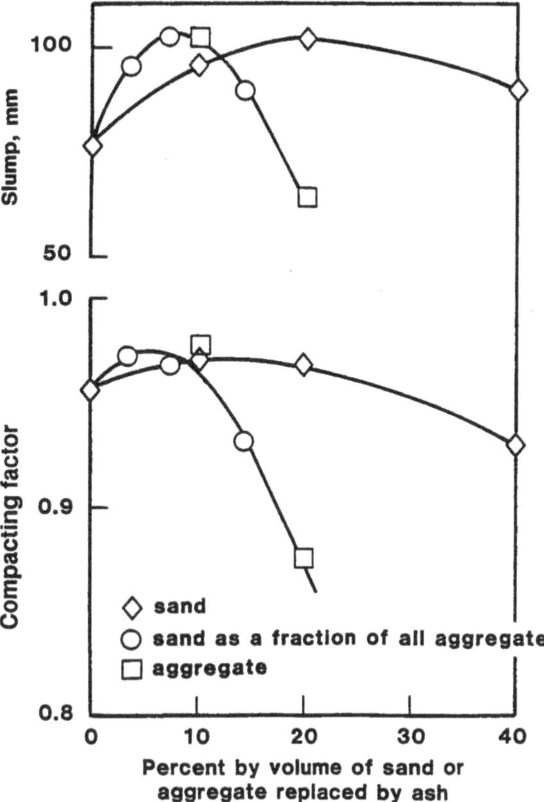

Fig. 2.11 Influence of replacement of aggregate by fly ash on the workability of concrete [42]

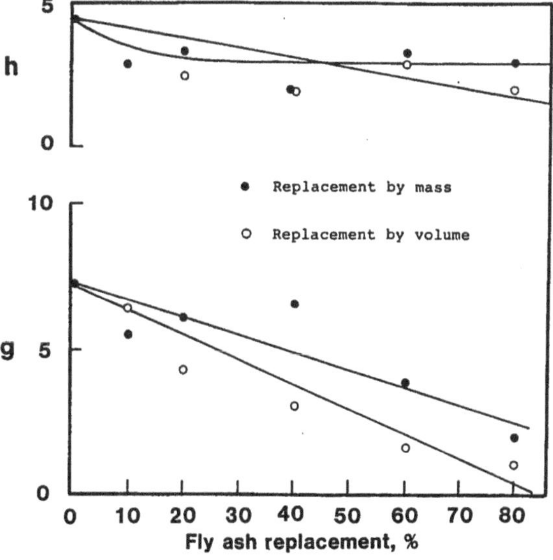

Fig. 2.12 Influence of partial replacement of cement by fly ash on the yield stress and plastic viscosity of various concrete mixtures [43]

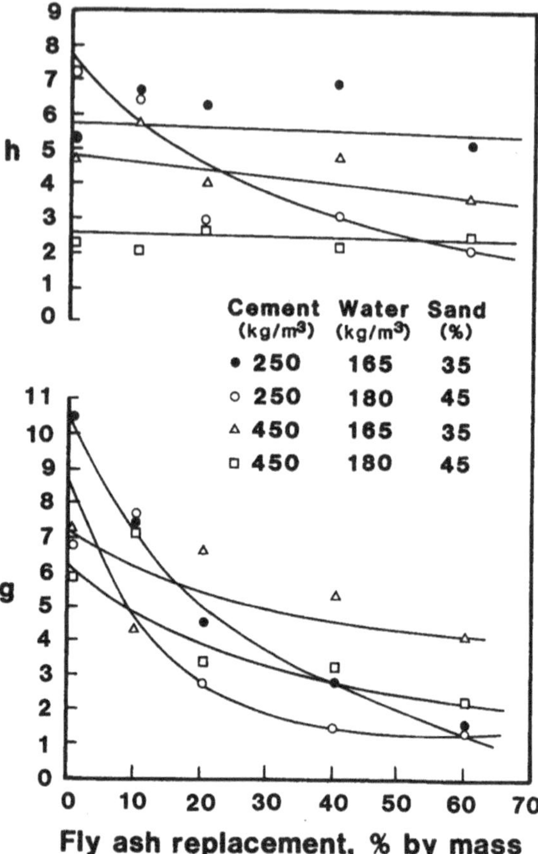

Fig. 2.13 Influence of partial replacement of cement by fly ash on the yield stress and plastic viscosity of concrete [43]

workability concretes. Figure 2.13 shows the effect of replacement by mass on mixtures of similar workability but with different cement contents and aggregate gradings. Both sets of data indicate that fly ash improves workability of concrete and, hence, reduces the water required for constant workability.

Copeland [44] reported that in the field the use of fly ash reduces bleeding in concretes made from aggregates that produce harsh mixes normally prone to bleeding. Johnson [45] reported that most concrete made in the Cape Town (South Africa) area suffers from excessive bleeding resulting from a lack of fines in the locally available dune sands. Johnson added that the problem can be overcome by using fly ash in the concrete to increase the overall paste volume.

In the CANMET study [34], it was found that 6 of the 11 ashes increased bleeding.

In line with the improved rheological properties and as a result of the fine-particular content, some fly ashes give a markedly improved finish when used as a replacement for either sand or cement, effects such as these make fly ash particularly valuable in lean mixtures and in concretes made with aggregates deficient of fines.

Effect of Fly Ash on Temperature Rise of Fresh Concrete

The hydration or setting of Portland cement paste is accompanied by an evolution of heat that causes a temperature rise in fresh concrete. Replacement of cement by fly ash results in a reduction in the temperature rise in fresh concrete. This is of particular importance in mass concrete, where cooling following a large temperature rise can lead to cracking. The first major use of fly ash in concrete was in the construction of a gravity dam, where it was employed principally to control temperature rise [46].

Data reported by Elfert [47] show the effects of fly ash and a calcined diatomaceous shale on the temperature rise of mass concrete (Fig. 2.14).

It has been estimated that the contribution of fly ash to early-age heat generation is 15–30 % of that of the equivalent of Portland cement [33].

Temperature rise, of course, depends on more than the rate of heat generation. It also depends on the rate of heat loss and the thermal properties of the concrete and its surroundings. Williams and Owens [48] presented an estimation (Fig. 2.15) of the effect of element size on the temperature rise in fly ash concretes.

Crown and Dunstan [50] reported the adiabatic temperature-rise data on concrete mixtures shown in Fig. 2.16. Whereas concrete containing 25 % low-calcium fly ash showed a reduced rate of heat generation, concrete with 25 % high-calcium fly ash produced as much heat (at a similar rate) as a Portland cement control.

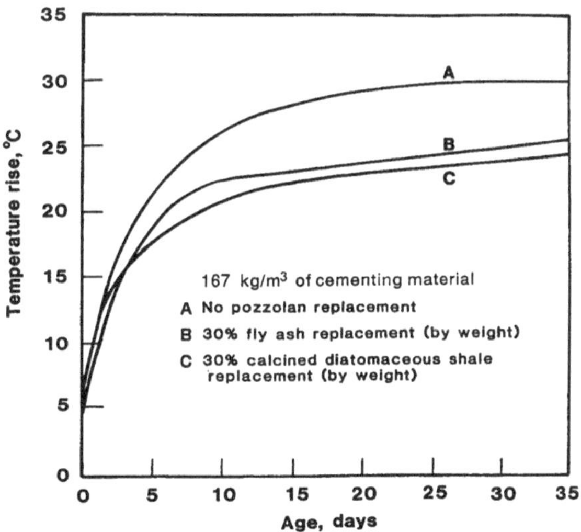

Fig. 2.14 Influence of pozzolans on the temperature rise of concrete [47]

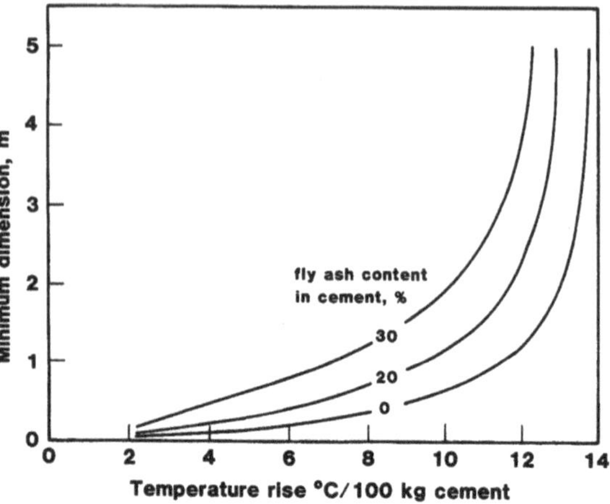

Fig. 2.15 Effect of unit minimum size on the temperature rise in fly ash and plain concrete [48]

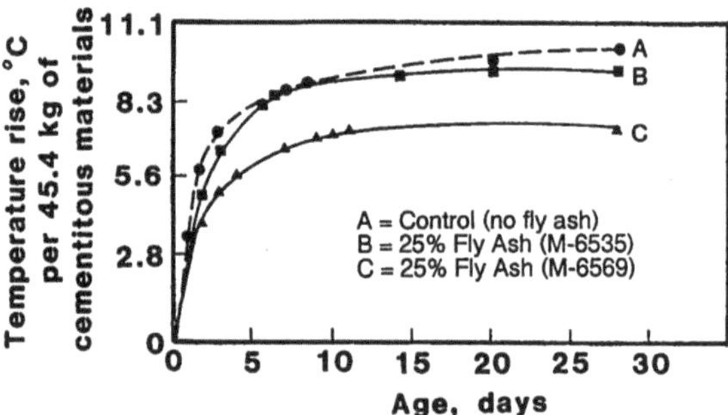

Fig. 2.16 Adiabatic temperature rise in concretes made with high-calcium and low-calcium fly ashes [49]

Effect on the Mechanical Properties of Hardened Concrete

Fly ash affects most of the properties of hardened concrete in one way or another. This chapter is concerned with the ways that the use of fly ash influences the following properties:

- Strength development
- Elasticity
- Creep, shrinkage, and thermal expansion.

The previous CANMET study [34] showed that concretes made under similar conditions, from the same cement and aggregates but incorporating different fly ashes, may develop strength at markedly different rates. Table 2.7 shows the strength development of the concrete mixtures. The different reactivities of the fly ashes are clearly seen from these date.

Table 2.7 Properties of hardened concrete [6]

Mixture no.	Compressive strength of 150 mm × 300 mm cylinders (MPa)				Flexural strength of 75 mm × 100 mm × 400 mm prisms (MPa)		
	7 days	28 days	91 days	365 days	14 days	28 days	91 days
Control	23.4	30.6	34.9	39.2	4.9	5.4	5.9
F1	18.4	35.7	31.4	38.3	4.4	4.4	5.4
F2	16.9	35.2	34.8	37.0	3.9	4.8	5.5
F3	14.4	21.0	27.6	34.4	4.0	5.0	5.3
F4	17.8	23.3	32.3	36.9	4.1	4.4	5.2
F5	20.1	28.0	33.9	44.3	3.5	4.4	5.3
F6	18.4	24.8	31.8	39.2	3.5	4.6	5.6
F7	16.7	24.1	29.1	35.7	3.9	4.5	5.4
F8	17.9	27.7	29.0	40.4	4.6	5.0	6.1
F9	16.7	24.9	31.1	35.6	4.3	4.2	5.7
F10	19.2	28.5	33.7	39.7	4.1	5.1	5.8
F11	21.1	29.4	35.3	40.1	4.8	5.3	6.6

Strength Development in Fly Ash Concrete

As discussed before, the main factors determining strength in concrete are the amount of cement used and water/cement. In practice, these are established as a compromise between the needs of workability in the freshly mixed state, strength and durability in the hardened state, and cost. The degree and manner in which fly ash affects workability are major factors in its influence on strength development. As was shown before, a fly ash that permits a reduction in the total water requirement in concrete will generally present no problems in the selection of mixture proportion any desired rate of strength development.

This chapter is concerned with the factors, other than mixture proportioning and workability that determine the rate of strength development in fly ash concrete. These are the characteristics of fly ash that mixture proportioning seeks to accommodate.

Many variables influence the strength development of fly ash concrete, the most important being the following:

- The properties of the fly ash
- Chemical composition
- Particle size
- Reactivity
- Temperature and other curing conditions.

Effect of Fly Ash Type on Concrete Strength

The first difference among fly ashes that should be recognized is that some are cementitious even in the absence of Portland cement. Frequently, these are the so-called ASTM Class C, or high-calcium, fly ashes usually produced at power plants burning subbituminous or lignitic coals.

In general, the rate of strength development in concretes tends to be only marginally affected by high-calcium fly ashes. A number of authors have noted that concrete incorporating high-calcium fly ashes can be made on an equal-weight or equal-volume replacement basis without any significant effect on strength at early ages [51–53].

Yuan and Cook [52] examined the strength development of concrete with and without high-calcium fly ash (CaO = 30.3 wt%). The data from their research are shown in Fig. 2.17. Using a simple replacement method of mixture proportioning (Table 2.8), they found the rate of strength development of fly ash concrete comparable to that of the control concrete, with or without air entrainment.

Raba and Smith [53] examined the concrete-making properties of a subbituminous fly ash (CaO = 20.0 wt%). Data on compressive strength gain attributed to the presence of high-calcium fly ash shown in Table 2.9. It should be noted that the mixture-proportioning approach replaced fine aggregate by volume. The mass of cement and coarse aggregate was kept constant for each series of determinations.

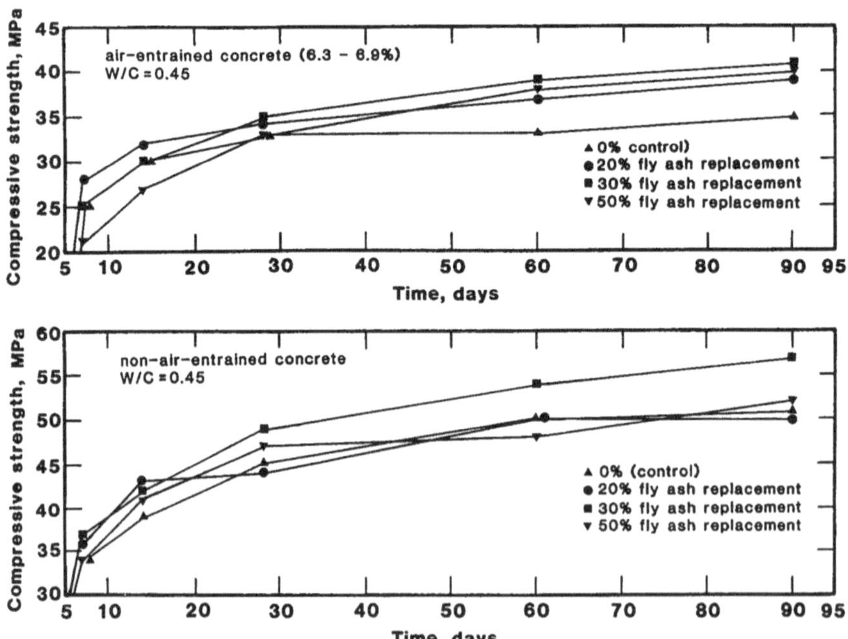

Fig. 2.17 Compressive-strength development of concrete containing high-calcium fly ash [52]

Table 2.8 Mixture designations, proportions, and properties of concretes incorporating high-calcium fly ash [52]

	Mixture designation			
	C1	C2	C3	C4
Proportion (kg/m^3)				
Cement	387	309	272	196
Fly ash	0	77	117	196
Cement + fly ash	387	386	389	392
Water	145	145	146	147
Coarse aggregate	1146	1144	1153	1160
Fine aggregate	701	690	678	654
Properties				
Slump	3	9	12	21
Air content	2.1	1.9	1.9	1.4
Unit weight	2377	2364	2364	2352
Fly ash as percentage of cement	0	20	30	50

Table 2.9 Compressive-strength gains attributed to the presence of high-calcium fly ash [53]

Cement (kg/m^3)	Fly ash (kg/m^3)	Compressive strength[a]		
		Percentage of control values		Gained (%)
		28 days	56 days	28–56 days
91	50	–	354	15.2
	59	–	393	16.0
	68	–	435	13.6
	77	–	494	10.5
136	50	210	228	10.2
	59	231	256	11.8
	68	245	269	11.1
	77	253	274	10.3
182	50	149	155	8.2
	59	153	167	12.4
	68	163	177	12.0
	77	187	190	6.2

[a] In these mixes, fly ash was used to replace fine aggregate

Low-calcium fly ashes, the so-called ASTM Class F ashes, usually a by-product of the burning of bituminous coals, were the first to be examined for use in concrete. Most of what has been written on the behavior of fly ash concrete examines using Class F ashes. In addition, the ashes used in much of the early work came from older power plants and were coarse in particle size, contained unburned fuel, and were often relatively inactive as Pozzolans. Used in concrete, proportioned by simple replacement, these ashes showed exceptionally slow rates of strength development. This has led to the general view that "fly ash reduces strength at all ages" [54]. Conversely, it has also resulted in considerable efforts to understand the factors affecting strength in fly ash concrete and ways to obtain desired rates of strength gain.

Gebler and Klieger [55] evaluated the effect of ASTM Class F and Class C fly ashes from 10 different sources on the compressive-strength development of concretes under different curing conditions, including the effects of low temperature and moisture availability. Their tests indicated that concrete containing fly ash had the potential to produce satisfactory compressive-strength development. The influence of the class of fly ash on the long-term compressive strength of concrete was not significant. In general, compressive-strength development of concretes containing Class F fly ash was more susceptible to low curing temperature than concretes with Class C fly ash or control concretes.

Gebler and Klieger concluded Class F fly ash concretes required more initial moist curing for long-term, air-cured compressive-strength development than did concretes containing Class C fly ashes or control concretes. At early ages, compressive strength of concretes containing fly ash, regardless of class, was essentially unaffected by moisture availability during curing, compared with concretes without fly ash. This showed the importance of proper curing for strength development of concrete with or without fly ash. Generally, at later ages, concretes without fly ash were less sensitive to unavailability of moisture than concretes containing fly ash.

Tikalsky et al. [56] examined concretes containing both Class C and Class F fly ashes ranging from 0 to 35 % by weight of Portland cement. They proportioned their mixtures to have similar slump and constant cementitious materials content by weight. As reported in other studies, they also found that compressive and flexural strengths of fly ash concretes were slightly lower at early ages than those of control concretes but exceeded those of concrete without fly ash at later ages. The creep of fly ash concretes was reduced for similar applied stress/strength, and shrinkage was similar or reduced under similar curing conditions when compared with control concrete.

Particle Size and Strength of Fly Ash Concretes

Particle size can influence strength development in two ways. First, as discussed before, particles >45 μm appear to influence water requirement in an adverse way. They counteract the proportioning methods used to compensate for the slow rate of reaction of fly ash at early ages. Secondly, cementing activity occurs on the surface of solid phases and through heterogeneous processes of diffusion and dissolution of materials in concentrated pastes. Surface area of particles must play a considerable role in determining the kinetics of such processes.

Results of studies of the influence of particle size on strength development are contradictory. On the one hand, in a study of 36 concrete mixtures, most containing fly ashes with a wide variety of properties, Crow and Dunstan [50] concluded that

> fineness of ash compared loosely with the pozzolanic activity, thus finer ashes reacted more readily with Portland cement. Fineness appeared more critical to the reaction of the low-calcium ashes than to those higher in calcium content.

This suggests that there was a relationship between the fineness of fly ash and strength development in concrete. On the other Cabrera et al. [57], in their study of the properties of 18 fly ashes produced from bituminous coals by power stations in the United Kingdom, showed there was no relationship between the 45 μm sieve retention values and the 28 days cube strengths of fly ash concretes. From their results they concluded that differences in the chemical and mineralogical compositions and physical of fly ashes did not influence the strength properties of fly ash concrete. They believed that a much wider range of ashes could be used in concrete than those allowed by British Standard BS3892.

Wesche and vom berg [58], from more than 340 tests on fly ashes from 14 sources, also found no correlation between fineness and compressive strength of mortars at 7 or 28 days, but they did find a minor correlation at 90 days.

In some respects, these results are predictable when samples from a large number of sources are examined in experiments designed to determine only one factor. Many fly-ash-related variables may influence strength development; poor correlation with particle size only indicates that particle size is not the dominant variable in fly ash reactivity. To establish a relationship with particle size, it is necessary either to limit all other variables to a minimum number or to perform a multi-variable experiment. To date, only the former, minimum-number approach has been taken.

Both Joshi [59] and Ravina [60] exploited the phenomenon of particle-size segregation that occurs in electrostatic precipitators. They used this to obtain fly ash fractions of different fineness from a single source.

When fly ash is collected in a multistage, electrostatic precipitator, segregation by particle size occurs: particles in the finer size categories are collected in the chambers farthest from the furnace. By taking ash from each chamber, one may obtain particles of different size distributions from the same source at the same time.

Unfortunately, size is not the only property that differs among fly ashes from different chambers. The chemical properties, density, and morphology of the particles also differ. Carbon is segregated with the larger particles, and alkali sulphates and chlorides collect on the surface of the finer particles in the cooler regions of the precipitator. Low-density particles are differentially distributed in the precipitator; cenospheres and irregularly shaped particles tend to precipitate from the gases in the first chamber and, thus, affect average density measurements. Examination of ashes from one source, segregated by size, would seem to be the simplest way of determining the influence of particle size on strength, and even then other, uncontrolled factors may influence the results.

Joshi [59] examined ash from different chambers of the precipitator of a modern power plant (Sundance, Alberta) burning subbituminous coal. The ashes had the following percentages of particles > 45 μm:

Ash	% retained on 45 μm sieve
1	5
2	16
3	32
4	38

Concretes were made with 10 and 20 % replacement of cement for each of the four ashes. Resulting compressive-strength data are shown in Fig. 2.18.

These experiments show that at both 10 and 20 % replacement, the finer fly ashes imparted greater strength. The coarse fraction appears not to have contributed significantly to the strength.

Ravina [60] reported similar results from an examination of the pozzolanic activity index of low-calcium ash from each of four hoppers (chambers) in one precipitator.

An alternative way to obtain specific sizes of ash from one source is by grinding. Monk [61] examined ashes in blended cements and cements interground with Portland cement. The ashes were from four base-load power plants the principal conclusions from this research were as follows:

- Portland fly ash cements prepared by intergrinding clinker, gypsum, and fly ash have a water requirement equal to, or lower than, that of equivalent cements produced by blending ordinary Portland cement and fly ash.
- Intergrinding did not reduce the workability of fly ash concrete. Breakdown of spherical particles was not observed, but the agglomerates of spheres in coarse ashes were separated.
- Compressive strengths at all ages for the interground cements were equal to, or higher than, those for the blended cements.

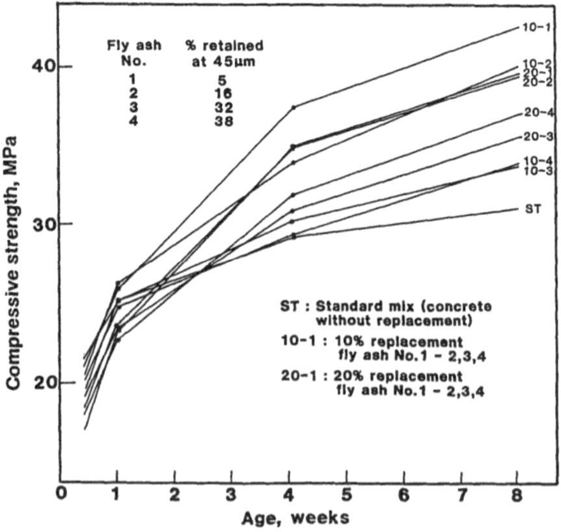

Fig. 2.18 Effect of coarse fractions of fly ash on compressive-strength development of concretes [59]

- Intergrinding improved the water-reducing properties and pozzolanic activity of coarse ashes, as indicated by compressive-strength improvement at later ages.

Matsufuji et al. [62] examined the properties of concrete containing ultra-fine particles produced from fly ash with ultra-high-temperature treatment. This treatment results in an increase of the specific surface area, from 20 to 130 m^2/g.

The specific surface area of the ultra-fine particles affects the mechanical properties of concretes made with these particles. In their experiment, Matsufuji et al. achieved a compressive strength of ~118 MPa (W/C = 0.25) with ash particles with a surface area of 71 m^2/g. The early-age compressive strength of concretes containing ultra-fine ash particles was lower than that of the plain concrete but higher at ≥28 days.

When experimental efforts are made to eliminate the effects of multi-variable interference, there is a well-defined correlation between strength development and particle size (or surface area). This suggests that for ash from one source, particle size is an important indication of reactivity. This viewpoint is reflected in almost all the quality control procedures recommended for selecting ash for use in concrete and has been incorporated in most standards.

Effects of Temperature and Curing Regime on the Strength Development in Fly Ash Concretes

When concrete made with Portland cement is cured at temperatures of >30 °C, there is an increase in strength at early ages but a marked decrease in strength in the mature concrete [63].

Concretes containing fly ash and control concretes behave significantly differently. Figure 2.19 shows the general way in which the temperature, maintained during early ages of curing, influences the 28 days strength of concrete.

Whereas ordinary Portland cement loses strength as a consequence of heating, fly ash concretes show strength gains. This is of great value in the construction of mass concrete or in concrete construction at elevated temperatures.

Kobayashi [64] reported that fly ash was used at a 25 % cement-replacement level in the concrete for an intake tunnel of the Kurobegwa No. 3 Power Station in Japan [65]. Because this tunnel is located in rock at temperatures of 100–160 °C, fly ash was used to combat the loss of strength that would have resulted had Portland cement concrete been used alone.

Bamforth [49] reported on the in situ and laboratory-observed effects of temperature on the strength development of mass concrete containing either fly ash or slag substituents.

Figure 2.20a shows the strength development of standard cubes made from the three concretes studied by Bamforth under standard laboratory curing conditions. Strength approximately equal to that of the control was reached at 28 days; in fact,

Fig. 2.19 Effect of temperature rise during on the compressive-strength development of concretes [48]

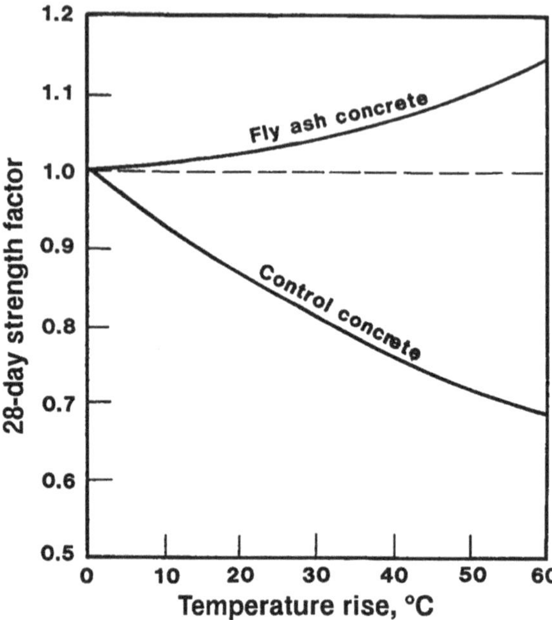

the strength of the fly ash concrete was close to that of the control concrete at earlier ages.

Figure 2.20b shows the effects of curing under temperature-matched conditions that simulated the effects of the early-age temperature cycle at the centre of the concrete mass. In all cases, early strength development was accelerated.

However, at 28 days, the strength of the fly ash concrete was enhanced by temperature, whereas the control concrete bad significantly less strength (30 % below that of the standard water-cured concrete).

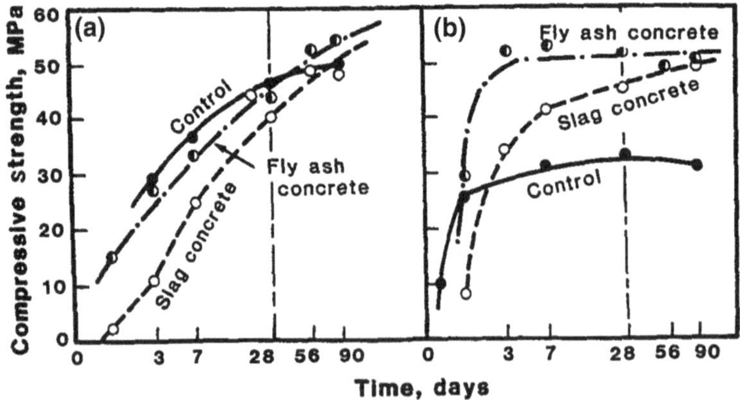

Fig. 2.20 Compressive-strength development of concretes **a** cured under normal conditions and **b** cured by temperature matching [49]

Ravina [66] discussed bow the effects of fly ash in concrete cured at moderately elevated temperatures may be advantageous in precast operations. Ravina's paper is of considerable interest for its examination of the way in which fly ash reacted in concrete cured under a controlled regime that included some exposure to beat.

Ravina examined concrete made with fly ashes of two size fractions from the same source:

- Coarse ash from the first precipitator field (with 30–35 % retained on a 45 μm sieve); and
- Fine ash from the third precipitator field (with 14–17 % retained on a 45 μm sieve).

Control mixtures contained no fly ash but had two cement contents. Ravina kept control specimens for 22 days at 23 °C prior to demolding and then placed them for 7 days in the fog room at 23 °C. He then stored the samples at 23 °C and 65 % RH before testing them.

Ravina kept thermally cured specimens at 23 °C for 2 days and then transferred them to a steam chamber, where be raised the temperature from 23 to 75 °C over a 2 days period and kept it at 75 °C for 4 days. He then discontinued beating and allowed the specimens to cool in the chamber. After 22 h, be demolded the samples and then stored them under the same conditions as those for the control specimens.

Ravina made two series of non-air-entrained mixtures (Table 2.10). In the first series, coarse ash alone replaced either all or 50 % (by volume) of the pit sand. In the second series, fine fly ash replaced 20 % of the cement (by weight) and coarse ash replaced 50 % (by volume) of the sand.

The compressive-strength data from Ravina's study are shown in Fig. 2.21. Ravina drew the following conclusions:

- Large quantities of fly ash in concrete cured at elevated temperatures significantly improve its compressive strength, but fly ash contributes less to the strength of concrete cured normally at ages ≤28 days.

Table 2.10 mixture designations and proportions of concretes cured at elevated temperatures [66]

	Mixture designation					
	1-A	1-B	1-C	2-A	2-B	3
Proportions (kg/m^3)						
Corse and medium aggregate	1165	1165	1125	1165	1165	1160
Quarry sand	535	535	520	535	535	535
Pit sand	380	190	–	370	180	430
Cement	300	300	300	240	240	240
Water	205	210	230	190	200	205
Coarse fly ash	–	155	305	–	150	–
Fine fly ash	–	–	–	60	60	–

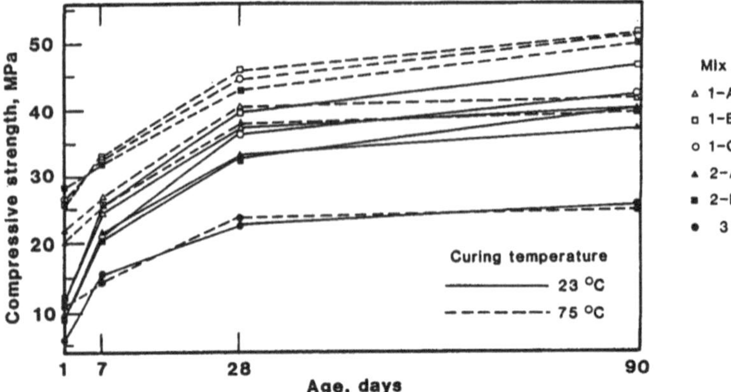

Fig. 2.21 Effect of curing temperature on the compressive-strength development of concretes incorporating fine and coarse fractions of fly ash [66]. *Note* For mix details, see Table 2.10

- Curing coarse and fine fly ash concrete at elevated temperatures bas a significant beneficial effect on the strength of the concrete at early and later ages.

Ravina's work and the observations of other authors [48] raise some significant issues regarding the nature of pozzolanic reactivity. The following would seem to warrant further investigation:

- The rate of reaction of fly ash–cement systems is clearly increased by temperature, as is the case for Portland cement. Yet, the products of hydration do not exhibit the poor mechanical properties associated with curing Portland cement at elevated temperatures. This suggests that the products of fly ash-cement hydration, their relative proportions, and their morphology are significantly different from those formed from thermally accelerated hydration of Portland cement alone.
- The rate of reaction of fly ash in cement systems is so significantly increased by temperature that the effects of particle size on pozzolanic behavior are largely overcome. This suggests that some pozzolanic activity tests that use thermal acceleration may give seriously misleading results.
- The pozzolanic reaction, once initiated by heat, appears to continue when the source of external heat is removed, even with coarse fly ash. This indicates the possibility of an activation effect similar to that observed for slags, which has not previously been associated with pozzolanic activity.

The sensitivity of fly ash to elevated temperature implies that it will also be sensitive to reduced temperature, with a consequent reduction in the rate of strength development. This has been observed in practice, and it is generally noted that fly ash concretes require more attention to curing in cold weather.

Curing regime is more important to the compressive-strength development of fly ash concrete, compared with plain concrete. Gopalan and Haque [67] reported the strength of normal and fly ash concretes cured in a fog room and in an uncontrolled environment. The strength of 91 days air-cured specimens was less than that of 7 days fog-cured specimens. On air curing, the percentage loss of strength increased both with an increase in fly ash content and the curing period. Haque et al. [68] also studied the effects of various curing regimes on the strength development of both plain concretes and two subbituminous fly ash concretes.

In situ strengths of low-grade 20 MPa fly ash concretes were 60–90 % of the fog-cured strengths. In situ strengths of high-grade 40 MPa fly ash concretes were consistently 90 % of the fog-cured strengths. The strength ratio obtained at 6 months appeared to be independent of the ash content.

Effect of Fly Ash on Elastic Properties of Concrete

Published data indicate that fly ash has little influence on the elastic properties of concrete.

Abdun-Nur [69] made the following observations on the early literature:

The modulus of elasticity of fly ash concrete is lower at early ages, and higher at later ages [1, 70]. In general, fly ash increases the modulus of elasticity of concrete when concretes of the same strength with and without fly ash are compared [71–73].

Lane and Best [36] stated the following:

Fly ash properties controlling the compressive strength of concrete also influence the modulus of elasticity, but to a lower extent. The modulus of elasticity, like compressive strength, is lower at early strength and higher at ultimate strength when compared with concrete without fly ash.

Figure 2.22 shows typical stress–strain relationships for concrete with and without fly ash.

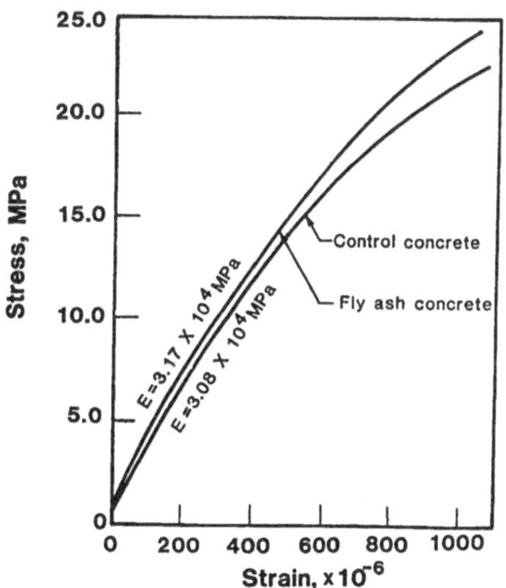

Fig. 2.22 Stress–strain relationship for concretes with and without fly ash [36]

Crow and Dunstan [50] who reported on an examination of the properties of 36 concretes, most containing fly ash at different levels and from different sources, drew the following conclusions about elastic properties:

> The elastic properties of concretes containing both Portland cement and fly ash are similar to those expected with Portland cement (alone). The modulus of elasticity and Poisson's ratio both increased with age, paralleling the compressive-strength development. The modulus of elasticity ranged from a low of 18.8 GPa at 28 days to a high of 39.6 GPa at 365 days. Most of the ashes (in concrete) had a 28-day Poisson's ratio in the range of 0.14–0.25.

Ghosh and Timusk [74] studied fly ash concretes, proportioned for equivalent 28 days strength, over a range of nominal strength values. They showed the relationship between strength and modulus of elasticity, reproduced in Fig. 2.23, and concluded that for all strength levels the modulus of elasticity of fly ash concrete is generally equivalent to that of the reference concrete. In all instances, the modulus was found to exceed that given by the ACI formula.

Nasser and Marzo [75] examined the effects of temperature on modulus of elasticity of concrete made from fly ash (from a Saskatchewan lignite source) and a sulphate-resistant, ASTM type V cement. They reported that over the temperature range of 21–232 °C, the modulus of elasticity was reduced by 910 % for specimens that were both sealed and unsealed to prevent moisture loss.

In a CANMET study [6], the 28 days Young's modulus of elasticity values for the fly ash concretes were 29–35.8 GPa, compared with the value of 33.5 GPa for the control concrete. The data obtained in this investigation showed no significant effect of fly ash or type of fly ash on the modulus of elasticity.

Nasser and Ojha [76], in a study of Saskatchewan lignite fly ash, reached a similar conclusion—the modulus of elasticity of concretes containing ≤20 % fly ash was similar to that of the control concrete. At higher percentages, fly ash reduces both the compressive strength of concrete and the modulus of elasticity.

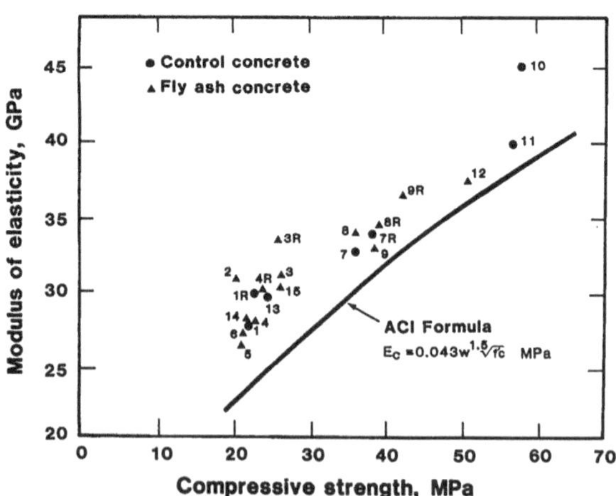

Fig. 2.23 Moduli of elasticity of fly ash concretes [74]

Effect of Fly Ash on Creep Properties of Concrete

Data on creep of fly ash concrete are limited. Lohtia et al. [77] reported the results of studies on creep and creep recovery of plain and fly ash concretes under stress/strength of 20 and 35 %. The concretes were made by replacing cement with equal weights of fly ash in the range of 0–25 %. From this work, they drew the following conclusions:

- Replacement of 15 % of cement by fly ash was found the optimum for strength, elasticity, shrinkage, and creep for the fly ash concretes studied.
- Creep versus time curves for plain and fly ash concretes were similar, with creep linearly related to the logarithm of time.
- Increase in creep with ~15 % fly ash content was negligible. However, slightly higher creep took place at fly ash contents of >15 %.
- The creep coefficients were similar for concrete with fly ash contents in the range of 0–25 %.
- Creep recovery was 22–43 % of the corresponding 150 days creep. For cement replacement >15 %, the creep recovery was smaller. No definite trend of creep recovery as a function of stress/strength was observed.

In another study, Ghosh and Timusk [74] examined bituminous fly ashes of different carbon contents and fineness values in concretes at nominal strength levels of 20, 35, and 55 MPa (water/cement of 1.0, 0.4, and 0.2, respectively). Each concrete was proportioned for equivalent strength at 28 days. Fly ash concretes showed less creep in the majority of specimens than the reference concretes showed. This was attributed to a relatively higher rate of strength gain after the time of loading for the fly ash concretes than for the reference concretes.

Yuan and Cook [52] reported the data in Fig. 2.24 from studies of high-strength concrete containing a high-calcium fly ash. The figure shows that concrete containing 30 and 50 % fly ash exhibited more creep than either the control concrete or a concrete with 20 % fly ash.

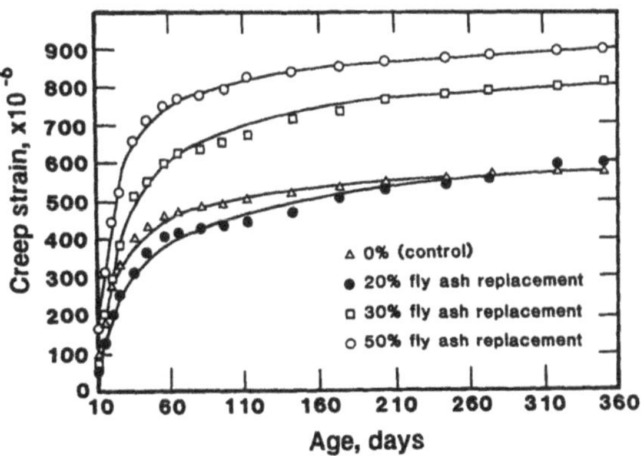

Fig. 2.24 Creep of fly ash concretes [45]

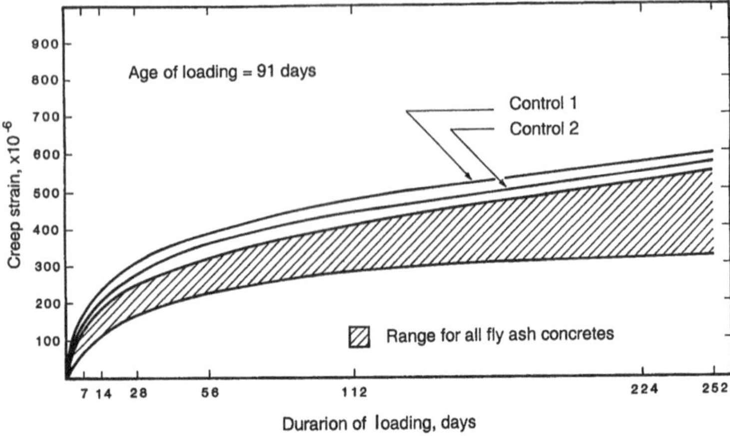

Fig. 2.25 Creep strains after initial moist curing of 91 days [6]

Gifford and Ward [78] examined lean mass concrete and concluded that fly ash reduces creep, as a result of a number of factors including the following:

- Fly ash increases the elastic modulus.
- Fly ash contributes to the total aggregate and reduces the volume of paste available to creep.

Investigation by Bamforth [49] on mass concrete showed that a reduction in creep of ~50 % can be obtained when cement is replaced by ~30 % fly ash. However, the results of Nasser and Al-Manaseer's work [79] showed that there was an increase in creep of—15 % in concretes containing 20 % fly ash and other admixtures. In another study, the same authors [80] examined the creep of sealed and unsealed concrete made with ASTM type I cement containing 50 % Saskatchewan fly ash. They tested the concrete specimens at different stress/strength and measured their creep for a maximum period of 112 days. The experimental results showed that the creep of concrete made with 50 % lignite fly ash was a linear function of stress/strength. The creep of this concrete was lower than that of plain concrete by—13 % for the unsealed and 39 % for the sealed specimens. In addition, they found that the ratio of creep values of sealed and unsealed concrete was about 2.44 for plain concrete and 3.67 for concrete with 50 % fly ash.

In a CANMET investigation, the creep-strain data for control and fly ash concretes were compared [6]. Figure 2.25 shows the creep strains after 91 days of initial moist curing. All fly ash concretes are shown to produce consistently lower creep strains than the control concrete. The strain reduction, which in most cases varies between 20 and 45 %, does not appear to be related to the type of ash.

Effect of Fly Ash on Volume Changes of Concrete

It has been generally reported that the use of fly ash in normal proportions does not significantly influence the drying shrinkage of concrete. Typical of the conclusions of most researchers in this area are those made by Davis et al. [1], who commented as follows:

For masses of ordinary thickness, such as are normally found in highway slabs and in the walls and frames of buildings, the drying shrinkage at the exposed surfaces of concrete up to the age of one year for fly ash cements is about the same as, or somewhat less than, that for corresponding Portland cements. At a short distance from the exposed surface the drying shrinkage up to the age of one year is substantially less for concretes containing corresponding Portland cements.

For very thin sections and for cements of normal fineness the drying shrinkage of concretes containing finely ground high-early-strength cements may be somewhat reduced by the use of fly ash.

Figure 2.26 shows data presented by Elfert [47] comparing the drying shrinkage and autogenous length change of fly ash concrete, plain concrete, and concrete made with other pozzolans. Typically, fly ash concrete performed better in these respects than the other concretes studied.

In their study, Ghosh and Timusk [74] showed that for the same maximum size of aggregate and for all strength levels, the shrinkage of concrete containing fly ash is lower than that for concrete not containing fly ash.

In their studies of concrete using high-calcium fly ash, Yuan and Cook [52, 81] concluded that the replacement of cement by fly ash has little influence on drying shrinkage. Their data are shown in Fig. 2.27. Similar conclusions may also be

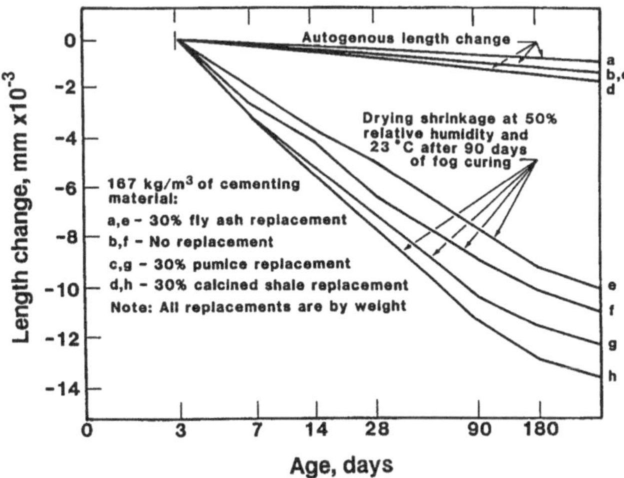

Fig. 2.26 Drying shrinkage and autogenous length change for concretes containing various pozzolans [47]

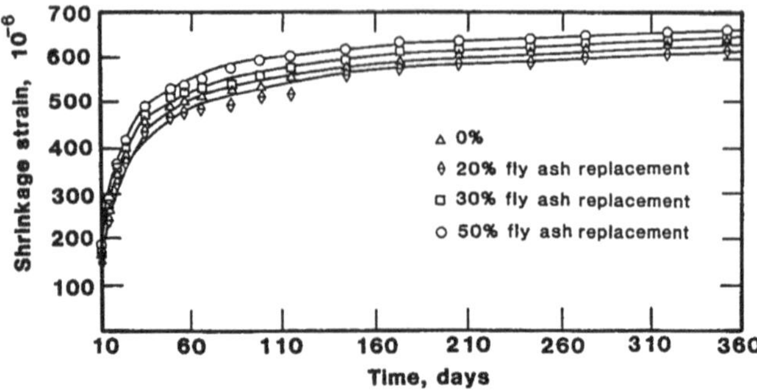

Fig. 2.27 Drying shrinkage of concretes incorporating high-calcium fly ash [52]

Table 2.11 Drying shrinkage of fly ash concretes [34]

Mixture no.	Shrinkage measurements			
	Initially cured for 7 days in water		Initially cured for 91 days in water	
	Moisture loss (%)	Drying shrinkage (10^{-6})	Moisture loss (%)	Drying shrinkage (10^{-6})
Control 2	55.0	442	53.7	453
F1	57.5	447	47.9	365
F2	57.3	364	45.4	280
F3	56.9	411	56.2	405
F4	54.7	379	49.2	387
F5	58.8	404	51.1	403
F6	60.6	475	56.4	454
F7	64.3	397	54.1	433
F8	56.3	400	–	327
F9	58.2	390	49.3	361
F10	58.4	642	55.2	500
F11	49.5	454	48.9	362

Notes Duration of drying, 224 days

drawn from the data on shrinkage obtained by CANMET [34] for concretes incorporating a range of fly ashes (Table 2.11).

Munday et al. [82] reported typical shrinkage-age relationships for fly ash concretes, as shown in Fig. 2.28, and concluded that a general relationship exists between drying shrinkage and equivalent cement content (Fig. 2.29). It was also found that wetting and drying of fly ash concrete results in a cumulative expansion over a number of cycles (Fig. 2.30). It was concluded that overall, the incorporation of fly ash does not significantly affect the drying shrinkage, wetting–drying expansion, or thermal expansion of concrete.

Drying shrinkage of fly ash concrete is affected by the initial moist-curing period. Ravindrarajah and Tam [83] in a study of properties of concrete containing low-calcium fly ash, reported that the drying shrinkage of fly ash concrete was

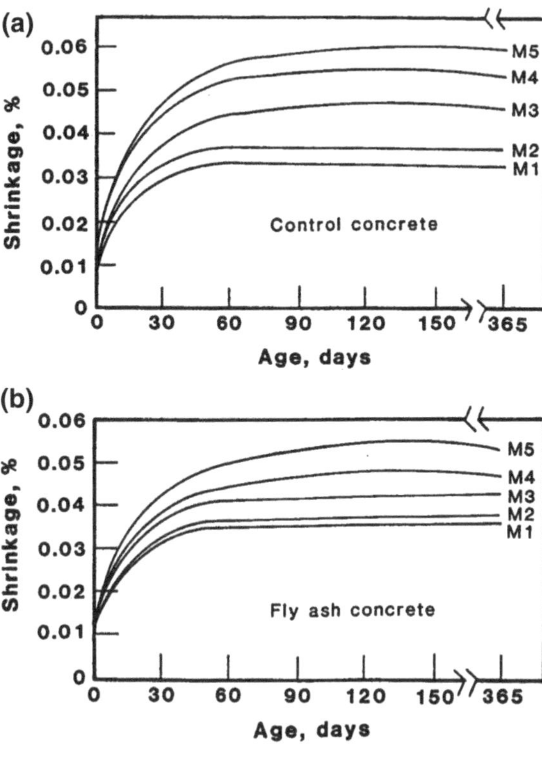

Fig. 2.28 Drying shrinkage of concretes incorporating low-calcium fly ash [82]

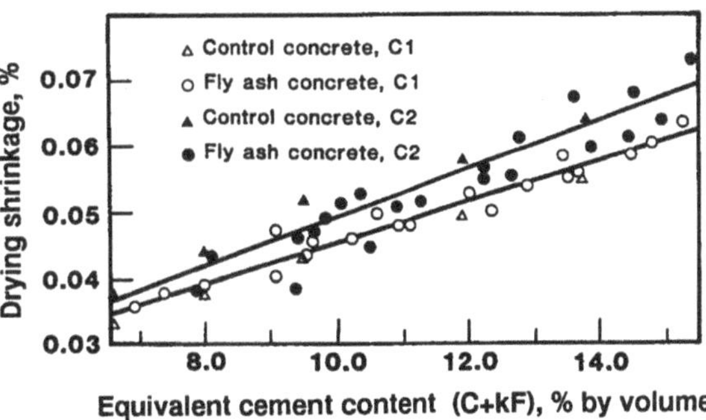

Fig. 2.29 Drying shrinkage of concretes versus equivalent cement content [82]

higher than that of control concrete when the initial moist-curing period was limited to 7 days. However, the difference in drying shrinkage between the two concretes dropped considerably when the initial moist-curing period was increased. In some cases, the drying shrinkage of fly ash concrete was slightly lower than that of the plain concrete.

In the CANMET study [6] of 11 Canadian fly ashes, the drying-shrinkage strain of concrete initially cured for 7 days in water was apparently little affected by the presence of fly ash in the concrete. The strain values for the control and most fly

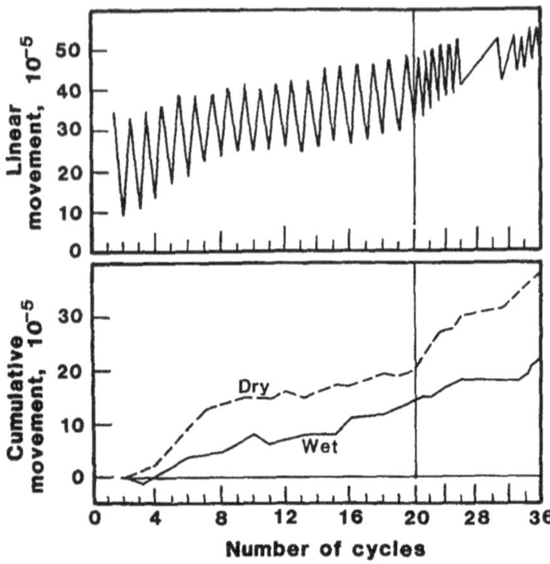

Fig. 2.30 Wetting and drying movements of fly ash concretes [82]

ash concretes were closely grouped together until after 224 days of drying (Fig. 2.31). The only significant deviation from this was for concrete made with lignite fly ash No. 10 (see Table 2.3), in which case the total shrinkage strain was found to exceed the average strain by ~50 %.

Thermal expansion of concrete is mainly affected by the thermal expansion of the coarse aggregate that constitutes its main component. Values for wet limestone and quartzite have been reported to be 4.0 and 11.7 micro-strain per degree Celsius, respectively, and that of cement paste is reported to be 11.0–20.0 micro-strain

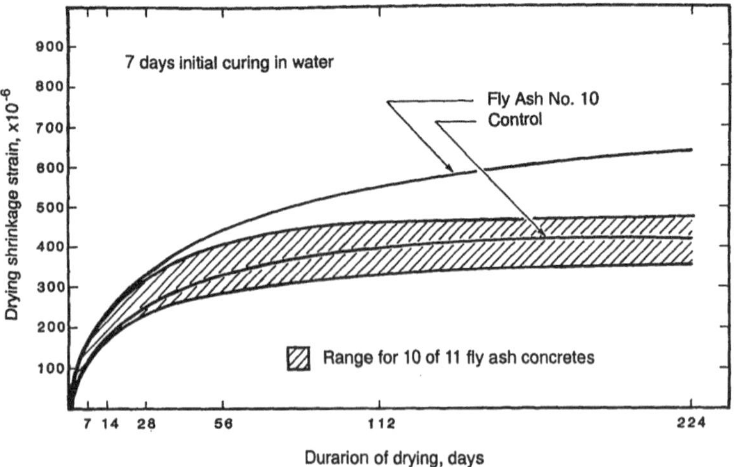

Fig. 2.31 Drying shrinkage strains after initial curing of 7 days in water [6]

per degree Celsius [78]. Gifford and Ward [78] quoted Dunstan as suggesting that fly ash slightly reduces thermal expansion. Their own data indicated an average reduction of 4 % at a fly ash level of 40 %.

Failure of concrete after a period less than the lifetime for which it was designed may be caused by the environment or by a variety of internal causes. External causes may be physical or chemical: weathering, extremes of temperature, Abrasion, or chemical action in the cement, aggregate, or reinforcement components. Internal causes may lie in the choice of materials or in inappropriate combinations of materials and may be seen as alkali-aggregate expansion or other forms of failure. Of all the causes of lack of durability in concrete the most widespread is excessive permeability. Permeable concrete is vulnerable to attack by almost all classes of aggressive agents. To be durable, Portland cement concrete must be relatively impervious.

Increasingly, concrete is being selected for use as a construction material in aggressive or potentially aggressive environments. Concrete structures have always been exposed to the action of sea water. In modern times, the demands placed on concrete in marine environments have increased greatly, as concrete structures are used in arctic temperate, and tropical waters to contain and support the equipment, people, and products of oil and gas exploration and production. Concrete structures are used to contain nuclear reactors and must be capable of containing gases and vapors at elevated temperatures and pressures under emergency conditions. Concrete is increasingly being placed in contact with sulphate and acidic waters. In all of these instances, the use of fly ash as concrete material bas a role, and an understanding of its effect on concrete durability is essential to its correct and economical application.

The following sections of this chapter seek to provide a general view of the present knowledge regarding the durability of fly ash concrete. The subject matter is vast, complex, and as yet incompletely understood. The reader desiring a more detailed treatment of the subject is invited to consult the cited literature.

Effects of Fly Ash on Permeability of Concrete

The movement of aggressive solutions into a concrete mass or the removal from concrete of dissolved reaction products must play a primary role in determining the rate of progress of concrete deterioration caused by chemical attack. Permeability of a concrete mass is, therefore, fundamental in determining the rates of mass transport relevant to destructive chemical action. It should be recognized that all the cementitious hydrates and some of the aggregates from which concretes are made are inherently subject to attack, not only by sulphates, chlorides, acids, and organic agents, but by water alone. That concrete survives aqueous environments at all is attributable to (a) the low equilibrium solubility of the hydrated components and (b) the low rate of mass transfer in well-compacted, cured concrete. Given any combination of cement and aggregate, it is generally observed that the

Table 2.12 Relative permeability of concretes with and without fly ash [84]

Fly ash		W/(C + F) by weight	Relative permeability (%)	
Type	% by weight		28 days	6 months
None	–	0.75	100	26
Chicago	30	0.70	220	5
	60	0.65	1410	2
Cleveland	30	0.70	320	5
	60	0.69	1880	7

Table 2.13 Mixture designations and proportions for concretes examined by Kanitaks [85]

Mixture	Constituents (per m^3)				
	Cement (kg)	Fly ash (kg)	Sand (kg)	Stone (kg)	Water (L)
N	400	–	586	1190	233
M	350	100	519	1213	227

less permeable the concrete. the greater will be its resistance to aggressive solutions or pure water.

A number of investigations have been made of the influence of fly ash on the relative permeability of concrete pipes containing fly ash substituted for cement in amounts of 30–50 %. Davis [84] examined the permeability of concrete pipes incorporating fly ash substituted for cement in amounts of 30-50 %. Permeability tests were made on 150 × 150 mm cylinders at the ages of 28 days and 6 months. The results of these tests are shown in Table 2.12.

It is clear from these data that the permeability of the concrete was directly related to the quantity of hydrated cementitious material at any given time, After 28 days of curing, at which time little pozzolanic activity would have occurred, the fly ash concretes were more permeable than the control concretes. At 6 months, this was reversed. Considerable imperviousness had developed, presumably as a result of the pozzolanic reaction of fly ash.

Kanitakis [85] used an Initial Surface Absorption Test to examine concrete with and without a low-calcium fly ash (CaO = 2.0 wt%) made using the mixture properties shown in Table 2.13. Absorption measurements were made at 7, 17, 28, and 56 days of curing, and the data shown in Fig. 2.32 were obtained.

The author concluded that at early ages, fly ash concrete behaves as a lean-mixture concrete and is, thus, permeable. At later ages, permeability is reduced as the pozzolanic action proceeds.

The rather limited observations on this very important aspect of fly ash concrete are consistent with the view expressed by Manmohan and Mehta [86] that the transformation of large pores to fine pores, as a consequence of the pozzolanic reaction between Portland cement paste and fly ash, substantially reduces permeability in cementitious systems.

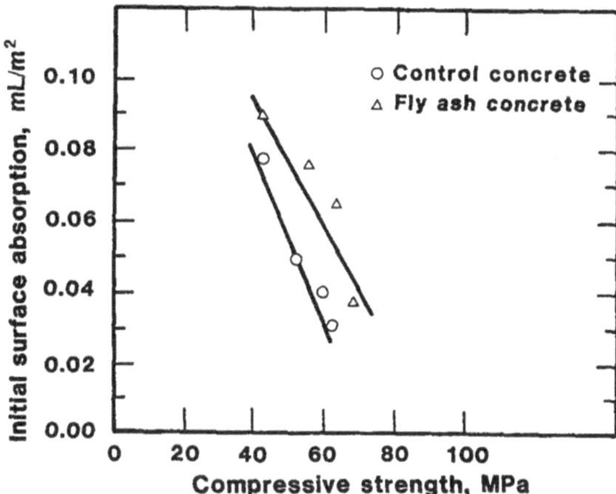

Fig. 2.32 Initial surface absorption of water versus compressive strength of concretes [85]

Diffusion of ions, such as chlorides, that are not specifically bound by the components of concrete is reasonably represented by Fick's diffusion equation [87, 88]

$$dc/dt = D_c\left(d^2C/dX^2\right) \tag{2.1}$$

where C is the ion concentration at a distance *x,* after a time *t;* and D, is the ion diffusion coefficient.

For concrete in offshore structures, Browne [86] reported values for D, of 1–50.0 × 10^{-9} cm^2/s for high-and normal-strength concretes, respectively.

Short and Page [89] reported on the diffusion of chloride ions in solution into Portland and blended cement pastes and found the following values of D, for different cement types:

Type of cement	D_c value (× 10^{-9} cm^2/s)
Normal Portland	44.7
Sulphate-resisting	100.0
Fly ash/Portland	14.7
Slag/Portland	4.1

It was concluded from these data that slag and fly ash cements were more effective in limiting chloride diffusion in pastes than were normal or sulphate-resisting cements.

Permeability to gases, in particular to air and carbon dioxide, is important for some aspects of concrete durability related to carbonation (see below). Kasai et al. [90] examined the air permeability of mortars (moist cured for 1, 3, and 3, **7** days) of blended cements made with fly ash and with blast-furnace slag and concluded the following:

- The air permeability of blended cement mortars is greater than that of Portland cement mortars.
- When early moist curing is extended, the permeability is reduced.

The considerable differences in the strengths at all ages of the mortars examined by these authors, with fly ash mortars being some 20–30 % weaker, suggest that the curing regimes adopted were inadequate to permit any conclusions relevant to the practical behavior of blended cement concretes in the field.

Thomas et al. [91] studied the effect of curing on the permeability to oxygen and water of fly ash concretes designed for equal workability and 28-day compressive strength. The results for oxygen permeability of concretes at 28-days are given in Tables 2.14 and 2.15 respectively. These results confirm the importance of curing, with reductions in curing period resulting in more permeable concrete. Despite exhibiting lower strengths, fly ash concretes moist cured for only 1 day were, generally, no more permeable to water and substantially less permeable to oxygen than similarly cured control concrete. As the period of curing increased, the fly ash concretes became considerably less permeable to water and oxygen than the control concrete. The authors argued that although the increased curing periods recommended in BS8110 for fly ash concrete were justified on the basis of concrete strength, fly ash concrete might require no more curing than control concrete to achieve equal durability, measured by oxygen and water permeability.

Table 2.14 Coefficient of oxygen permeability of concretes after 28 days of storage [91]

Mixture	PFA	Coefficient of oxygen permeability ($\times 10^{-17}m^2$)						
		20 °C, 65 % RH[a]			20 °C, 40 % RH[a]			28 days of water storage
		Curing period			Curing period			
		1 days	3 days	7 days	1 days	3 days	7 days	
A	0 %	24.3	14.8	11.9	27.2	23.8	18.7	3.97
B	15 % M367	21.3	11.6	11.2	12.9	9.50	8.10	2.02
C	30 % M367	9.67	5.85	4.01	15.3	12.2	7.50	2.05
D	50 % M367	18.0	5.26	5.26	16.1	9.53	4.69	0.64
E	30 % M368	6.37	4.64	1.77	–	–	–	1.08
F	30 % M369	18.6	7.51	3.21	–	–	–	1.61

[a] Storage conditions following curing

Table 2.15 Coefficient of water permeability of concretes after 28 days of conditioning [91]

Mixture	PFA	Coefficient of water permeability ($\times 10^{-13}$m/s)						
		20 °C, 65 % RH[a]			20 °C, 40 % RH[a]			28 days of water storage
		Curing period			Curing period			
		1 days	3 days	7 days	1 days	3 days	7 days	
A	0 %	3800	500	320	2200	910	280	1.79
B	15 % M367	3100	270	33	1800	66	60	0.80
C	30 % M367	3800	1100	100	2300	1200	130	0.70
D	50 % M367	3800	130	130	7000	1900	160	0.89
E	30 % M368	1800	36	36	1021	62	5.6	0.63
F	30 % M369	2400	52	52	2000	790	9.5	0.60

[a] Storage conditions following curing

Effects of Fly Ash on Carbonation of Concrete

Calcium hydroxide and, to a lesser extent, calcium silicates and aluminates in hydrated Portland cement react in moist conditions with carbon dioxide from the atmosphere to form calcium carbonate. The process, termed carbonation, occurs in all portland cement concretes. The rate at which concrete carbonates is determined by its permeability, the degree of saturation with water, and the mass of calcium hydroxide available for reaction. Well-compacted and properly cured concrete, at a low water/cement, will be sufficiently impermeable to resist the advance of carbonation beyond the first few millimeters.

If carbonation progresses into a mass of concrete, two deleterious consequences may follow: shrinkage may occur; and carbonation of the concrete immediately adjacent to steel reinforcement may reduce its resistance to corrosion.

In 1968, results from long-term investigations of concrete in Japan [92, 93] indicated that concrete made with blended cements was subject to more rapid carbonation than normal Portland cement concrete. Other investigations [94–96] did not show any appreciable differences in this regard, provided that the strengths of the concretes being compared were equal.

In 1980, Tsukayama et al. [97] reported data from experiments conducted over a long period on fly ash concretes exposed in the field in Japan. They found that the depth of carbonation (termed neutralization by the authors) was related to the quality of the concrete in the following ways:

- A linear relationship was found between water/cement (excluding fly ash) and depth of neutralization (Fig. 2.33).
- At identical water/cement, when fly ash was added, depth of carbonation was found to be slightly decreased (see Fig. 2.33).
- The period of curing in water, after casting, was found to have a substantial influence on concrete exposed indoors. Concrete exposed outdoors was considered adequately cured after about 1 week in water (Fig. 2.34).

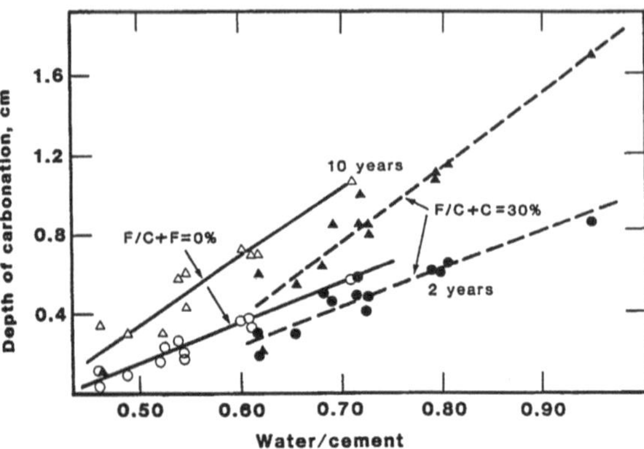

Fig. 2.33 Depth of neutralization (carbonation) versus water/cement of concretes [97]

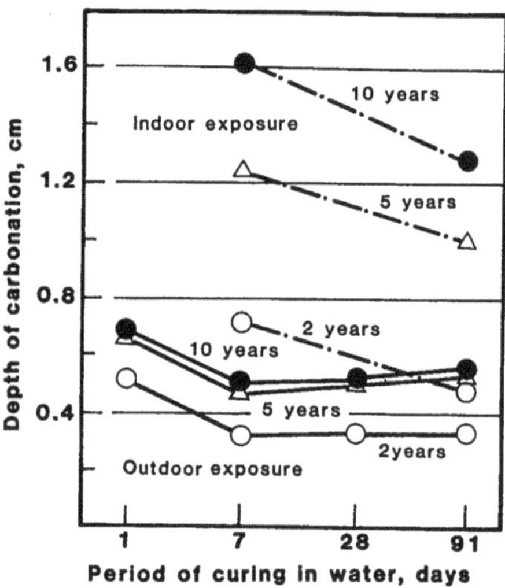

Fig. 2.34 Depth of neutralization (carbonation) versus period of curing of fly ash concretes [97]

Gebauer [98] examined slabs of steel-reinforced concretes with the properties and mixture proportions shown in Table 2.16 the slabs, 100 cm × 50 cm × 7 cm, were formed by compaction and vibration. They were cured for 7 days at 20 °C and 95 % RH. Subsequently, one slab of each composition was placed in an outdoor testing station; control slabs were kept under moist curing conditions (indoor slabs). The compressive strengths of companion prisms were measured at 28 and 365 days.

Tables 2.17 and 2.18 summarize the data from indoor and exposure, respectively.

Gebauer reported that carbonation depth increased as

- Strength and pulse velocity decreased (regardless of composition);
- Water/cement increased;
- Cement content decreased;

Table 2.16 Mixture proportions and properties of concretes [98]

Mix no.	Cement type	Cement (kg/m^3)	W/C	Slump (cm)	Compressive strength (MPa)	
					28 days	1 year
1	Portland #1	300	0.50	3.7	36.7	47.7
2	Portland #1	300	0.55	10.5	32.8	42.3
3	Portland #1	300	0.60	16.0	28.5	38.6
4	Portland #2	300	0.50	2.7	43.2	52.9
5	Portland #2	300	0.55	8.3	37.9	48.0
6	Portland #2	300	0.60	15.0	33.8	43.6
7	Blended #1 10 % ash	300	0.50	5.5	46.4	46.4
8	Blended #1 10 % ash	300	0.55	13.5	42.9	42.9
9	Blended #1 10 % ash	300	0.60	16.0	38.9	38.9
10	Blended #1 20 % ash	300	0.50	8.0	45.2	45.2
11	Blended #1 20 % ash	300	0.55	14.0	39.8	39.8
12	Blended #1 20 % ash	300	0.60	20.0	35.5	35.5
13	Portland #3	350	0.47	3.8	54.8	54.8
14	Blended #3 20 % ash	350	0.47	3.5	55.8	55.8
15	Portland #1	350	0.45	3.3	53.4	53.4
16	Blended #1 20 % ash	350	0.43	4.0	54.8	54.8
17	Blended #1 20 % ash	350	0.46	4.8	54.5	54.5

Table 2.17 Carbonation of concrete slabs stored indoors [98]

Mix no.	Strength (MPa)	Pulse velocity (m/s)	Depth of carbonation (mm)	
			Bottom	Top
1	51.8	4983	0.5	1.0
2	51.5	4885	1.0	2.0
3	46.7	4779	1.0	2.5
4	60.5	5046	0.5	1.0
5	55.5	4970	0.5	1.0
6	50.0	4802	1.0	2.0
7	47.9	4983	1.0	1.5
8	50.3	4897	1.0	1.5
9	42.0	4756	2.0	3.0
10	41.8	4909	1.0	1.5
11	37.5	4837	2.0	2.0
12	35.9	4722	3.0	2.0
13	59.9	5020	0.5	0.5
14	51.7	4995	1.5	1.0
15	62.2	5046	0.5	0.5
16	56.5	5020	1.5	2.0
17	47.5	4958	1.5	2.5

Table 2.18 Carbonation of concrete slabs stored indoors [98]

Mix no.	Strength (MPa)	Pulse velocity (m/s)	Depth of carbonation (mm)	
			Bottom	Top
1	62.0	4945	0.5	1.5
2	57.6	4825	1.0	3.0
3	50.6	4699	1.0	5.0
4	63.1	4921	0.5	1.0
5	61.0	4861	0.5	2.5
6	55.0	4779	1.0	3.0
7	52.7	4897	1.0	3.0
8	48.1	4707	1.0	2.5
9	40.2	4341	2.0	5.0
10	46.7	4873	1.0	5.5
11	40.2	4734	2.0	5.0
12	39.4	4623	3.0	5.5
13	63.9	4945	0.5	1.5
14	53.4	4921	1.5	2.5
15	65.4	4970	0.5	1.5
16	51.6	4995	1.5	2.5
17	51.9	4897	1.5	2.0

- Fly ash content increased.

The primary correlation was between carbonation depth and compressive strength (determined on cores of the slabs, not on test prisms), as is seen in Figs. 2.35 and 2.36.

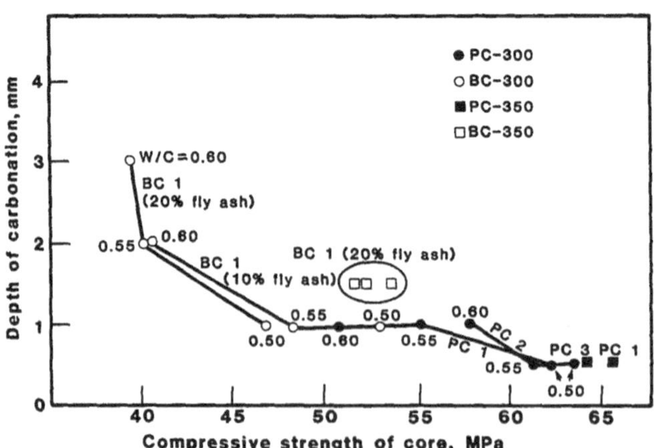

Fig. 2.35 Depth of carbonation versus compressive strength of fly ash concretes (indoor exposure) [98]

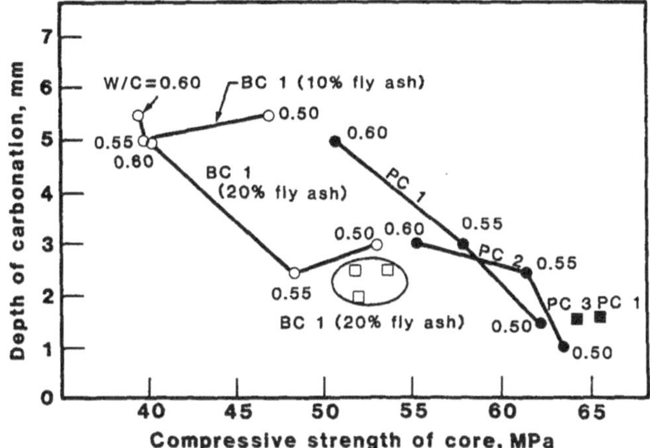

Fig. 2.36 Depth of carbonation versus strength of fly ash concretes (indoor exposure) [98]

The rate of strength development in thin slabs was different for the fly ash concretes and the control concretes. This was not unexpected, because fly ash was incorporated by simple replacement.

Ho and Lewis [99] examined the rates of carbonation of three types of concrete (plain, water-reduced, and fly ash) at equal slump. Accelerated carbonation was induced by storing specimens in an enriched CO_2 atmosphere (4 %) at 20 °C and 50 % RH for 8 weeks. The authors noted that 1 week under these conditions is approximately equivalent to 1 year in a normal atmosphere (0.03 % CO_2).

The data from this study are presented in terms of depth of carbonation versus 28-day compressive strength in Fig. 2.37.

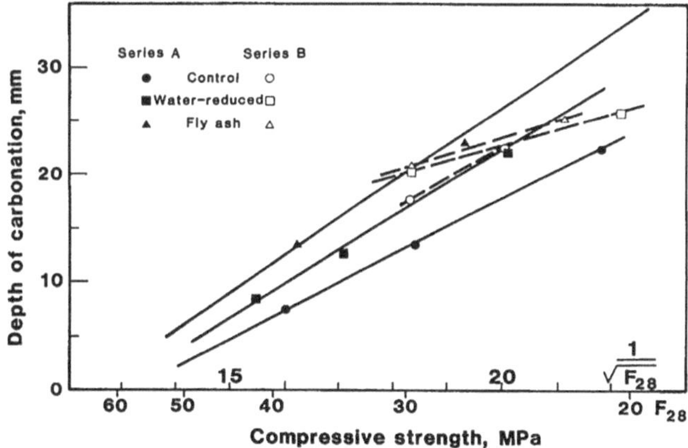

Fig. 2.37 Depth of carbonation versus 28-day compressive strength of fly ash concretes (accelerated testing) [99]

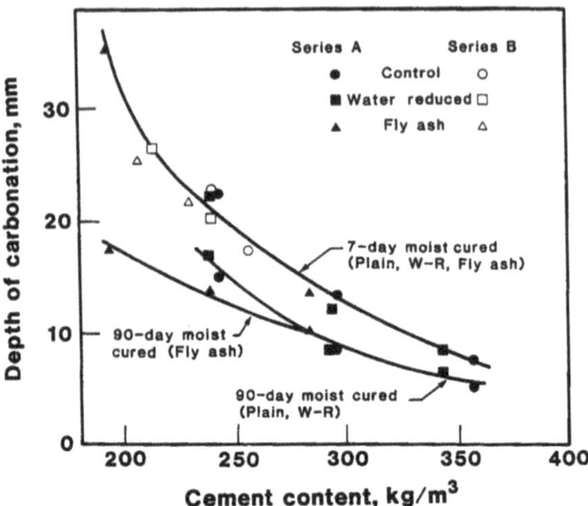

Fig. 2.38 Depth of carbonation versus cement content of fly ash concretes (accelerated testing) [99]

It is clear that there is an inverse relationship between strength and the depth of carbonation and that the fly ash concrete more readily carbonated than the non-fly ash concrete, especially at lower strength (<30 MPa).

Depth of carbonation versus cement content is shown in Fig. 2.38 as a function of the time of moist curing.

The authors concluded as follows:

- Concretes having the same strength and water/cement do not necessarily carbonate at the same rate.
- Based on a common 28-day strength, concrete containing fly ash showed a significant improvement in quality when curing was extended from 7 days to 90 days. such improvement was much greater than that achieved for the plain concrete.
- The depth of carbonation is a function of the cement content for concretes moist cured for 7 days. However, with a further curing to 90 days, concrete containing fly ash showed a slower rate of carbonation than plain and water-reduced concretes.

There is no doubt that these conclusions are consistent with the reported observations. However, the approach taken in this research has a major weakness, the influence of which must not be disregarded. In the accelerated test, concrete specimens that have been moist cured for 7 days and then conditioned in the laboratory for 21 days at 20 °C and 50 % RH are exposed to CO_2 (at elevated pressure) for 8 weeks. The age of the concrete at the start of the test is 28 days. Its maturity, however, is considerably less than an equivalent 28-day moist-cured concrete (although it is arguable that it is closer to the condition of real concrete in most construction situations). The concrete is exposed to carbonation at a rate at least five times that of atmospheric exposure while at the same time being kept at

50 % RH [99]. The disparity between the rate of carbonation and the rate of maturing is extreme and becomes greater the longer the experiment proceeds.

As Butler et al. [100] noted, the accelerated test may be suitable for different cements with different mixture proportions. It is not applicable when comparing concretes made with Portland cement with those made with blended cements because of the slow rate of pozzolanic reactions (Table 2.19).

Nagataki et al. [101] reported the long-term results of experiments carried out since 1969 to investigate the depth of carbonation in concrete with and without fly ash. Table 9.8 shows the depth of carbonation of different concrete mixtures at the ages of 2, 5, 10, and 15 years. The authors concluded that the initial curing period affects the carbonation of concrete cured indoors; hence, it is necessary for fly ash concrete to have a longer curing period in water at early ages. The carbonation of concrete cured outdoors is not affected by the initial curing period in water, provided it is cured in water for a period of 7 days.

Table 2.19 Average depth of carbonation [101]

C + F (kg/m^3)	F/(C + F) (%)	Age (years)	Average depth of carbonation (mm)					
			Outdoors[a]				Indoors[a]	
			1 days	7 days	28 days	91 days	7 days	91 days
250	0	2	6.4	3.9	3.4	3.9	7.9	5.2
		5	7.0	5.3	5.9	5.7	13.6	11.7
		10	8.6	6.5	7.0	7.4	18.2	15.2
		15	10.7	7.3	7.6	7.4	22.1	18.3
	30	2	8.9	6.4	5.9	6.2	11.1	7.9
		5	13.1	10.3	10.4	10.8	21.0	16.5
		10	14.1	11.0	11.5	11.8	28.4	21.8
		15	14.9	13.3	12.7	14.6	34.5	27.6
290	0	2	5.0	1.9	2.3	2.7	6.1	4.1
		5	4.3	2.0	2.2	3.5	6.2	6.4
		10	4.1	2.0	2.1	3.8	11.4	10.2
		15	3.7	2.2	4.4	5.4	14.6	13.4
	30	2	6.7	5.1	5.5	4.1	9.3	6.6
		5	7.1	5.9	5.5	5.6	14.5	10.3
		10	7.3	7.1	6.0	5.7	20.2	15.7
		15	9.9	8.2	6.3	6.6	22.3	19.0
330	0	2	0.9	0.2	0.1	0.5	2.2	1.4
		5	2.7	1.2	1.6	1.5	5.7	5.0
		10	1.6	0.8	1.2	1.1	6.2	4.4
		15	0.6	0.2	1.0	1.8	9.0	7.9
	30	2	2.8	1.6	2.1	1.6	5.7	3.8
		5	5.7	2.5	4.3	4.6	10.2	9.4
		10	3.6	2.0	2.3	2.0	12.8	9.8
		15	3.0	2.2	3.3	3.6	15.2	13.9

[a] Initial curing period in water

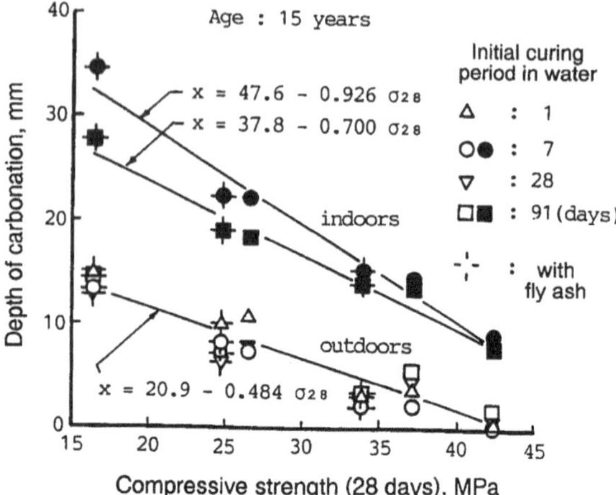

Fig. 2.39 Relation between compressive strength and depth of carbonation [101]

Nagataki et al. [101] found a linear relationship between the compressive strength at 28 days and the depth of carbonation, regardless of fly ash replacement (see Fig. 2.39). This relationship is not valid under changing exposure conditions or initial curing periods in water.

The rate of carbonation can be affected by the type of cement, the environmental conditions, the curing conditions, the water/cement, and the microstructure of concrete. Other published data show that the rate of carbonation is proportional to the water/cement ratio and to the square root of exposure time [102, 103].

Using the results in their work, Nagataki el al. [101] obtained the equation shown below for the rate of carbonation of concretes incorporating fly ash:

$$x = AB\,[\alpha(W/C) - \beta]\,t^{\gamma} \tag{2.2}$$

where x is depth of carbonation (cm); A is the correcting factor; B is the factor for initial curing period in water; ex, α, β are factors for fly ash; W/C is water/cement; t is age (years); and γ is the factor for environmental conditions.

The values of A, B, α, β, and γ, determined under various conditions, are shown in Table 2.20. It is indicated that γ has different values depending on environmental conditions and whether the concrete is cured indoors or outdoors.

Ramezanianpour [104, 105] examined the depth of carbonation of mortars containing some natural and artificial pozzolans under accelerated and long-term

Table 2.20 Coefficients of equation [101]

Exposure condition	F/(C + F) (%)	A	B	α	β	γ
Outdoors	0	0.55		4.211	1.831	
	30	0.77	1	3.311	1.777	0.25
Indoors	0	1.50	1.00 (for 7 days)	1.656	0.622	
	30	1.61	0.78 (for 91 days)	1.445	0.619	0.5

conditions. He found that under all curing and storage conditions the differences between the depth of carbonation of normal Portland cement and that of fly ash mortar mixtures were insignificant. He derived models for the rate of long-term carbonation that took into account the porosity of mortar mixtures and the square root of exposure time.

Ohga and Nagataki [106] investigated the effect of replacement ratio of fly ash, initial curing period in water, and air content on the carbonation phenomenon in concrete. They reported that the carbonation of concrete with fly ash was affected by initial curing conditions and increased with an increase in the replacement ratio of fly ash. They also found that the depth of carbonation of concrete was in proportion to the square root of exposure duration in the accelerated carbonation test.

In general, it appears that good-quality fly ash concrete is comparable to plain concrete in its resistance to carbonation. If concrete is placed at a low cement factor, with insufficient curing (either lack of moisture or low temperature), it should come as no surprise to find that it is not resistant to all forms of chemical and physical aggression, including carbonation [107].

Effects of Fly Ash on the Durability of Concrete Subjected to Repeated Cycles of Freezing and Thawing

It is now generally accepted, other criteria also being met, that air entrainment renders concrete frost resistant. Fly ashes, in common with other finely divided mineral components in concrete, tend to cause an increase in the quantity of admixture required to obtain specified levels of entrained air in concrete. In some instances, the stability of the air or the rate of air loss from fresh concrete is also affected. In general, the observed effects of fly ash on freezing and thawing durability support the view expressed by Larson [108]

> Fly ash has no apparent ill effects on the air voids in hardened concrete. When a proper volume of air is entrained, characteristics of the void system meet generally accepted criteria.

Gebler and Klieger [109] extended their study of air entrainment of fly ash concrete to include an examination of the air-void parameters of hardened concretes cast after initial mixing and after 30, 60, and 90 min. From these experiments, the authors concluded as follows:

> Spacing factors ($\overline{L}$) of specimens cast over a period of 90 min were essentially constant for the majority of concretes containing fly ash. In addition, the initial spread of results of specific surface and voids per inch was essentially similar for concretes containing Class F or Class C ash. However, when measured on specimens cast at 90 min, concretes with Class F ash exhibited greater variability of results for these air-void parameters than concretes with Class Cash.

Sturrup et al. [110] related the freezing and thawing performance of low-calcium fly ash concretes to carbon content. Accelerated freezing and thawing tests (ASTM C 666, procedure A) and outdoor exposure tests were conducted on concrete specimens containing low-calcium ashes of 5.4, 12.3, and 23 % carbon at 15, 30, 45, and 60 % replacement of cement by weight. Water/(cement + fly ash) was kept constant at 0.6; air content was kept at 6.5 ± 1 %, with the exception of a specimen with the fly ash containing 23 % carbon (air content = 3.6 %).

It was reported that correlation between durability factor, as determined by resonant frequency, and carbon content was poor; correlation with weight loss resulting from freezing and thawing cycling was more definite. This was taken to indicate that surface scaling, rather than internal damage, was the result of frost action on the specimens [111]. This was confirmed by observation of outdoor-exposed specimens [112].

Yuan and Cook [52] reported on the freezing and thawing resistance of concrete incorporating high-calcium fly ash. They examined two series of concrete specimens (non-air-entrained, air-entrained) with 0, 20, 30, and 50 % replacement of cement by weight. The freezing and thawing durability as determined by the relative dynamic modulus is shown in Figs. 2.40 and 2.41 and as determined by weight loss for air-entrained concrete is shown in Fig. 2.42.

- Yuan and Cook made the following resistance of air-entrained concrete is evident with or without fly ash replacement.
- The concrete with 20 % fly ash was found to be more durable than the control.
- As the quality of fly ash in air-entrained concrete was increased to 50 %, more scaling damage was noted after 400 cycles.

Ramakrishnan et al. [37] also examined the freezing and thawing durability of concrete containing a high-calcium fly ash (CaO = 20.1 wt%). Their data for concretes with ASTM type III cement are shown in Figs. 2.43 and 2.44. Weight

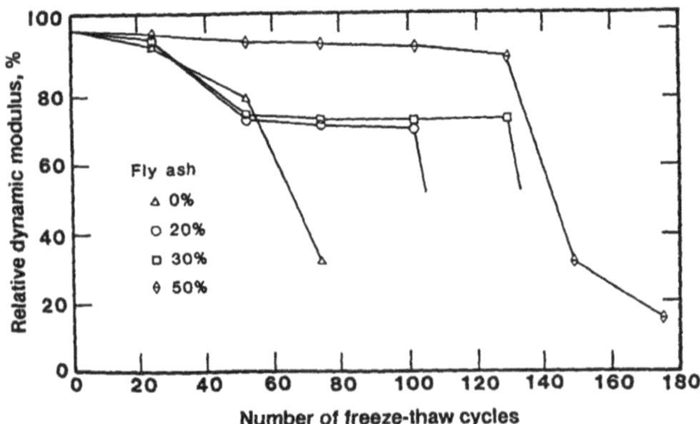

Fig. 2.40 Relative dynamic modulus of elasticity versus number of freezing and thawing cycles for air-entrained fly ash concretes after 14 days of curing [52]

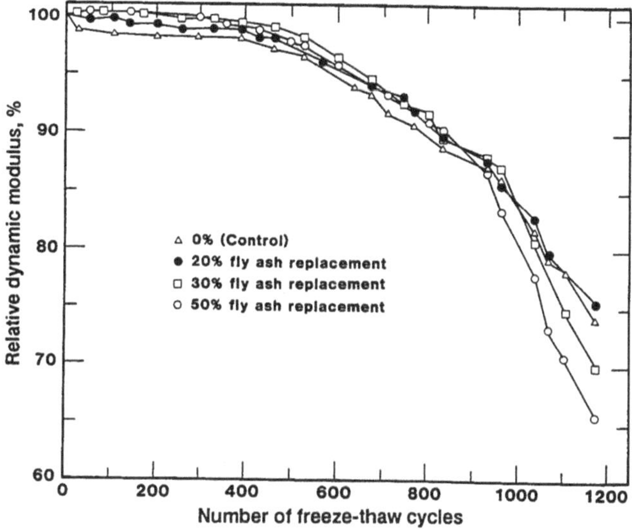

Fig. 2.41 Relative dynamic modulus of elasticity versus number of freezing and thawing cycles for air-entrained fly ash concretes after 14 days of curing [52]

Fig. 2.42 Weigth loss versus number of freezing and thawing cycles for air-entrained fly ash concretes after 14 days of curing [52]

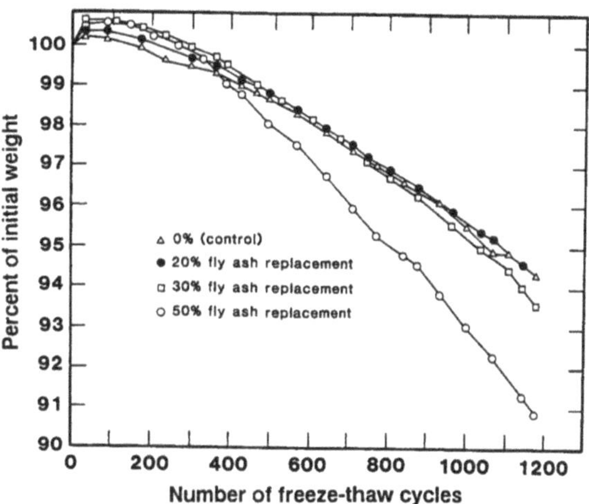

change, pulse velocity development, and dynamic modulus were similar for concretes with and without fly ash.

The freezing and thawing resistance of fly ash concretes incorporating both low- and high-calcium fly ashes was determined in the CANMET study [6]. Table 2.21 shows the durability factors found for specimens made from the concrete described elsewhere in this report.

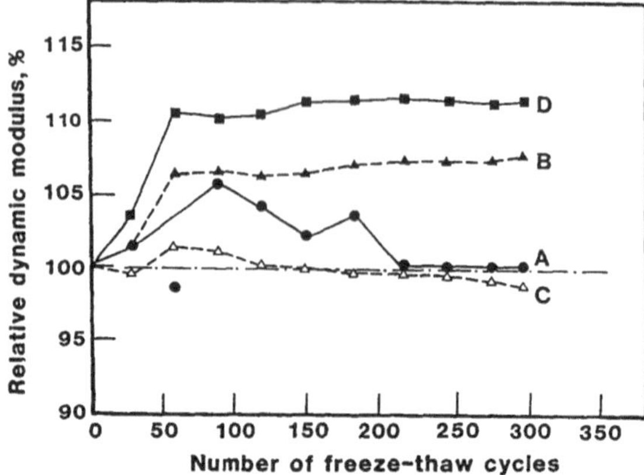

Fig. 2.43 Relative dynamic modulus changes of fly ash concrete under rabid freezing and thawing cycling [37]. *A*, control mix (freeze–thaw cure); *B*, fly ash mix (standard moist cure); *C*, fly ash mix (freeze–thaw cure); *D*, control mix (standard moist cure)

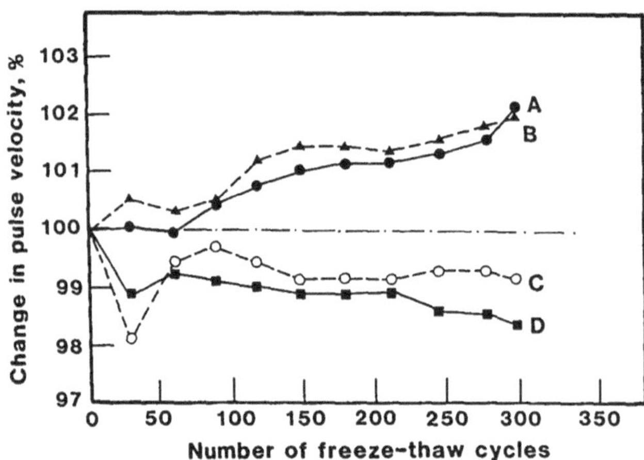

Fig. 2.44 Pulse velocity changes of fly ash concretes under rapid freezing and thawing cycling [37]. *A*, control mix (standard moist cure); *B*, fly ash mix (standard moist cure); *C*, fly ash mix (freeze–thaw cure); *D*, control mix (freeze–thaw cure)

Virtanen [115] compared the freezing and thawing resistance of fly ash, condensed silica fume, and slag concretes. Virtanen's observations on AEA requirement were discussed before. An estimate was made of the relative theoretical service life for each type of concrete as a function of air content (Fig. 2.45), and Virtanen made the following observations:

- The air content has the greatest influence on the freeze–thaw resistance of concrete.

Table 2.21 Freezing and thawing durability factors for fly ash concretes [34]

Mixture no.	Air content (%)	Durability for fly ash concretes[a]
Control 1	6.5	97.7
Control 2	6.4	98.1
F1	6.2	96.4
F2	6.2	98.8
F3	6.2	96.8
F4	6.3	98.8
F5	6.4	97.2
F6	6.5	96.8
F7	6.1	97.6
F8	6.2	96.9
F9	6.4	97.6
F10	6.5	97.2
F11	6.6	95.8

[a] Determined in accordance with ASTM 666

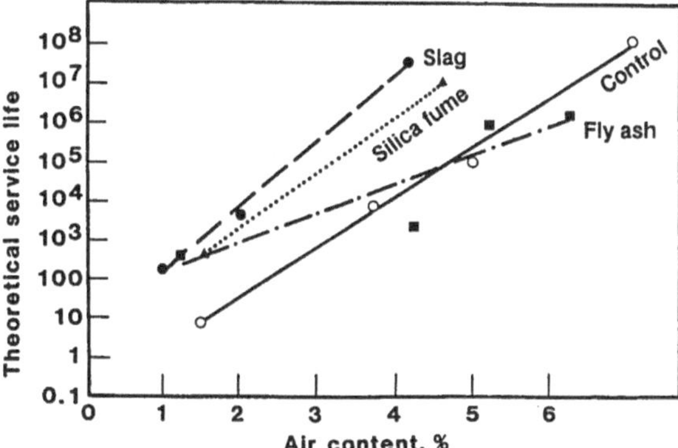

Fig. 2.45 Relationship between theoretical service life and air content of fresh concrete [115]

- Addition of fly ash has no major effect on the freeze–thaw resistance of concrete if the strength and air content are kept constant.
- The addition of blast-furnace slag fly ash may have a negative effect of the freeze–thaw resistance of concrete when a major part of the cement is replaced by them.

Larson [108], discussing some of the difficulties of interpreting the findings of much of the early works on freezing and thawing resistance of fly ash concrete, made the following observation:

Fly ash concrete durability characteristics are influenced and obscured by all the factors operating on ordinary concrete. They are also related to variations in the fly ash itself and perhaps to the associated phenomenon of increased air-entraining-agent requirement. When valid comparisons are made with equal strengths and air contents, however, there are no apparent differences in the freezing and thawing durability of fly-ash and non-fly-ash concretes.

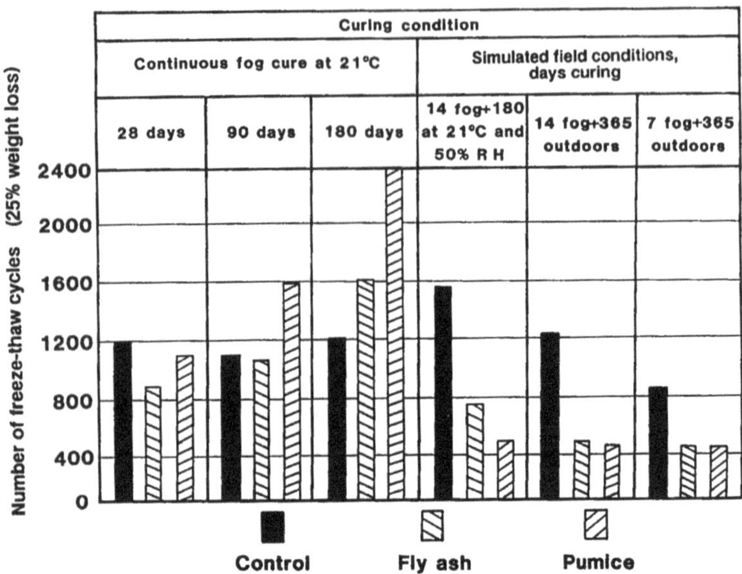

Fig. 2.46 The effect of curing conditions in freezing and thawing durability of concrete containing pozzolans [47]

Elfert [47] noted that the type of curing of specimens used to evaluate freezing and thawing resistance greatly influences the results obtained with fly ash and pozzolan concretes. Figure 2.46 illustrates the differences in results obtained under moist-curing and simulated field condition.

Another aspect of the freezing and thawing testing procedure was criticized by Brown et al. [113] who made the following comments on the freezing and thawing testing of blended cements:

- When blended cements are tested according to ASTM C 666-73, the standard method for measuring the freezing and thawing durability of Portland cement concretes, inferior resistance is usually observed. This is probably because test initiation after only a short curing period does not make proper allowance for the generally lower rate of strength development of blended cements.
- Freezing and thawing studies, when initiated after longer curing periods, have indicated that blended cements, due to development of strengths equivalent or superior to those of Portland cements, also develop superior resistance to freezing and thawing.

These points should certainly be considered when one is reviewing reports of all aspects of the durability of fly ash concrete, not merely its frost resistance.

Johnston [114] studied the freezing and thawing durability of concretes containing up to 42 % ASTM Class C fly ash by weight of cement with various ratios of water to cementitious material. Johnston concluded the following: [116]

- Replacement of cement by Class C fly ashes does not detract from freezing-thawing performance in ASTM C 666, Procedure A, provided the dosage of air-entraining admixture is increased to the point needed to achieve an air void-spacing factor of less than 0.25 mm, and provided water-to-cementitious materials ratio does not exceed about 0.53. At W/C + F values of 0.6 or higher, DF_{300} values show a definite tendency to decrease to below 90 % of their values for equivalent concretes without fly ash.
- For concretes with Class C fly ash, satisfactory performance in ASTM C 666, Procedure A, is not necessarily accompanied by satisfactory scaling resistance in ASTM C 672. They exhibit more severe scaling than control concretes without fly ash at W/C + F as low as 0.53, despite apparently adequate air void spacing factor. However, under field conditions, concrete with the same fly ash, but a lower W/C + F of 0.4, has not scaled noticeably after eight winters of exposure to heavy traffic and deicing salt. Therefore, it appears that the safe upper limit of W/C + F with respect to scaling is no greater than the 0.45 recommended for concrete without fly ash.

Nasser and Lai [117] studied the effects of a lignite fly ash on the resistance of concrete to freezing and thawing. Their results showed that the use of high percentages of fly ash in concrete (35–50 %) reduced its resistance to freezing and thawing even though it contained ~6 % air and was moist cured for 80 days. Nasser and Lai found that concrete containing 20 % fly ash had a satisfactory performance, provided its air content and strength were comparable to those of control concrete.

Klieger and Gebler [118] also evaluated the durability of concretes containing ASTM Class F and Class C fly ashes. Their test results indicated that air-entrained concretes, with or without fly ash, that were moist cured at 23 °C generally showed good resistance to freezing and thawing. For the specimens cured at a low temperature (4.4 °C), air-entrained concretes with Class F fly ash showed slightly less resistance to freezing and thawing than similar concretes made with Class C fly ash.

Klieger and Gebler also carried out the Deicer scaling test of air-retained concretes in the presence of a 4 % NaCl solution. Their results showed that the performance of air-entrained concretes without fly ash was better than that of those with fly ash, regardless of the type of curing. Air-entrained concretes made with either class of fly ash and air cured at 23 °C showed equal results when subjected to NaCl solution during freezing and thawing tests.

Bilodeau et al. [119], in an investigation carried out at CANMET, determined the scaling resistance of concrete incorporating fly ashes. Water/(cement + fly ash) of 0.35, 0.45, and 0.55 were used, and concrete without fly ash and those containing 20 and 30 % fly ash as replacement by mass for cement were made.

Their results showed that the concrete containing S $\leq$ 30 % fly ash performed satisfactorily under the scaling test with minor exceptions (see Tables 2.22 and 2.23).

However, the performance of the fly ash concrete was more variable than that of the control concrete. The satisfactory scaling resistance of concrete (scaling residue < 0.8 kg/m^2) is attributed to the adequate air-void system in all cases, even

Table 2.22 Mass of scaling residue after 50 freezing and thawing cycles-series I [140]

Time of moist curing (days)	Time of air drying (weeks)	Mass of scaling residue (kg/m^2)								
		w/(C + F) = 0.35			w/(C + F) = 0.45			w/(C + F) = 0.55		
		Percentage of fly ash			Percentage of fly ash			Percentage of fly ash		
		0	20	30	0	20	30	0	20	30
3	3	0.195	0.504	0.184	0.149	0.160	0.206	0.123	0.282	0.321
	4	0.122	0.237	0.208	0.178	0.200	0.680	0.126	0.281	0.638
	5	0.076	0.128	0.143	0.091	0.243	0.634	1.131	0.734	0.354
	6	0.100	0.074	0.158	0.105	0.306	0.263	0.129	0.255	0.226
7	3	0.154	0.047	0.371	0.135	0.212	0.362	0.160	0.335	0.426
	4	0.147	0.092	0.265	0.158	0.448	0.209	0.172	0.312	0.885
	5	0.098	0.108	0.151	0.114	0.180	0.199	0.119	0.238	0.396
	6	0.192	0.038	0.223	0.169	0.268	0.177	0.118	0.370	0.562
14	3	0.139	0.670	1.094	0.188	0.264	0.409	0.517	0.895	0.705
	4	0.144	0.158	0.449	0.126	0.198	0.202	0.131	0.636	0.625
	5	0.1	0.066	0.493	0.135	0.319	0.839	0.162	0.811	0.613
	6		0.064	0.189	0.117	0.293	0.463	0.286	0.728	0.814

Table 2.23 Mass of scaling residue after 50 freezing and thawing cycle-series II, III, IV, V [140]

Time of moist or membrane curing (days)	Time of air drying (weeks)	Mass of scaling residue (kg/m^2)											
		Mixture series II Mixture nos. 10–12 (membrane cured)			Mixture series III Mixture nos. 13–15 (membrane cured)			Mixture series IV Mixture nos. 16–18 (moist cured)			Mixture series V Mixture nos. 19–21 (membrane cured)		
		Percentage of fly ash			Percentage of fly ash			Percentage of fly ash			Percentage of fly ash		
		0	20	30	0	20	30	0	20	30	0	20	30
3	3	0.129	0.105	0.100	0.072	0.070	0.233	0.194	0.497	0.671	0.215	0.292	0.407
	4	0.069	0.084	0.098	0.062	0.073	0.052	0.225	0.250	0.485	0.253	0.158	0.223
	5	0.087	0.069	0.127	0.057	0.094	0.117	0.105	0.371	0.289	0.266	0.366	0.112
	6	0.062	0.072	0.096	0.097	0.128	0.128	0.176	0.389	0.359	0.240	0.116	0.160
7	3	0.241	0.154	0.111	0.073	0.085	0.097	0.240	1.197	0.527	0.264	0.248	0.203
	4	0.092	0.266	0.136	0.087	0.105	0.164	0.215	0.318	1.041	0.185	0.271	0.374
	5	0.095	0.158	0.174	0.081	0.062	0.157	0.150	0.323	0.491	0.297	0.496	0.174
	6	0.080	0.092	0.201	0.078	0.166	0.162	0.338	0.119	0.392	0.187	0.338	0.325
14	3	0.031	0.096	0.132	0.052	0.205	0.239	0.770	0.157	2.063	0.142	0.350	0.321
	4	0.081	0.119	0.103	0.046	0.064	0.269	0.199	0.549	0.559	0.247	0.450	0.103
	5	0.057	0.103	0.127	0.085	0.099	0.139	0.164	0.318	1.962	0.126	0.130	0.152
	6	0.043	0.154	0.100	0.057	0.258	0.098	0.195	0.337	0.404	0.128	0.273	0.633

Note W/(C + F) = 0.45

for fly ash mixtures that have a water/cementitious materials of 0.55 and are moist cured for 3 days.

Carette and Langley [120] studied the performance of fly ash concrete subjected to 50 freezing and thawing cycles in the presence of de-icing salts. They concluded that the incorporation of fly ash in concrete mixtures with ≤30 % replacement of Portland cement did not show significant difference in salt-scaling resistance in the presence of a 4 % calcium chloride solution when examined by visual rating of surface deterioration. In the measurement of weight loss due to surface deterioration, which they believed was a meaningful way to assess surface deterioration, concretes containing fly ash showed greater weigh t loss than control concrete.

Carette and Langley [120] stated that the surface scaling appeared not to be sensitive to the length of time that specimens were moist cured or air dried subsequent to initial moist curing, at least within the periods investigated.

Effects of Fly Ash on the Durability of Concrete Exposed to Elevated Temperatures

The influence of elevated temperatures on the strength of concrete during curing was discussed at some length in the strength section. In recent years, the use of concrete in structures required to withstand elevated temperatures under some circumstances (such as nuclear reactor containment structures) has generated renewed interest in the effects of high temperatures on fly ash and other concretes.

Nasser and Lohtia [121, 122] and Nasser and Marzouk [75, 123] studied plain and fly ash concretes at temperatures ≤230 °C. Carette et al. [124] studied concretes with normal Portland cement, slag, and fly ash at sustained temperatures ≤600 °C. Data from this research are shown in Fig. 2.47. In general, the incorporation of fly ash appears not to influence the behavior of concrete at elevated temperatures. Loss of strength and changes in other structural properties occur at approximately the same temperatures for both types of concrete.

Abrasion and Erosion of Fly Ash Concrete

Under many Circumstances, concrete is subjected to wear by attrition, scraping, or the sliding action of vehicles, ice, and other objects. When water flows over concrete surfaces, erosion may occur. In general, regardless of the type of test performed, the abrasion resistance of concrete is usually found to be proportional to its compressive strength [63]. Similarly, at constant slump, resistance to erosion improves with increased cement content and strength. It may be anticipated that fly ash concrete that is incompletely or inadequately cured may show reduced resistance to abrasion.

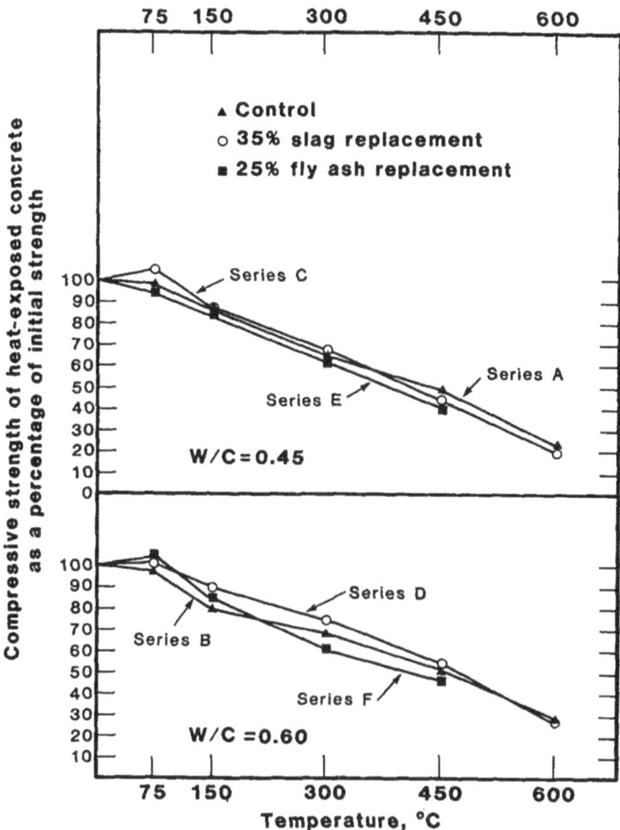

Fig. 2.47 Compressive strength of concretes after 1 month of exposure to various elevated temperatures [124]

Abdun-Nur [69] cited three publications indicating that abrasion resistance may be reduced in fly ash concrete. However, in all of these there is no indication that attempts were made to compare fly ash concrete and plain concrete at equal strength or equal maturity.

Liu [125] examined the abrasion-erosion resistance of concrete using a newly developed underwater abrasion test. One of the concrete mixtures examined by Liu incorporated fly ash at a 25 vol% replacement for Portland cement. Details of the type and origin of the fly ash used were not reported. Performance of the fly ash concrete, cured for 90 days to an average compressive strength of 49 MPa, was compared with that of a concrete of similar mixture proportions containing no fly ash and cured for 28 days to an average compressive strength of 47 MPa.

Little difference in abrasion resistance was found between the two concretes for test periods ≤36 h. At prolonged test times, the performance of the fly ash

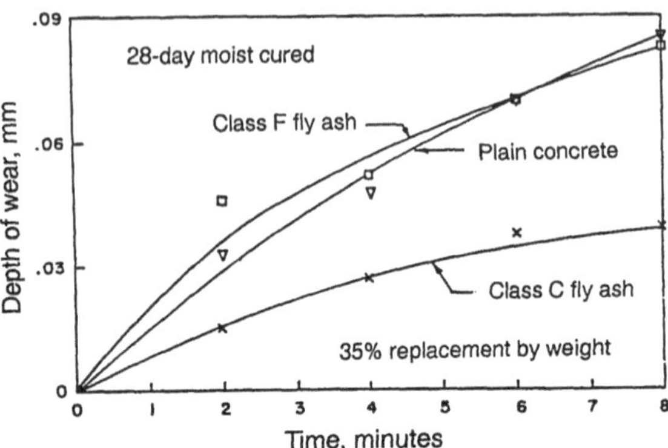

Fig. 2.48 Abrasion resistance of concrete [126]

concrete was inferior to that of the control. After 72 h, the fly ash concrete had lost ~25 % more mass (7.6 % loss) to abrasion-erosion than the control (61 % loss).

Carrasquillo [126] examined the abrasion resistance of concretes containing no fly ash, 35 % ASTM Class C fly ash, or 35 % ASTM Class F fly ash. Specimens tested were cast from concretes having similar strengths, air contents, and cementitious materials contents. As shown in Fig. 2.48, the resistance to abrasion of concrete containing Class C fly ash was greater than that of the concrete containing Class F fly ash or no fly ash. The letter two exhibited approximately equal abrasion resistance; measurement was based on the depth of wear.

Naik et al. [127] carried out an investigation of the compressive strength and abrasion resistance of concrete containing ASTM Class C fly ash. They proportioned concrete mixtures to have cement replacement in the range of 15–70 wt% fly ash. The water/cementitious materials varied from 0.31 to 0.37. Their results showed that the abrasion resistance of concrete containing ≤30 % fly ash was similar to that of the control concrete. However, abrasion resistance of concretes containing ≤40 % fly ash was lower than that of control concrete without fly ash.

Effects of Fly Ash on the Durability of Concrete Exposed to Chemical Attack

Introducing fly ash as a component of concrete has been shown to influence the concrete's resistance to chemical attack. Leaching of calcium hydroxide, acidic dissolution of cementitious hydrates, the action of atmospheric and dissolved carbon dioxide, and the reactivity of cement components to ions in solution are the main causes of deterioration of concrete exposed to chemical action.

Biczok [128] enumerated four conditions related to concrete quality and the constituents of concrete on which the destructive effects of aggressive waters depend:

1. type of cement used and its chemical and physical properties;
2. quality of concrete aggregates and their physical properties and gradation;
3. method used for preparing concrete, the water/cement, the proportion of cement, and the placement; and
4. condition of the surface exposed to the water.

Of these, condition 1 relates strictly to the nature of the cementitious binder used, whereas conditions 2, 3, and 4 apply to one or more aspects of the permeability of concrete.

With regard to cement type, two factors are influential in determining the relative durability of fly ash concrete:

- The chemical composition of the cement, vis-a-vis the cementitious components produced during hydration, has a pronounced influence on resistance to chemical action. The most notable example is the use of low-C_3A (AS 1 M type V) cements as a means of controlling attack by sulphate ions.
- A combination of chemical composition and physical properties, notably fineness, determines the rate at which cement hydration proceeds and, at least for the early life of a structure, must influence its permeability.

Fly ash used as a replacement for Portland cement has an indirect influence on both factors. At early ages, fly ash serves only as an inert component. At later ages, it contributes to the formation of cementitious components but, as Kovacs [129] showed, it does so in a manner that changes the relative proportions of the usual hydrate materials. Finally, it converts some of the calcium hydroxide, which is produced when cement hydrates, to less reactive calcium silicates and aluminates through the pozzolanic reaction. The removal of free calcium hydroxide by reactive combination with pozzolans was shown by Lea [130] to progress as is illustrated in Fig. 2.49, in which the quantities of free $Ca(OH)_2$ in mortars made with and without pozzolan are compared as a function of age.

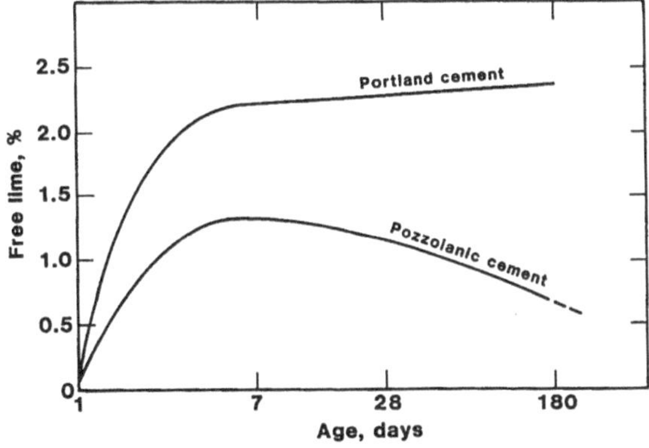

Fig. 2.49 Free lime content of 1:3 cement/sand mortars [130]

It is generally agreed that in concrete this process leads to long-term gains in watertightness, strength, and resistance to aggressive environments.

In recent years, research has been directed at the role played by fly ash in changing the chemical balance of the cementitious components of concrete, either as a factor in concrete durability or with respect to the development of test methods.

Effects of Fly Ash on Sulphate Resistance of Concrete

In 1937, Davis et al. [1] reported that some fly ashes increased the resistance of concrete to sulphate attack, others were ineffective, and some were deleterious and caused increased sulphate deterioration.

In 1967, Dikeou [131] reported the results of sulphate-resistance studies on 30 concrete mixtures made with Portland cement, Portland -fly ash cement, or fly ash. From this work it was concluded that all of the 12 fly ashes tested greatly improved sulphate resistance. The relative order of improvement is shown in Fig. 2.50.

Kalousek et al. [132] reported studies on the requirements of concrete for long-term service in a sulphate environment. From their study, they drew the following conclusions:

- 84 % of the AS1 M types V and II cement concretes without pozzolan showed a life expectancy of <50 years.
- Certain pozzolans very significantly increased the life expectancy of concrete exposed to 2.1 % sodium sulphate solution. Fly ashes meeting present-day specifications were prominent among the group of pozzolans showing the greatest improvements.

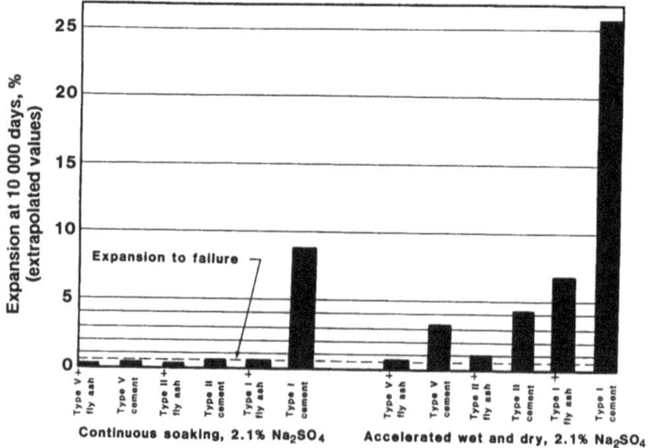

Fig. 2.50 Sulphate expansion of concretes containing 30 % fly ash [131]

- Concretes for long-term survival in a sulphate environment should be made with high-quality pozzolans and a sulphate-resisting cement. The pozzolan should not increase significantly, but preferably decrease, the amount of water required.
- Cement to be used in making sulphate-resisting concrete with pozzolan of proven performance should have a maximum C_3A content of 6.5 % and maximum C_4AF content of 12 % Restrictions of cements to those meeting present-day specifications for type V cement does not appear justified.

The fly ash samples examined by Dikeou [131] and those examined by Kalousek et al. [132] all originated from bituminous coals.

In 1976, Dunstan [133] reported the results of experiments on 13 concrete mixtures made with fly ashes from lignite or subbituminous coal sources. On the basis of this work, he concluded that lignite and subbituminous fly ash concrete generally exhibited reduced resistance to sulphate attack.

Dunstan's work was extended, and in 1980 he published a report summarizing the results of a 5-year study on sulphate attack of fly ash concretes [134]. This report includes a theoretical analysis of sulphate attack and its causes. The basic postulate of Dunstan's thesis is that CaO and Fe_2O_3 in fly ash are the main contributors to the resistance or susceptibility of fly ash concrete to sulphates.

Dunstan noted that as the calcium oxide content of ash increases above a lower limit of 5 % or as the ferric oxide content decreases, sulphate resistance is reduced. To select potentially sulphate-resistant fly ashes (or, more important, fly ashes that can improve the sulphate resistance of concrete), Dunstan proposed the use of a resistance factor *(R)*, calculated as follows:

$$R = (C - 5)/F \tag{2.3}$$

where C is the percentage of CaO; and *F* is the percentage of Fe_2O_3.

Figures 2.51 and 2.52 show the results from two types of laboratory sulphate-resistance tests on samples of concretes containing high-calcium fly ashes with the properties shown in Table 2.24. Figures 2.53 and 2.54 show the results from similar tests on samples of concretes containing low-calcium fly ashes with the properties also shown in Table 2.24. The influences (positive and negative) of fly ash are clearly seen from these data.

The findings of Dunstan's work were summarized in terms of the selection of fly ashes for sulphate-resistant concrete as follows:

R limits[a]	Sulphate resistance
<0.75	Greatly improved
0.75–1.5	Moderately improved
1.5–3.0	No significant change
>3.0	Reduced

[a] At 25 % cement replacement

[b] Relative to ASTM type II cement at a water/cement of 0.45

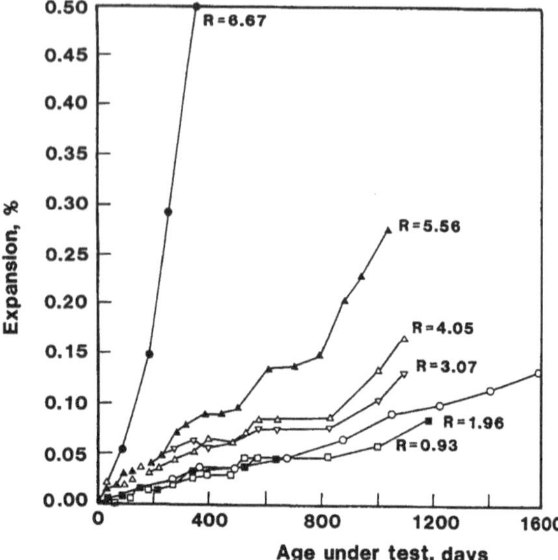

Fig. 2.51 Sulphate expansion of concretes containing high-calcium fly ash (soak test) [134]

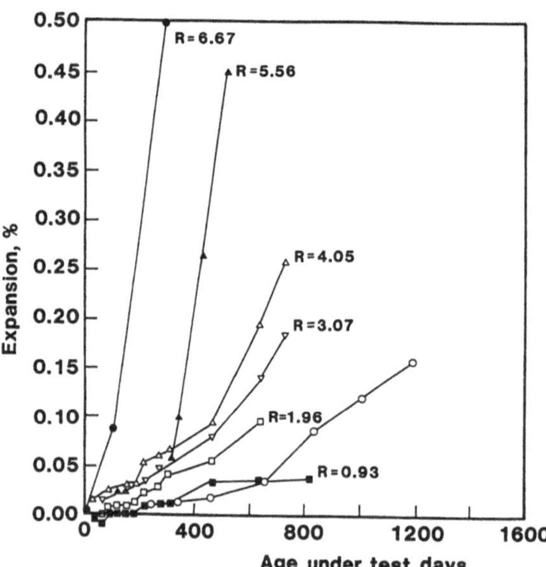

Fig. 2.52 Sulphate expansion of concretes containing low-calcium fly ash (wet-dry test) [133]

The U.S, Bureau of Reclamation incorporated a more conservative version of Dunstan's limits in its revised reprint of the Concrete Manual [135, 136], details of which are given in Table 2.25.

Table 2.24 Characteristics of fly ashes examined by Dunstan [126, 134]

Fly ash no.	Composition (mass %)							
	SiO_2	Al_2O	Fe_2O	CaO	MgO	SO_3	Alkali	LO1
M-6498	46.1	19.0	18.6	8.2	1.3	1.6	0.72	2.0
M-6498	28.1	20.0	4.1	32.0	6.4	3.8	0.68	0.2
M-6498	34.7	24.8	4.2	26.1	5.2	1.4	0.81	0.2
M-6498	37.2	15.6	5.6	24.3	11.3	0.9	0.07	0.3
M-6498	37.1	11.8	7.3	21.8	5.6	2.6	4.23	0.3
M-6498	31.1	17.1	7.9	25.3	8.1	3.3	1.35	1.1
M-6498	51.8	27.2	2.0	10.7	2.1	0.7	0.86	1.2
M-6498	49.6	25.7	3.0	11.3	2.1	0.7	1.20	1.3
M-6498	61.4	23.4	3.7	7.0	1.2	0.5	0.81	2.5
M-6498	32.8	19.6	4.1	28.0	5.5	3.4	1.54	0.5
M-6498	36.9	18.1	4.7	24.0	4.8	2.8	1.86	0.6
M-6498	41.1	17.9	4.9	20.2	4.4	2.2	2.21	0.8
M-6498	45.7	18.4	5.3	15.5	3.8	1.6	2.83	0.9
M-6498	51.5	24.5	5.7	10.2	2.1	0.9	0.96	1.2
M-6498	34.7	24.8	4.2	26.1	5.2	1.4	0.81	0.2

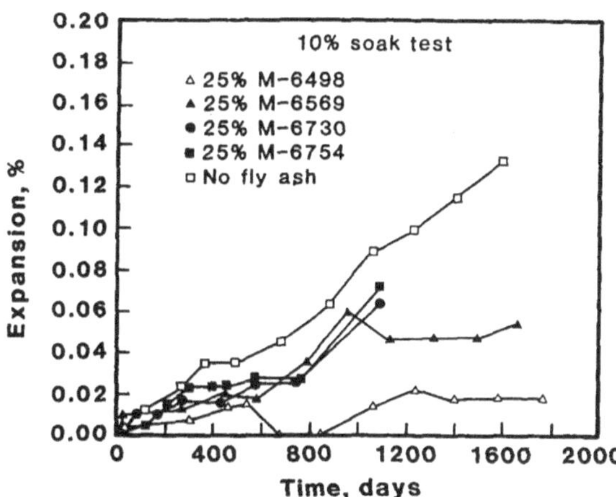

Fig. 2.53 Sulphate expansion of concretes containing low-calcium fly ash (10 % soak test) [133]

Dunstan [137] examined the *R* value as an advance indicator of potential sulphate resistance of concretes containing fly ash after >12 years of testing. The sulphate-resistance test results indicated that the *R* value remained a good indicator of potential sulphate resistance of fly ash concretes. For concrete containing 15–25 % fly ash with an *R* value of <3, the sulphate resistance would be as good as or better than that of a concrete with the same cement and without fly ash.

Tikalsky et al. [138] reported the effect of 24 different fly ashes on the sulphate resistance of fly ash concretes. They found that except for some fly ashes, Dunstan's R factor could provide a good indication of the effect of fly ashes on the sulphate resistance of concrete. Some fly ashes that contained calcium oxide at

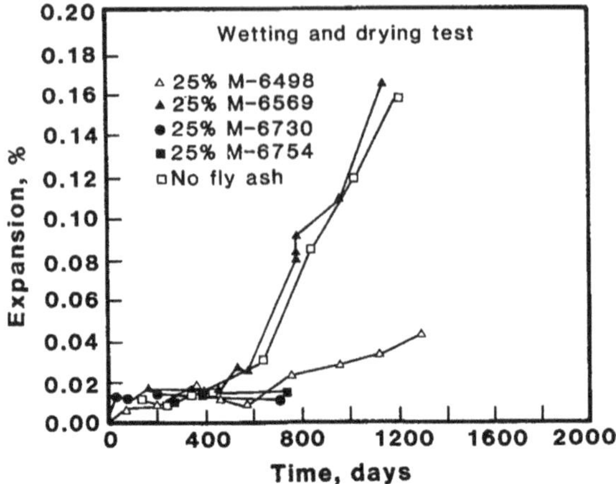

Fig. 2.54 Sulphate expansion of concretes containing low-calcium fly ash (wet-dry test) [133]

Table 2.25 US Bureau of Reclamation cementitious materials options for sulphate resistance [135]

I Positive sulphate attack (0.1–0.2 %, or 150–1500 ppm)
A Type II cement
B Type II cement plus class N, F, or C pozzolan with R < 2.5
C Type IP (MS) cement with R < 2.5
II Sever sulphate attack (0.2–2 %, or 1500–10000 ppm)
A Type V cement
B Type V cement plus class N, F, or C pozzolan with R < 2.5
C Type II cement plus class N, F, or C pozzolan with R < 2.5
D Type IP (MS) cement
with R < 2.5 if C_3A is < 5.0
with R < 1.5 if C_3A is < 5.0–8.0
III A Type V cement plus class N, F, or C pozzolan with R < 1.5
B Type II cement plus class N, F, or C pozzolan with R < 0.75
C Type IP (MS) cement
with R < 1.5 if C_3A is < 5.0
with R < 0.75 if C_3A is < 5.0–8.0

10–25 wt% did not conform to the general trend defined by Dunstan's R factor. In their investigation, the effect of a given ash replacement level of cement within the range of 25–45 vol%.

Mather [139] reported data from two studies at the laboratories of the U.S. Army Corps of Engineers. Various pozzolans were being investigated for their influence on sulphate resistance of concrete. The data presented were obtained from exposure of mortars to 0.352 M Na_2SO_4 solution. Care was taken in the experiments to expose the mortar bars to the sulphate solution only after they had reached approximately equal maturity as determined by measurements of compressive strength on companion mortar cubes.

Three non-sulphate-resisting cements, with C_3A contents of 14.6 % (cement RC-756), 13.1 % (cement RC-714), or 9.4 % (cement RC-744), were used. Ten pozzolans were examined: one condensed silica fume, one volcanic glass, three fly ashes of subbituminous origin, one bituminous fly ash, and four fly ashes from lignite coals. It is unfortunate that the chemical and physical properties of these materials were not recorded in this preliminary report of the study.

The pozzolans were incorporated in mortars by replacement of 30 vol. % of portland cement.

Figures 2.55, 2.56, 2.57 and 2.58 show the results obtained from the experiments reported by Mather [139]. The ranking of effectiveness in reducing sulphate expansion was found to be, from best to worst, as follows: condensed silica fume, volcanic glass, subbituminous fly ash (three examples), bituminous fly ash, and lignite fly ash (four examples).

Mather summarized the findings of the study as follows:

> What seems to be suggested (by the results) IS that a pozzolan of high fineness, high silica content and highly amorphous silica is the most effective pozzolan for reducing expansion due to sulphate attack on mortars made with non-sulphate resisting cements…. The pozzolans that resulted in poorer performance… were in 6 of 7 cases fly ashes produced by the combustion of lignite.

It might be added that the ranking of subbituminous ashes as better than the one bituminous ash examined is in direct contradiction with the findings of Dunstan [133, 134]. However, in the absence of specific properties of the ashes examined

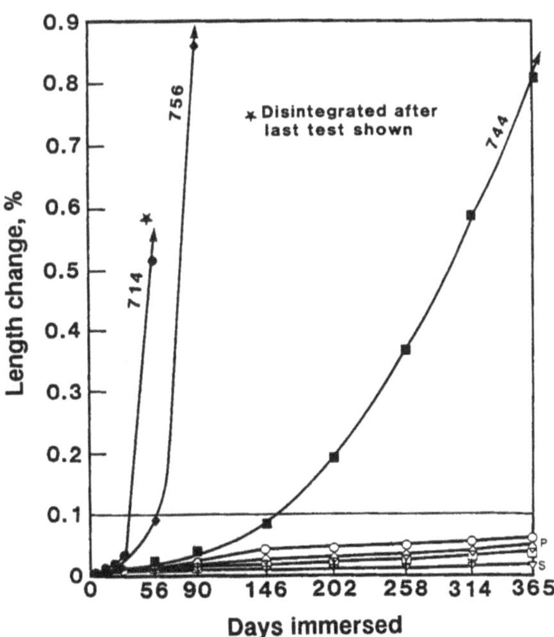

Fig. 2.55 Effect of condensed silica fume and volcanic glass pozzolan on the expansion of mortar prisms incorporating non-sulphate-resisting cement in sulphate solution [139]. *P*, replacement of cement by pozzolan; *S*, 30 % replacement of cement by condensed silica fume; *714*, cement with calculated $C_3A = 13.1$ %; *756*, cement with calculated $C_3A = 14.6$ %; *744*, cement with calculated $C_3A = 9.4$ %

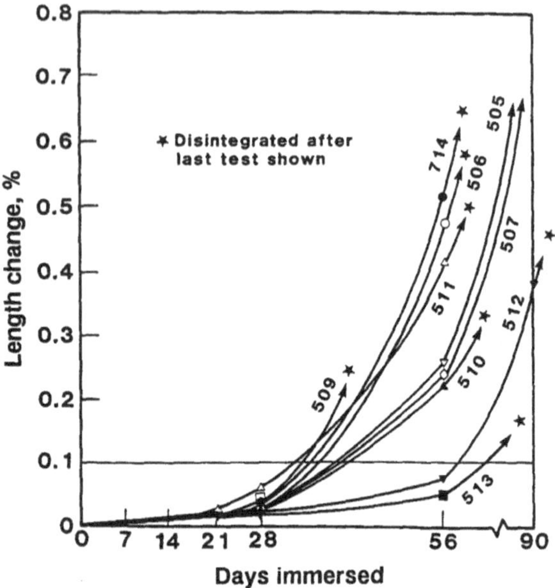

Fig. 2.56 Effect of fly ash on the expansion of mortar prisms incorporating non-sulphate-resisting cement in sulphate solution [139]. *Curve 714* represents cement with calculated $C_3A = 13.1$ %; remaining curves represent replacement of cement *714* by 30 % of eight different, but unidentified, fly ashes

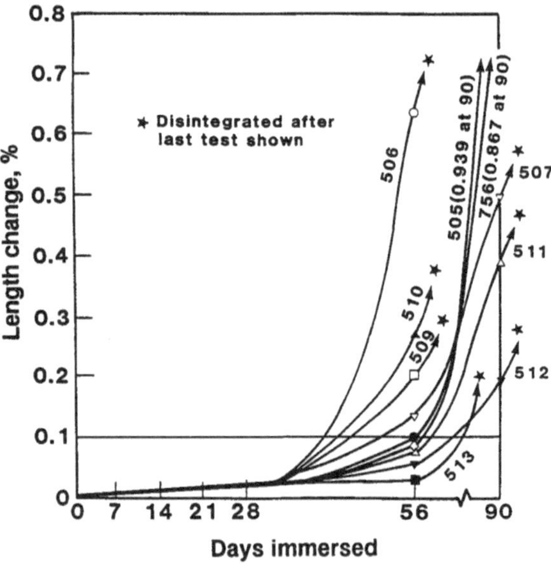

Fig. 2.57 Effect of fly ash on the expansion of mortar prisms incorporating non-sulphate-resisting cement in sulphate solution [139]. *Curve 756* represents cement with calculated $C_3A = 14.6$ %; remaining curves represent replacement of cement *756* by 30 % of eight different, but unidentified, fly ashes

and without comparable data from mortars made with other fly ashes, it would be inadvisable to draw further conclusions from this report. Clearly, the influence of fly ash on the sulphate resistance of concrete is not completely understood, and much more research is needed to establish guidelines on this important aspect of concrete durability.

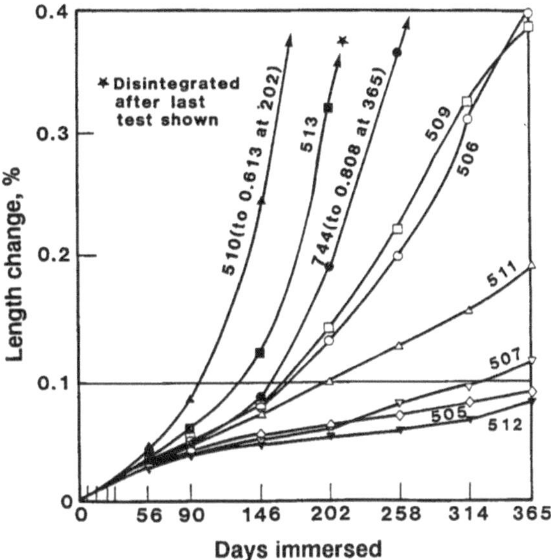

Fig. 2.58 Effects of fly ash on the expansion of mortar prisms incorporating non-sulphate-resisting cement in sulphate solution [139]. *Curve 744* represents cement with calculated $C_3A = 9.4\ \%$; remaining curves represent replacement of cement *744* by 30 % of eight different, but unidentified, fly ashes

Effects of Fly Ash on Alkali-Aggregate Reaction in Concrete

Shortly after Stanton [140] discovered that alkali-aggregate reactions (AAR) caused expansion and damage in some concretes, he reported that the effects could be reduced by adding finely ground reactive materials for the concrete mixture. Subsequently, a variety of natural and artificial pozzolans and mineral admixtures, including fly ash, were found to be effective in reducing the damage caused by AAR. As discussed below, the effectiveness of fly ash (and other mineral admixtures) in reducing expansion due to AAR appears to be limited to reactions involving siliceous aggregates. A form of AAR, known as alkali-carbonate reaction, has been reported and has been shown to be relatively unresponsive to the addition of pozzolans.

The damage caused by AAR. As discussed below, the effectiveness of fly ash (and other mineral admixtures) in reducing expansion due to AAR appears to be limited to reactions involving siliceous aggregates. A form of AAR, known as alkali-carbonate reaction, has been reported [141] and has been shown to be relatively unresponsive to the addition of pozzolans [142].

It is not within the scope of this review to consider numerous complex aspects of AAR; these matters are the subject of much current research and are poorly understood. Rather, consideration here is limited to some aspects directly relevant to the selection of fly ash as a means to reducing expansion caused by alkali-silica reaction.

Table 2.26 Minimum percentage replacement of fly ash for the effective control of expansion [143]

Replacement material	Minimum replacement for effectiveness (vol%)		
	14 days	6 months	Average
Fly ash I	46	36	41
Fly ash II	48	36	42
Fly ash III	52	36	44
Fly ash IV	45	34	40

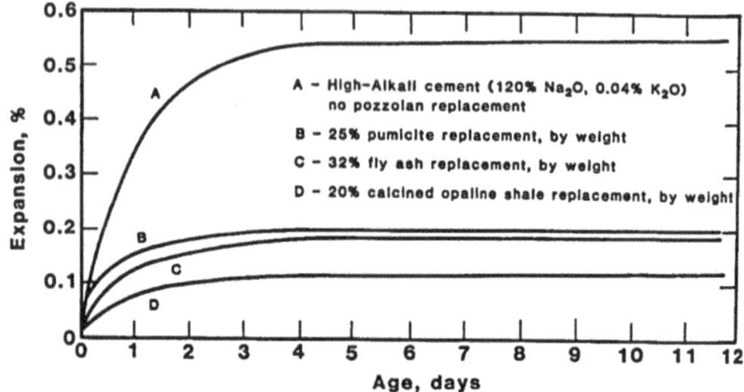

Fig. 2.59 Effect of pozzolans on reducing expansion of mortars made with a high-alkali cement and crushed Pyrex™ glass [47]

The effectiveness of fly ash in the control of the alkali-silica reaction has been widely reported. Pepper and Mather [137] reported the minimum percentage replacement (by volume) of cement by fly ash required to reduce expansion in test specimens by 75 %. Their results are shown in Table 2.26. Elfert [47] reported data of a similar nature from work carried out at the U.S. Bureau of reclamation (Fig. 2.59).

While it is clear that some fly ashes are effective in reducing expansion due to AAR, it is questionable whether the early strength losses caused by replacement of 40-50 % of the cement by low-calcium fly ash would be tolerable for more than a limited number of applications. Similar levels of replacement using high-calcium fly ashes may be more acceptable.

During a study of alkali-reactive aggregates in Nova Scotia, Duncan et al. [144] showed effective suppression of expansion of expansion by replacement of moderate-alkali cement (0.71 % as Na_2O) with a little as 25 % fly ash. Some data are shown in Fig. 2.60.

In other studies, the following factors have been identified as particularly important:

- The concentration of soluble alkali in the system;
- The type of aggregate; and
- The quality of fly ash used.

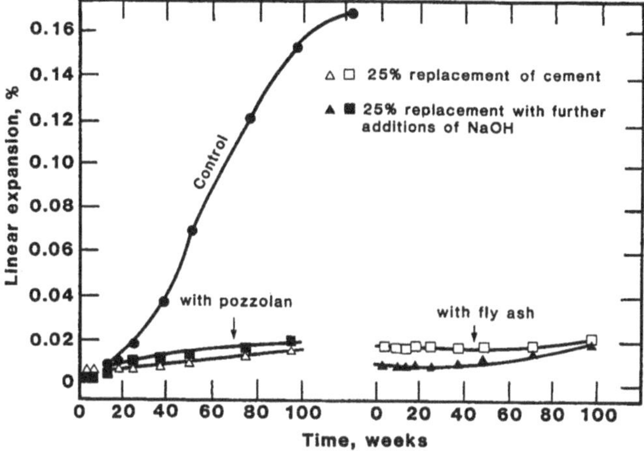

Fig. 2.60 Expansion of concrete prisms made with metagreywacke aggregate and 25 % cement replaced with pozzolan [144]

It is now generally accepted that, with regard to alkali-silica reactivity, it is the concentration of soluble alkali, rather than the total alkali content of the system, that affects expansion. Figure 2.61 shows the relationship between water-soluble alkali content and the time required for cracking in concrete containing Beltane opal as aggregate, as reported by Hobbs [145].

In general, Hobbs estimated that the lower limit of alkali concentration at which mortar test specimens exhibit excessive cracking was 3.4 kg/m^3 as acid-soluble alkali [146]. This is equivalent to 2.5 kg/m^3 as water-soluble alkali [145].

The source of alkali (Na_2O or K_2O) is not considered important. Thus, soluble alkali from fly ash is regarded as being as deleterious as that from Portland cement. High-calcium fly ashes, for example, contain large amounts of soluble alkali sulphates, which have been reported to increase the rate of deterioration through "the alkali-silica reaction" [147].

Not all aggregates are susceptible to the alkali-silica reaction, nor do all susceptible aggregates behave in the same way. The alkali-silica reaction is a long-term process that has been found to occur most commonly with aggregates that contain non-crystalline or cryptocrystalline silica.

Aggregates and their mineralogical constituents known to react with alkalis in concrete include the following [148]:

- silica materials -opal, chalcedony, tridymite, and cristobalite;
- zeolites, especially heulandite;
- glassy to cryptocrystalline rhyolites, dacites, andesites, and their tuffs; and
- certain phyllites.

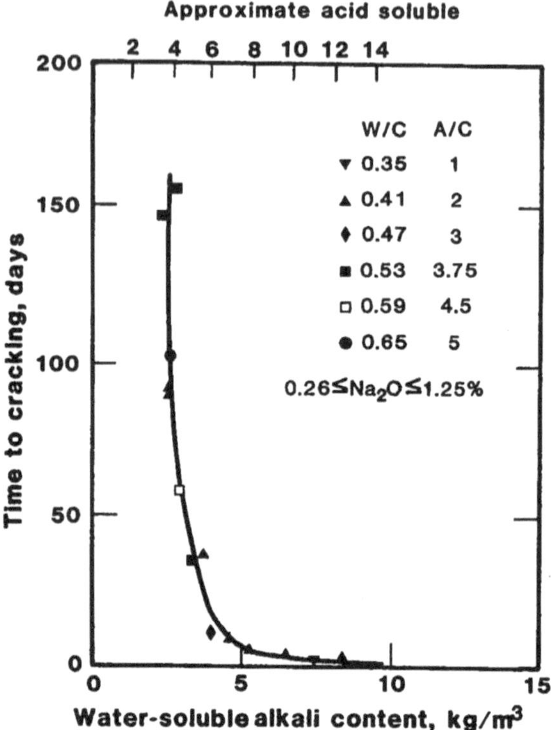

Fig. 2.61 Variation in time to cracking with water-soluble alkali content at the most critical alkali/Beltane opal ratio [145]

In Britain, the main reactive component of aggregates susceptible to alkali-silica reactivity has been found to be chert (flint) [149]. In South Africa, studies have been made on hornfels of the Malmesbury Group [150].

Much of the experimental work reported in the literature has been directed to the study of very reactive, porous, opaline aggregates, such as the Beltane opal from California and Pyrex™ glass. Both aggregates are generally more reactive than many of the natural aggregates encountered in practice, and this should be considered when the reported data are evaluated for practical applications.

As with most other aspects of fly ash use, ashes from different sources have significantly different effects on alkali-silica reactivity. As was noted above, some high-calcium fly ashes have been found in the laboratory to be ineffective or deleterious in relation to alkali reactivity.

In studies using Beltane opal, Hobbs [151] reported the expansion data shown in Fig. 2.62 for fly ashes with the chemical compositions summarized in Table 2.27.

The following conclusions were drawn by Hobbs from these experiments:

- The partial replacement of a high-alkali cement by fly ash reduced the long-term expansion due to alkali-silica reaction but, even when 30 or 40 % of the cement was replaced, most of the blended cement mortars cracked at earlier or similar ages as the Portland cement mortars.

Fig. 2.62 Variation in expansion with age for specimens where part of the aggregate was replaced with fly ash [151]

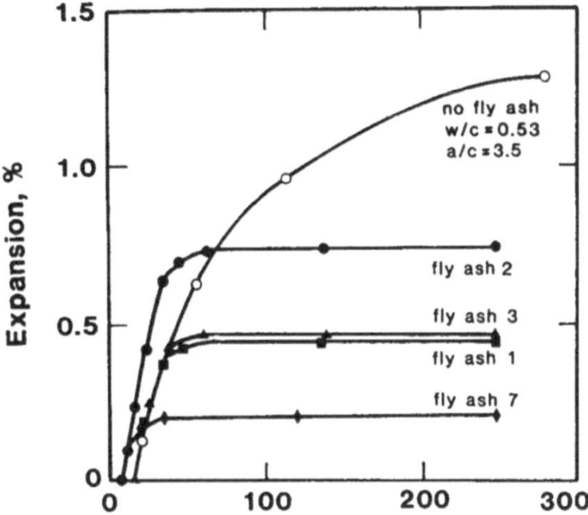

Table 2.27 Composition and properties of fly ashes examined by Hobbs [151]

Composition (wt%)	Fly ash no.			
	1	2	3	7
SiO_2	50.02	51.48	46.58	49.72
Fe_2O_3	9.02	8.70	14.24	5.22
Al_2O_3	26.83	28.08	25.22	32.45
CaO	1.48	1.27	4.10	2.77
MgO	0.93	0.93	0.95	2.41
SO_3	0.79	1.15	1.29	0.53
LOI	3.43	1.74	1.84	3.24
Na_2O (total)	0.88	1.13	0.80	0.38
Na_2O (water sol.)	0.07	0.10	0.08	0.02
K_2O (total)	3.90	3.85	2.35	1.40
K_2O (water sol.)	0.07	0.11	0.04	0.02

- The effectiveness of the fly ashes in reducing the long-term expansion varied widely. It is suggested that the effectiveness of the (fly ashes) may be dependent upon their alkali content or fineness.
- Where part of the cement was replaced by fly ash, the lowest mortar alkali content, expressed as equivalent Na_2O, at which cracking was observed was 2.85 kg/m^3. This figure relates only to the acid soluble alkalis contributed by the Portland cement and compares with a figure of 3.5 kg/m^3 for a Portland cement mortar.
- If it is assumed that fly ash acts effectively like a cement with an alkali content of 0.2 % by weight, the lowest alkali content at which cracking was observed was 3.4 kg/m^3.

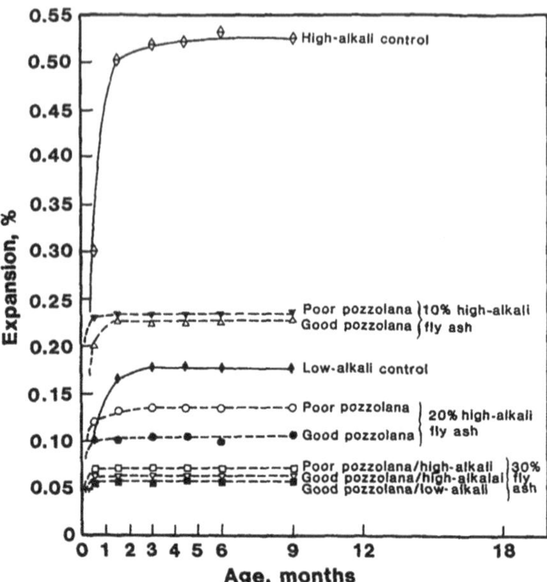

Fig. 2.63 Expansion of mortar bars containing high-alkali cement and fly ash with Pyrex aggregate [149]

- Both fly ashes and granulated blast-furnace slags act as alkali diluters, slags being more effective in reducing damage due to alkali-silica reaction than fly ashes.
- From the above it may be concluded that, when the aggregate to be used contains a reactive constituent, and when the concrete is to be exposed to external moisture, damage due to alkali-silica reaction is unlikely to occur if the acid-soluble equivalent Nap content of the concrete is below 3 kg/m^3. In calculating the alkali content of the concrete, granulated blast· furnace slags may be assumed to contain no available alkalis but fly ash should be assumed to have an available alkali content of 0.2 % by weight.

After conducting experiments using Pyrex glass, Nixon and Gaze [149] presented the data shown in Fig. 2.63 and drew the following conclusions:

- When Pyrex glass is used as the reactive aggregate, the partial replacement of a high alkali Portland cement by fly ash or by granulated blast-furnace slag produces a significant reduction in expansion of mortar bars at all replacement levels tested (10, 20 and 30 percent fly ash). The reductions are greater than could be accounted for by simple dilution of the alkali content of the Portland cement.
- Weight for weight the fly ashes are more effective (than granulated slag) in reducing expansion....
- Only small differences were found between the effectiveness of different fly ashes. These differences could best be correlated with a measure of the pozzolanicity of the ash. The ashes with lower alkali content did, on the whole,

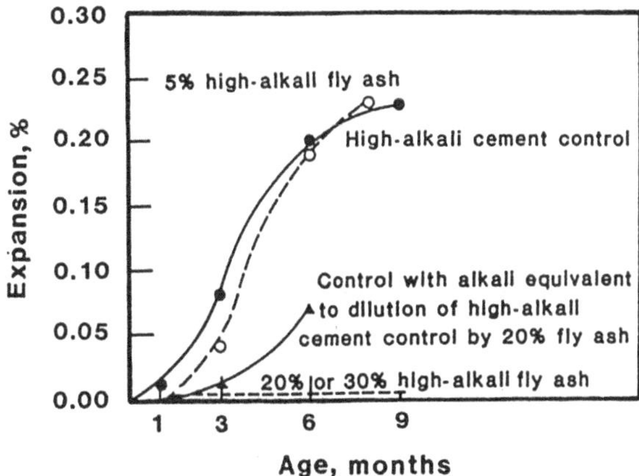

Fig. 2.64 Expansion of concrete prisms containing high-alkali cement and fly ash with 30 % flint or quartz aggregate [152]

seem to perform slightly better than those with high alkali (content) but this effect was secondary to the pozzolanicity. The available alkali content of the ashes gave no better correlation with the observed expansions than did the total alkali content….

Nixon and Gaze [152] also reported on studies using chert aggregate in fly ash concrete and published the data shown in Fig. 2.64.

Oberholster and Westra [150] studied alkali-silica reactivity of the Malmesbury Group aggregates and examined, among many additives, a low-calcium fly ash. Oberholster and Westra reported the fly ash to be more effective in reducing expansion than would be expected from a simple dilution of alkali content.

The study by Oberholster and Westra [150] included an examination of concrete prisms in which the fly ash effectively suppressed expansion at cement replacement levels of ≥20 % on an equal volume basis.

Stanton [140], Porter [153], and Pepper and Mather [143], in early studies of the use of fly ash to reduce expansion caused by alkali-silica reactions, noted that small additions of fly ash to mortars containing an opal aggregate may increase expansion, whereas lager amounts may reduce expansion. The general form of the relationship between ash quantity and expansion observed by these researchers is illustrated in Fig. 2.65.

In another study, Hobbs [154] reported that replacing 5 wt% of Portland cement by four fly ashes, a ground granulated slag, or limestone fines had little effect on the expansion of mortar bars tested at a critical alkali/silica.

Dunstan [148] examined 17 fly ashes of both bituminous and subbituminous origin, with Pyrex^{TM} glass as an aggregate. Dunstan suggested that the expansion–fly ash replacement relationship takes the form shown in Fig. 2.66 and that the amount of fly ash corresponding to the pessimum point (the point of maximum expansion) is related to the CaO content: as CaO increased with fly ash replacement, so the pessimum point increased. This suggests that high-calcium fly ashes would show an increased contribution to expansion at the levels of replacement

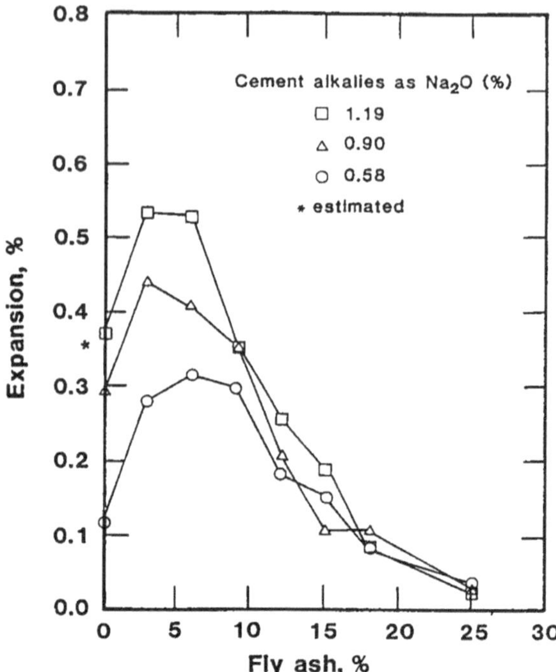

Fig. 2.65 Effect of cement alkali and fly ash on alkali-aggregate reaction [148]

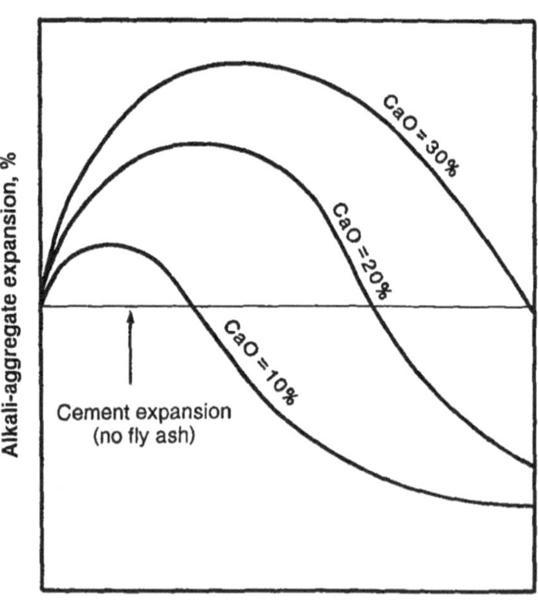

Fig. 2.66 Theoretical expansion due to alkali-aggregate reaction [148]

normally used with low-calcium fly ashes and would only become effective in retarding expansion caused by alkali-silica reactions at higher replacement levels.

Farbiarz and Carrasquillo [155] investigated the effectiveness of using fly ash in concrete to reduce the damage to concrete caused by alkali-aggregate reaction. Their results showed that the addition of fly ash can effectively reduce alkali-aggregate expansions. Nevertheless, for fly ashes with >1.5 % alkali content there was a pessimum limit for the percentage of fly ash below which no beneficial effect was achieved. This limit is inherent to each particular fly ash.

Carrasquilo and Snow [156] reported the alkali-aggregate reaction of mortar mixtures made with ASTM Class C and Class F fly ashes having available alkali contents of 0.57–4.35 %. Based on their test results, the following general conclusions were drawn:

- For mixture proportions with or without fly ash, mortar-bar expansion increases as the alkali content of the cement increases.
- The replacement of a portion of cement with fly ash effectively reduces the expansion in mortar bars caused by alkali-aggregate reaction, regardless of aggregate reactivity and the chemical composition of the cement, provided an adequate replacement with fly ash is used.
- The available alkalis in fly ash do participate in the alkali-aggregate reaction in concrete. In general, it was found that for similar mixture proportions, mortar-bar expansion increases as the alkali content of the fly ash increases.
- The CaO content of fly ash does not seem to have a significant effect on alkali-aggregate reaction in concrete.
- When the available alkali content of fly ash is <1.5 %, its beneficial effect in preventing expansion due to alkali-aggregate reaction increases as the percentage of cement replaced increases, regardless of the class of the fly ash, the alkali content of the cement, the alkali content of the fly ash, and the aggregate reactivity.
- When the available alkali content of fly ash is >1.5 %, there is a minimum percentage of cement replacement below which the fly ash causes expansion greater than that caused by a mixture without fly ash and above which the fly ash causes smaller expansions. This minimum is known as the pessimum limit.
- Minimum percentage of cement replacement rather than a maximum allowable available alkali content of fly ash could provide an adequate guideline for the use of fly ash to prevent damage of concrete due to alkali-aggregate reaction.

Nagataki et al. [157] examined eight different ASTM Class F fly ashes for controlling alkali-aggregate reaction of concretes made with Pyrex™ as an aggregate and with a replacement ratio of fly ash of 0–30 wt% of cement. The expansion of their mortar prisms depended on the type and the replacement ratio of fly ash. The expansion value was independent of equivalent sodium oxide content in fly ash but correlated with the concentration of soluble alkali ions in fly ash. Nagataki et al. suggested that the use of finer silicon dioxide and fly ash containing a higher amount of amorphous silica, together with a higher replacement ratio of fly ash, can control the alkali-aggregate reaction. They proposed a method to

Table 2.28 Mix proportions, properties of fresh concrete, and 28-day strength results [158]

Mix series	Total cementitious content (kg/m^3)	Fly ash (%)	Alkali concrete (eq Na_2O)		Reactive sand content[a] (%)	Total w/b[b]	Wet density (kg/m^3)	CF[c]	28-d Strength (MPa)
			Total (kg/m^3)	From cement (kg/m^3)					
A	741	–	7.9	7.9	20	0.30	2444	0.69	68.6
B	735	5	8.7	7.5	20	0.30	2411	0.73	71.0
C	737	20	11.3	6.3	20	0.29	2398	0.68	67.7
D	740	30	13.1	5.5	20	0.28	2391	0.67	70.9
E	600	–	6.4	6.4	20	0.36	2421	0.86	65.7
F	599	–	6.4	6.4	30	0.36	2420	0.89	67.7
G	598	20	9.2	5.1	30	0.34	2394	0.91	66.5
H	603	30	10.6	4.5	30	0.34	2398	0.92	65.8

[a] As percentage of total aggregate
[b] Water to binder ratio
[c] Compacting factor

evaluate the expansion of mortar containing fly ash based on amorphous silicon dioxide content, replacement ratios, and mean diameter of fly ash.

Thomas et al. [158] studied the effect of fly ash on alkali–silica reaction in concrete containing natural UK aggregate. Mixture proportions and properties of fresh and hardened concretes are given in Table 2.28. Petrographic examinations showed that concrete containing either no fly ash or 5 % fly ash exhibited considerable expansion and cracking caused by alkali-silica reaction (see Fig. 2.67). However, concretes containing 20–30 % fly ash showed little expansion (<0.03 %) and no cracking after 7 years of exposure.

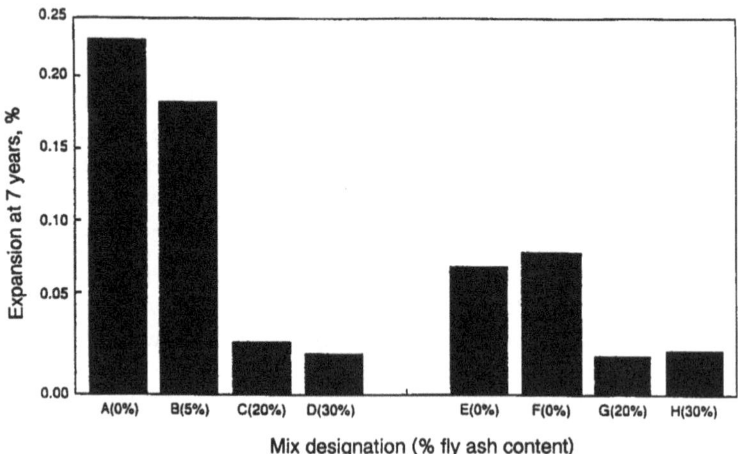

Fig. 2.67 Expansion of concrete prisms (75 mm × 75 mm × 200 mm) after 7 years of external exposure [158]

In summary, the following points may be made:

- There are substantial published data to show that low-calcium fly ashes are effective in reducing expansion caused by alkali-silica reactions when the fly ashes are used at a replacement level in the range of 25–30 %.
- The use of high-calcium ashes has received less attention; hence, the background information relevant to their use is less well developed. If they are to be used, there is some indication that effective replacement levels may be higher than those for low-calcium ashes.
- The mechanism and details of control of expansion caused by alkali-silica reactions are not fully understood, and there remains much research to be carried out before a satisfactory understanding can be developed.

Effects of Fly Ash on the Corrosion of Reinforcing Steel in Concrete

Recently, an issue of concern has been the corrosion of steel reinforcement in fly ash concrete structures exposed to chloride ions from deicing salts or sea water.

If the concrete cover over steel reinforcement is sufficiently thick and impermeable, it will normally provide adequate protection against corrosion. The protective effect of the concrete cover is of both a physical and a chemical nature and functions in three ways:

- It provides an alkaline medium in the immediate vicinity of the steel surface.
- It offers a physical and chemical barrier to the ingress of moisture, oxygen, carbon dioxide, chlorides, and other aggressive agents.
- It provides an electrically resistive medium around the steel members.

Under alkaline conditions (pH higher than ~11.5), a protective oxide film will form on a steel surface, rendering it passive to further corrosion.

When concrete carbonates (see Effects of Fly Ash on Carbonation of Concrete in this chapter) and the depth of carbonation reaches the steel–concrete boundary, passivation may be reduced and corrosion may occur if sufficient oxygen and moisture reach the metal surface. Chlorides or other ions may also undermine the protective effect of passivation and encourage corrosion.

The Reunion international des laboratories d'essais et de recherches sur les materiaux et les constructions (RILEM) Technical Committee on Corrosion of Steel in Concrete [159] made the following statements, which give perspective to this issue:

- The efficiency of the (concrete) cover in preventing corrosion is dependent on many factors which collectively are referred to as its "quality". In this context, the "quality" implies impermeability and a high reserve of alkalinity which satisfies both the physical needs and chemical requirements of the concrete

cover. If the concrete is permeable to atmospheric gases or lean in cement, corrosion of the reinforcement can be anticipated and good protection should be attempted by the use of dense aggregate and a well compacted mix with a reasonably low water/cement.
- If chloride corrosion is excepted, it is now usually agreed that carbonation of concrete cover is the essential condition for corrosion of reinforcement.

As discussed in Effects of Fly Ash on Carbonation of Concrete in this chapter, the issue of carbonation of fly ash concrete has received some attention in recent years. However, it is our opinion that the carbonation of fly ash concrete is not a matter of concern, provided attention is paid to obtaining adequate impermeability in the concrete mass.

In 1950, the question was raised [160] whether the sulphur-containing components of fly ash could corrode the reinforcing steel in fly ash concrete. Gilliland [161] noted that most of the sulphur in fly ash is present as sulphate and, therefore, would have an effect similar to the sulphate components in Portland cement. Further, he pointed out that corrosion of steel is greatly affected by pH; at the high pH prevailing in concrete, corrosion rates would be expected to be slow. Ryan [162] presented further information on the same point and drew the following conclusions:

- Sulphur compounds in fly ash are usually so limited by specifications that they are not materially different in the concrete, whether fly ash is used or not. Moreover, the alkaline condition in the concrete is unfavourable to a sulphate attack on steel.
- Carbon in fly ash would appear by theoretical considerations to be much more significant in concrete than is sulphur. The actual effect should be investigated. However, if it is kept under 3 percent in the fly ash, its percentage in the concrete becomes so small that if it is well dispersed, its effect on the electrical conductivity of the concrete, and therefore upon the corrosion of the steel, should be quite minor.

These conclusions seem to be generally acceptable in the light of reported research that has shown that fly ash concrete does not decrease the corrosion protection of steel reinforcement, compared with normal concrete [163–165]. Larsen et al. [166, 167] found that the inclusion of fly ash in concrete increased corrosion protection.

In regard to the quality of the concrete cover over steel and to the points raised by the RILEM Technical Committee on Corrosion of Steel in Concrete (see above) [159], fly ash may influence both permeability and the alkalinity of the system.

The permeability of fly ash concrete and the related issue of carbonation were discussed in the first two sections of this chapter. It is sufficient here to recall that properly proportioned fly ash concrete subjected to adequate curing should in general be less permeable at later ages than the corresponding plain concrete. The danger of permeability lies in the premature exposure of fly ash concrete to

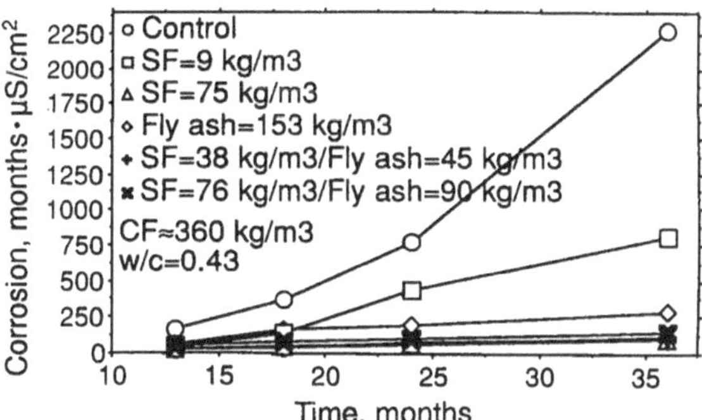

Fig. 2.68 Total corrosion versus time for reinforcing steel in concretes at W/C = 0.43, exposed to 3 % NaCl, as a function of pozzolan additions [169]. *SF*, Silica fume; *CF*, Cement factor

aggressive agents, as a result of either inadequate proportioning, incomplete curing, or poor fly ash quality.

It has been suggested that because the pozzolanic reaction consumes calcium hydroxide as it progresses, this may cause a decrease in the pH of the pore water in fly ash cement paste. The pore solutions in hydrated cement are highly alkaline; as Diamond [168] has shown, this results from the presence of sodium and potassium ions rather than from the presence of calcium hydroxide. In studies of two fly ash cement systems, Diamond [168] showed that the alkalinity is determined almost entirely by the dissolution of sodium and potassium salts from cement; at quite early ages, the concentration of calcium in solution is reduced to very low levels. In the samples studied by Diamond, pore-solution pH was reduced from 13.75 in a control system to ~13.55 in the presence of fly ash.

Berke et al. [169] reported the long-term effects of fly ash and silica fume on chloride ingress, electrical resistivity, microstructure, and corrosion of steel reinforcement in concrete. Figure 2.68 shows the corrosion of steel reinforcement in concrete with 0.43 water/cementitious materials after 3 years of exposure to 3 % NaCl. It can be seen that the steel reinforcement in control concrete and in concrete with 9 kg/m^3 silica fume is corroding after 3 years. Concretes containing fly ash were effective in lowering the corrosion rates. However, at a higher water/cementitious materials (W/C = 0.52), fly ash concretes showed higher corrosion rates than the other concretes.

Effects of Fly Ash on Concrete Exposed to Sea Water

The deterioration of concrete in marine environments is a complex subject and cannot be treated adequately here. This discussion has been limited to an examination of some of the aspects of using fly ash in marine concrete, focusing on the rather limited, directly applicable published data and some extrapolations from many reports on the behavior of plain concrete in the sea.

Exposure of concrete to the marine environment subjects it to an array of severely aggressive factors, including most of those discussed in the preceding sections of this chapter.

Concrete in the tidal zone is the most severely attacked, subjected as it is to alternating wetting and drying; wave action; abrasion by sand and debris; frequent freezing and thawing cycles; and corrosion of reinforcement all occurring in a chemically aggressive medium. Permanently immersed concrete is less severely affected [170, 171].

Very little direct observation of fly ash concrete in sea water has been reported in the literature, although research in this area is in progress [172].

In 1978, CANMET [173, 174] started a long-term project on the performance in a marine environment of concretes incorporating supplementary cementing materials. The test specimens are exposed to repeated cycles of wetting and drying and to ~100 freezing and thawing cycles per year.

Even under exposure to severe marine conditions, concretes incorporating 25 % fly ash from a bituminous source were in satisfactory condition after 10 years. The only exceptions were the specimens with a water/(cement + fly ash) of 0.60. It was concluded that fly ash concrete at 25 % cement replacement level (by mass) can be satisfactory under such conditions of exposure, provided the water/cementitious materials is ≤0.50.

Ohga and Nagataki [175] evaluated the effectiveness of fly ash for controlling alkali-aggregate reaction in the marine environment. They measured the expansion of mortars made with high-and low-alkali cements. The alkali content in the mixture was adjusted by adding NaCl or NaOH. The specimens were then exposed to humid air, distilled water, and NaCl solution. Figure 2.69 shows that the

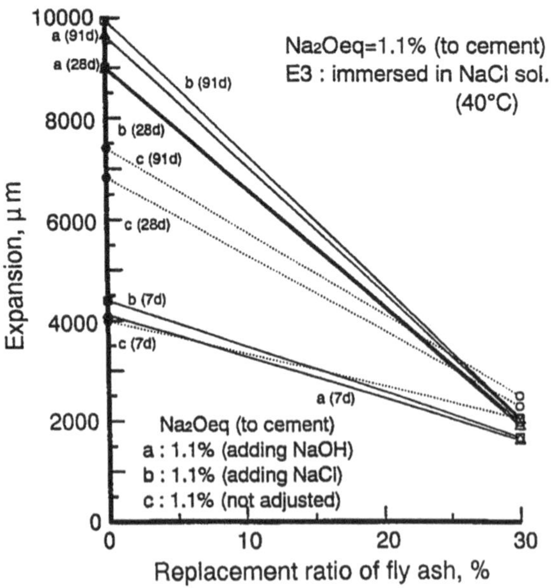

Fig. 2.69 Effect of replacement ratio of fly ash on the expansion of mortar at 40 °C [175]. Concrete in the sea is subjected to the aggressive influences discussed above

Table 2.29 Potential influence of aggressive on fly ash concretes in sea water

Form of attack	Influence of fly ash	Conditions
Wetting–drying	None	If proportioning and curing are adequate
Freezing-thawing	None	If proportioning, air entrainment, and curing are adequate
Sulphate	Improved by low-calcium ash	
Alkali-aggregate reaction	Expansion reduced	
Corrosion of steel	None	If proportioning and curing are adequate
Permeability	Improved at late ages	If proportioning and curing are adequate
Magnesium salts	Not known	

expansion due to alkali-aggregate reaction can be controlled, even in a marine environment, if an adequate amount of a suitable fly ash is used.

Table 2.29 gives a profile of the ways in which fly ash concrete might be affected by exposure to sea water.

Whereas permeability is considered the major factor affecting the durability of concrete in sea water, it is evident that fly ash has the potential to contribute to a number of aspects of concrete durability in the marine environment. It is clear also that this is an aspect of fly ash concrete behavior that is greatly in need of research.

Recent Development on the Durability of Fly Ash Concretes

In recent years most of the researches have been concentrated on the durability of fly ash concretes in severe environments. Corrosion of reinforcement due to chloride penetration and carbonation of concretes containing fly ash has been investigated at longer ages.

The capacity of binding chloride ions in fly ash concrete under marine exposure was investigated [176]. The free and total chloride contents in concrete were determined by water and acid-soluble methods, respectively. In order to study the effects of W/B ratios, exposure time, and fly ash contents on chloride binding capacity of concrete in a marine site, a class F fly ash was used as a partial replacement of Type 1 Portland cement at 0, 15, 25, 35, and 50 % by weight of binder. Water to binder ratios (W/B) were varied at 0.45, 0.55, and 0.65. Concrete cube specimens of 200 mm were cast and placed into the tidal zone of a marine environment in the Gulf of Thailand. Consequently, acid-soluble and water-soluble chlorides in the concrete were measured after the concrete was exposed to the tidal zone for 3, 4, 5, and 7 years. It was found that the percentage of chloride binding capacity compared to total chloride content increased with the increase of fly ash in the concrete. The percentage of chloride binding capacity significantly decreased within 3–4 years after the concrete was exposed to the marine environment, and then its value was almost constant. The research also showed that the W/B ratio does not noticeably affect the chloride binding capacity of concrete.

In another study concrete specimens were produced by using fly ash at 0, 15, 30 and 45 % ratios replaced by CEM I 42.5 R cement. The specimens were moist cured for 28 and 56 days and then remained in two different conditions which are air and water. Physical, mechanical, and durability properties of the produced concrete were determined. Chloride ion permeability test was carried out on concrete specimens. Reinforced concrete specimens were evaluated under accelerated corrosion technique. It was seen that corrosion currents decreased and deterioration occurrence time took longer due to the moist curing at series which were cured for 28 and 56 days. The lowest initial corrosion current values and the longest deterioration occurrence times were obtained concretes containing 30 and 45 % fly ash replacement and cured in water. Results showed that specimens having no fly ash additive had high chloride permeability. It was found that all of the series had low chloride permeability due to use of 15 % fly ash replacement. The use of 30 and 45 % fly ash showed great reductions in chloride permeability of specimens cured for 28 and 56 days in water. Based on their results, it was concluded that 15 % replacement of fly ash with cement was an optimum value for durability enhancement against corrosion. Durability of reinforced concretes containing fly ash significantly increased when 56 day moist curing was applied [177].

Recently a research was focused on investigating the durability of concretes containing fly ash and silica fume exposed to combined mode of deterioration. For this purpose, the chloride ion diffusivity of concrete was evaluated before and after 300 freeze–thaw cycles [178].

It was found that the coefficient of chloride ion diffusivity increased as water to cementitious material ratio and air content increased. When a proper curing was provided, all concrete showed good durability factor regardless of amount of air content.

Test results clearly showed that the coefficient of chloride ion diffusivity for all concretes increased after freeze–thaw cycles. However, fly ash concrete showed good resistance to chloride ion diffusivity before and after freeze–thaw cycles when low w/cm as well as a proper curing and air content were provided. This was attributed to the adsorption of chloride ion to C–S–H layers and the improved microstructure by the pozzolanic reaction. Higher coefficient of chloride ion diffusivity were observed for fly ash concrete than plain concrete when w/cm was 0.6 and air content was 2 %.

In order to study the effect of fly ash on alkali silica mitigation of concrete, 45 concrete blocks (915 mm × 915 mm × 815 mm or 350 mm cubes) containing alkali-silica reactive aggregates, and various levels of high-alkali cement and fly ash were placed on an outdoor exposure site in S.E. England for a period of up to 18 years. The reactive aggregates used included a variety of flint sands and a crushed greywacke combined coarse and fine aggregate. Length-change measurements were conducted periodically throughout this period. All concrete blocks without fly ash showed excessive expansion and cracking within 5–10 years of production and in many cases the ultimate expansion exceeded 1.0 % after 15–18 years. Fly ash used at replacement levels of 25 and 40 % was effective in significantly reducing expansion and cracking with all three flint aggregates at all levels of alkali. Of the 27 blocks containing fly ash and flint sand only two blocks showed evidence of damage after 16–18 years. The expansion of these blocks was significantly lower than similar blocks with the same Portland cement content without fly ash. None of the blocks with greywacke aggregate and fly ash exhibited cracking (expansion data were not available for these blocks). Collectively the data confirm that fly ash, when used at levels of 25–40 %, does not effectively contribute alkalis to the alkali-silica reaction [179].

Durability of special concretes such as high strength and high performance concretes, self consolidating concretes containing fly ash have been investigated by a number of researchers. In a recent study, durability properties of high strength concrete utilizing high volume Class F fly ash sourced from Western Australia have been investigated [180]. Concrete mixtures with fly ash as 30 and 40 % of total binder were used to cast the test specimens. The compressive strength, drying shrinkage, sorptivity and rapid chloride permeability of the fly ash and control concrete specimens were determined. The 28-day compressive strength of the concrete mixtures varied from 65 to 85 MPa. The fly ash concrete samples showed less drying shrinkage than the control concrete samples when designed for the same 28-day compressive strength of the control concrete. Inclusion of fly ash reduced sorptivity and chloride ion permeation significantly at 28 days and reduced further at 6 months. In general, incorporation of fly ash as partial replacement of cement improved the durability properties of concrete.

In an experimental program properties of self consolidating concrete (SCC) made with Class F fly ash were investigated. The mixtures were prepared with five percentages of class F fly ash ranging from 15 to 35 %. Properties investigated were self consolidating parameters such as slump flow, J-ring, V-funnel, L-box and U-box, strength properties such as compressive and splitting tensile strengths, and durability properties (deicing salt surface scaling, carbonation and rapid chloride penetration resistance). SCC mixes developed 28 day compressive strength between 30 and 35 MPa and splitting tensile strength between 1.5 and 2.4 MPa. The carbonation depth increased with the increase in age for all the SCC mixes. Maximum carbonation depth was observed to be 1.67 mm at 90 days and 1.85 mm at 365 days for SCC with 20 % fly ash content. Also, the pH value for all the mixes was observed to be greater than 11. Deicing salt surface scaling weight loss increased with the increase in fly ash content except for the mixture containing 15 % fly ash. At 365 days age, the weight loss was almost consistent for all percentages of fly ash. SCC mixes made with fly ash exhibited very low chloride permeability resistance less than 700 and 400 Coulombs at the age of 90 and 365 days, respectively [181].

Application of Fly Ash in Concrete

For most civil engineering applications, the decision to use fly ash in concretes will depend on the availability of materials, local concrete practice, and, most importantly, economics. However, some types of special concretes require fly ash or other mineral by-products to attain specific properties. In particular, high-strength concretes (>60 MPa at 28 days) and roller-compacted mass concretes both depend on the use of fly ash for their structural and economic success.

High-Strength Concrete

High-strength concretes can be classified broadly into three groups:

- concretes with a compressive strength of 50–70 MPa, used over the
- past 10 years in a number of construction applications;
- concretes with a compressive strength of 70–100 MPa; and
- concretes with a compressive strength of >100 MPa.

Although field-placeable concrete in the strength range of 70–120 MPa is now commercially available, the major documented uses of high-strength concrete in Building applications have been for 85 MPa concretes in the major cities of Canada and the United States. Thus, for the present time, use of fly ash in high-strength concrete must be restricted to the lower end of the above strength range.

The most important reasons for using high-strength concrete are the following:

- to minimize the size of concrete structural members in buildings;
- to obtain more rapid production cycles in the precast and prestressed
- concrete industry; and
- to obtain high strength and high modulus of elasticity in structural members built to withstand large stresses.

In general, the production of high-strength concrete requires the use of mixtures with low water/cement but with sufficient workability to permit placement in a heavily reinforced structure. To meet these requirements, the following factors are usually considered important [182–184];

- use of high cementitious contents (550 kg/m^3) at slumps of 75–100 mm;
- stringent selection of cement and aggregates;
- use of low water/cement, attained through water-reducing agents and
- superplasticizers and
- use of pozzolans, such as fly ash and silica fume, and blast-furnace slag.

Fly ash in high-strength concrete provides at least two benefits:

- a good-quality fly ash generally permits a reduction in the water
- content of a concrete mixture, without loss of workability

Table 2.30 Mixture proportions and properties of fresh high-strength concretes [182]

	Quantity/m^3	
	Water tower place	River plaza
Materials		
Cement (kg)	383.7	385.6
Fine aggregate (kg)	464.9	471.7
Stone (kg)	1403.5	1300.7
Water (kg)	136.1	149.7
Water-reducing admixture (ml)	751.1	1271.5
Fly ash (kg)	45.4	45.4
Properties		
Slump (mm)	115	115
Air content (%)	–	1.5
Unit weight	2433	2383

Table 2.31 Compressive strengths of high-strength concrete [182]

Age (days)	Compressive strength (MPa)					
	Water tower place			River plaza		
	Air[a]	Moist	Cores	Yard	Job site	Cores
7	–	52.7	–	–	50.4	46.5
28	6301	64.8	–	71.3	64.9	55.8
56	–	72.9	–	77.5	72.4	73.7
90	64.9	–	–	80.2	78.7	72.1
180	63.5	–	–	91.1	–	–
368	66.9	–	–	–	–	–
730	61.8	–	79.7	–	–	–

[a] 7 days moist curing followed by curing at 50 % RH, 21 °C

- fly ash produces increased strength at later ages of curing, which cannot be achieved through the use of additional Portland cement.

Typical mixture proportions for high-strength concrete used in two structures built in the Chicago area during the 1970s are given in Table 2.30, and the compressive-strength values obtained with these mixtures are given in Table 2.31. Concrete mixtures with strengths of 55 MPa and incorporating fly ash and using similar proportions have been used in construction in the Toronto area [88].

Cook [185] reported on a comprehensive investigation of concretes in the strength range of 50–75 MPa (28 days) made using a high-calcium fly ash (CaO = 30.3 wt%). Tables 2.32 and 2.33 illustrate how fly ash can contribute to the strength development of high-strength concretes and the flexibility in the selection of mixture proportions that can be used to obtain essentially similar concretes. Table 2.32 presents the concrete properties attained by using different

Table 2.32 Mixture proportion and properties of fresh and hardened high-strength concrete at a constant ash/cement ratio [185]

Nominal cement content (kg/m^3):	279	335	390	446	502
Material					
Cement (kg/m^3)	282	341	396	449	501
Fly ash (kg/m^3)	71	85	99	112	131
Limestone (kg/m^3)	1144	1147	1141	1130	1121
Sand (kg/m^3)	735	643	578	513	454
Water (kg/m^3)	148	157	161	169	179
W/R admixture (mL/m^3)	696	851	967	1083	1238
Properties of fresh concrete					
Slump (mm)	102	102	114	102	95
Unit weight	2379	2373	2376	2373	2374
W/(C + F)	0.42	0.37	0.33	0.30	0.28
F/C	0.25	0.25	0.25	0.25	0.26
Compressive strength (MPa)					
7 days	44.3	48.1	53.1	55.1	59.3
28 days	54.6	62.7	67.4	72.7	70.2
56 days	61.0	68.0	75.7	75.9	78.0
90 days	63.3	68.9	75.0	75.7	75.0
180 days	67.8	76.9	83.6	86.6	85.3

Table 2.33 Mixture proportions and properties of fresh and hardened high-strength concretes at various ash/cement ratios [185]

	Mixture no.			
	1	2	3	4
Materials				
Cement	320	281	279	245
Fly ash	80	120	71	105
Limestone	1127	1127	1127	1127
Sand	698	702	769	720
Water	151	142	148	145
Admixture[a]	–	–	–	–
Mix properties				
Slump (mm)	102	95	83	108
W/(C + F)	0.38	0.36	0.42	0.41
F/(C + F)	0.20	0.30	0.20	0.30
Compressive strength (MPa)				
7 days	44.9	44.0	42.0	40.5
28 days	56.6	55.8	52.7	52.3
56 days	58.6	61.9	56.8	58.8
180 days	68.9	74.4	70.5	71.6

[a] Quantities not reported

quantities of cementitious materials at a constant ash/cement ratio (0.25). As the cement factor was increased, sand content and the water/cementitious materials were reduced. Strength at 28 days was controlled over a range of 55–70 MPa. The data in Table 2.33 show the effect on strength of increasing fly ash content from 20 to 30 % for two basic concrete mixtures.

Naik et al. [186] examined three concrete mixtures incorporating fly ash to achieve nominal strengths of 70 MPa. In one of the mixtures, they used ASTM Class C fly ash as a partial replacement for cement. Water-reducing, retarding admixture and a superplasticizer were also added to the concrete to lower the water/cementitious materials and to achieve a slump of 150 mm. Details of the mixture proportions and the result of the compressive-strength results at various ages are given in Table 2.34.

Naik et al. [186] also showed that high-strength concretes containing mineral admixtures had a dense matrix and, hence, a high resistance to chloride-ion penetration, especially at later ages.

The use of three classes of "classified fly ash" (classified by air-separation method as having maximum particle diameters of 20 pm) as concrete admixtures was investigated by Ukita et al. [187]. The compressive strength of concrete mixtures with 30 % replacement of cement with fly ash reached 92 MPa for a water/cementitious materials of 0.27 %. In their research, Ukita et al. showed that the use of classified fly ash reduced water requirement, improved workability, enhanced strength and water tightness, and increased resistance to alkali–silica reaction in the concrete [188].

Table 2.34 Concrete mixture proportions and strength data for high-strength concrete [186]

Nominal strength (MPa)	70
Cement (ASTM type I) (kg/m^3)	355
Fly ash (ASTM Class C) (kg/m^3)	207
Water (kg/m^3)	180
Water/cementitious materials	0.3
Sand (saturated surface dry) (kg/m^3)	712
Coarse aggregate (12.5 mm max. crushed limestone) (kg/m^3)	978
Slump (mm)	150
Entrapped-air content (%)	0.3
Retarding admixture (ASTM type A) (kg/m^3)	1
Superplasticizer (ASTM type F) (kg/m^3)	7.3
Compressive strength on 100 mm × 200 mm cylinders (MPa)	
7 days	58.0
28 days	69.6
91 days	83.8
365 days	93.9

Roller-Compacted Concrete

In the 1970s, a method for the construction of dams, termed roller compaction, was proposed [189–192]. The method, which in many ways is more related to the procedures used in geotechnical engineering than to conventional concrete practice, depends on the placement of layers of a low-workability concrete in the interior of a dam and its compaction using vibrator rollers.

The ACI [193] described roller-compacted concrete (RCC) as a dry concrete material that is consolidated by external vibration by vibratory rollers. It differs from conventional concrete in its required consistency: for effective consolidation, RCC must be dry enough to support the weight of the placement equipment but fluid enough to permit distribution of the paste throughout the mass during mixing and compaction.

Roller-compacted concrete must satisfy four requirements:

- It must have a high density, with a minimum of air voids.
- The layers of concrete must bond together [186].
- The generation of heat in the dam must be minimized.
- To resist thermal cracking, the hardened concrete must have a high capacity to withstand tensile strains.

Three types of materials have been investigated and used for RCC:

- cement-stabilized, soil-like materials;
- lean concrete, with a cementitious-materials content of 100–150 kg/m^3, of which 30 % may be fly ash; and
- rich concrete, with a cementitious-materials content of 180–270 kg/m^3, of which 60–80 % may be fly ash.

RCC can be made from any of the basic types of Portland cement in combination with pozzolans. In regard to proportioning, the principal difference between the selection of the relative quantities of the cementitious components in RCC, compared with more conventional concretes, is the use of large quantities of fly ash. The principal function of fly ash in RCC is to provide a large volume of fine material to occupy the space between larger particles, which could otherwise be accomplished only with the use of additional cement.

RCC normally needs to have a high paste content. Dunsan [194] showed that as paste/mortar falls below 0.35–0.38, the density of the compacted mass is significantly reduced. Below this level of paste content, some of the voids in the fine aggregate are not being filled. To obtain maximum density, the paste content must be increased; however, to attain this by the addition of Portland cement results in two serious disadvantages:

- The rate of heat evolution increases, and the possibility of thermal cracking becomes greater.
- The cost of the concrete may not be economical.

Table 2.35 Mixture proportions for some roller-compacted concretes [97]

Source	Max. aggregate size (cm)	Mix data (kg/m^3)				
		Cement	Pozzolan	Water	Fine aggregate	Coarse aggregate
1	7.6	56	77	77	660	1649
2	11.4	139	0	80	618	1774
3	7.6	139	0	86	683	1691
4	7.6	139	0	83	676	1602
5	7.6	42	78	83	676	1602
6	3.8	75	164	89	745	1426
7	3.8	45	178	84	727	1438
8	3.8	116	139	103	657	1438

Table 2.36 Properties of some roller-compacted concretes [97]

Source	Age (days)	Compressive strength (MPa)	Shear strength (MPa)	
			Mass	Joint
1	138	23	4	2
2	72	26	5	1
3	66	23	6	–
4	120	23	6	3
5	120	16	4	1
6	90	26	–	3
7	90	18	2	2
8	90	41	–	–

To obtain maximum density in RCC, the desired approach is to use large volumes of fly ash to increase the paste content.

Pozzolanic activity is somewhat secondary: strength in RCC can develop over long periods after placement. ACI Committee 207 reported the following [193]:

Where there is a deficiency in fines [in RCC] a pozzolan does not have to be highly reactive to be effective. Thus, many fly ashes whose reactivity, due to insufficient particle fineness, would not meet....ASTM specifications would be suitable for most roller compaction applications.

Typical RCC mixture proportions and the corresponding strength data, as reported by ACI Committee 207, are shown in Tables 2.35 and 2.36.

As in more conventional mass concrete, fly ash helps reduce the rate of heat evolution in RCC and, hence, extent of temperature rise.

Permeability of RCC used in the Upper Stillwater Dam was reported to be equal to, or less than, that of conventional mass concrete [193]. Both AEA and water-reducing admixtures have been used at normal dosages in RCC. ACI Committee 207 [187] stated that these admixtures are effective in reducing the vibration time for fill consolidation. However, the effectiveness of AEA in RCC and the appropriate dosage rates are as yet not fully established.

The use of RCC is not restricted to mass structures such as dams. It has been employed to replace riprap for erosion control of floodway sill in Alaska, as a foundation rock protection, in a lock floor, and in other structures [196].

As Joshi [197] pointed out, RCC are closely related to the family of stabilize materials commonly used for pavements and other applications. The use of RCC and lean concrete containing large amounts of fly ash has been extensively studied for pavement construction in the United Kingdom. Sherwood and Potter [198] reported studies of ash-modified lean concrete for use in road bases. They suggested that ash-modified lean concrete performs as well as conventional lean concrete, provided the development of thermal and shrinkage cracks was similar, and that it could withstand repeated traffic-induced stress at early ages.

Dunstan [194] reported on field applications of lean concretes containing high amounts of fly ash and drew the following conclusions:

- A considerable material cost saving (>20 %) was achieved.
- Good-quality control over the batching process is required to maintain consistency of material.
- It is difficult to obtain satisfactory levels of air entrainment in concrete incorporating large quantities of fly ash.
- Very little early age cracking was seen in the base materials.
- A fly ash that did not conform to British Standard specifications was found to be adequate, with no deleterious effects.

The Electric Power Research Institute (EPRI) of the United States began a program in 1984 involving six new demonstration projects. The demonstration projects were structured to show the environmental and technological acceptability of ash use in road construction in a controlled and monitored segment of a highway. The project of most interest to the concrete-design community was the one completed in North Dakota, in which a Class C fly ash was used to replace 70 % of the cement in a concrete pavement.

Naik et al. [199] reported the results of research performed in developing and using high volumes of ASTM Class C and Class F fly ash concrete mixtures for pavement construction. They concluded that the high-volume fly ash concrete incorporating ASTM Class F or Class C fly ash can be used to produce high-quality pavements with excellent performance.

With regard to air entrainment, Oliverson and Richardson [195] noted that research on the freezing and thawing durability of RCC is needed. Certainly, given the well-known difficulties associated with air entrainment of some fly ash concretes, considerable research in this area is required if RCC is to find use under the climatic conditions to which most pavement concrete is exposed in Canada.

Torii and Kawamura [200] studied the engineering properties and durability of dry, lean-rolled concrete containing fly ash, and they concluded as follows:

- The use of fly ash with cement in dry, lean-rolled concretes contributed to the improvement of the compaction of the concrete mixtures.

- Dry, lean-rolled concretes with fly ash showed relatively high strength development at later ages, compared with those without fly ash, when they were cured in water. Therefore, the curing conditions significantly influenced the strength development of dry, lean-rolled concretes made using fly ash.
- Low drying shrinkage of dry, lean-rolled concretes with fly ash was advantageous in the prevention of cracks in pavements.
- Resistance of dry, 1ean-rolled concretes with fly ash to freezing—thawing cycles was less than, or equal to, that of dry, lean concretes without fly ash.

References

1. R.E. Davis, R.W. Carlson, J.W. Kelly, H.E. Davis, Properties of cements and concretes containing fly ash. J. Am. Concr. Inst. **33**, 577–612 (1937)
2. O.E. Manz, United nations economic commission for Europe, Geneva, Switzerland, Report EP/SEM, 7/R, 1980, p. 51
3. O.E. Manz, Worldwide production of coal ash and utilization in concrete and other products, in *Proceedings of 10th International Ash Use Symposium*, Orlando, FL, Jan 18–21, 1993 (American Coal Ash Association, Washington, DC)
4. Canadian Standards Association, Supplementary cementing materials and their use in concrete construction. CSA, Rexdale, ON, CAN-A23.5-M82, 1982
5. American Society for Testing and Materials, Specification for fly ash and raw of calcined natural pozzolan for use as a mineral admixture in Portland cement concrete, ASTM, Philadelphia, PA, ASTM C618-78, 1978
6. G.G. Carette, V.M. Malhotra, Characterization of Canadian fly ashes and their relative performance in concrete, in *Energy, Mines and Resources Canada*, Ottawa, ON, CANMET Report 86-6E, 1986
7. V.M. Malhotra, G.G. Vallace, A new method for determining fineness of cement, in *Energy, Mines and Resources Canada*, Ottawa, ON, Mines Branch Investigation Report 63-119, 1993
8. K. Wesche, *Fly Ash in Concrete Properties and Performance*. International Union of Testing and Research Laboratories, Paris, France, RILEM Report 7 (Chapman and Hall, London, 1990)
9. P.K. Mehta, Testing and correlation of fly ash properties with respect to pozzolanic behavior. Electric Power Research Institute, Palo Alto, CA, Report CS3314, 1994
10. J.G. Cabrera, C.J. Hopkins, The effect of PFA on the rheology of cement pastes, in *Proceedings of International Symposium on the Use of PFA in Concrete*, University of Leeds, Leeds, UK, Apr 14–16, ed.by J.G. Cabrera, A.R. Cusens, Department of Civil Engineering, University of Leeds, Leeds, UK, 1982, pp. 323–343
11. E.E. Berry, Fly ash for use in concrete. Part 1: A critical review of the chemical, physical, and pozzolanic properties of fly ash. Energy, Mines and Resources Canada, Ottawa, ON, CANMET MSL Report 76-25, 1976
12. ACI Committee 226. Fly ash in concrete. ACI Mater. J. **84**, 381–409 (1987)
13. P.K. Mehta, Pozzolanic and cementitious by-products as mineral admixtures for concrete—A critical review, in *Proceedings of 1st International Conference on the Use of Fly Ash, Silica Fume, Slag, and Other Mineral By-products in Concrete*, Montebello, PQ, July 31–Aug 5, 1983, ed. by V.M. Malhotra (American Concrete Institute, Detroit, MI, Special Publication SP-79, 1983)
14. S. Diamond, Symposium N. Annual Meeting of the Materials Research Society, Boston, MA, Nov 1981, pp. 12–23

15. O.E. Manz, G.J. McCarthy, R.J. Stevenson, B.A. Dockter, D.G. Hassett, A. Thedchanamoorthy, Characterization and classification of North American lignite fly ashes for use in concrete, in *Proceedings of 3rd CANMET/ACI International Conference on the Use of Fly Ash, Silica Fume, Slag, and Other Mineral By-products in Concrete*, Trondheim, Norway, June 18–23, 1989, Supplementary Papers, pp. 16–32
16. P.K. Mehta, Pozzolanic and cementitious by-products in concrete Another look. in *Proceedings of 3rd CANMET/ACI International Conference on the Use of Fly Ash, Silica Fume, Slag, and Other Mineral By-products in Concrete*, Trondheim, Norway, June 18–23, 1989, Special Publication SP-1 14, vol 1, ed. by V.M. Malhotra (American Concrete Institute, Detroit, MI, 1989), pp. 1–43
17. J.D. Watt, D.J. Thorne, Composition and pozzolanic properties of pulverized fuel ashes: Composition of fly ashes from some British power stations and properties of their component particles. J. Appl. Chem. **15**, 585–594 (1965)
18. S. Diamond, F. Lopez-Flores, On the distribution between physical and chemical characteristics between lignitic and bituminous fly ashes, in *Proceedings of Symposium on Effects of Fly Ash Incorporation in Cement and Concrete* ed. by S. Diamond (Materials Research Society, Boston, MA, 1981), pp. 34–44
19. K. Takemoto, H. Uchikawa, Hydration of pozzolanic cement, in *Proceedings of 7th International Congress on the Chemistry of Cement*, Paris, France, Editions Septima, Paris, France, Sub-Theme IV-2, 1980, pp. 1–28
20. A. Ghosh, P.L. Pratt, Studies of the hydration reaction and microstructure of cement fly ash paste. in *Proceedings of Annual Meeting of the Materials Research Society*, Boston, MA, Nov 1981
21. H. Uchikawa, Effect of blending components on hydration and structure formation, in *Proceedings of 8th International Congress on the Chemistry of Cement*, Rio de Janeiro, Brazil, Sept 22–27, 1986, Special Report, vol 1
22. H. Uchikawa, S. Uchida, Influence of pozzolana on the hydration of C3A, in*Proceedings of 7th International Congress on the Chemistry of Cement*, Paris, France, Editions Septima, Paris, France, Sub-Theme IV, 1980, pp. 24–29
23. K. Mohan, H.F.W. Taylor, Paste of Iricalcium silicate with fly ash analytical electron microscopy, trimethylsilitation and other studies, *Effect of fly ash incorporation in cement and concrete*. Proceedings of Annual Meeting of the Materials Research Society, Boston, MA, Nov 1981
24. J. Skalny, J.F. Young Mechanism of Portland cement hydration, in *Proceedings of 7th International Congress on the Chemistry of Cement*, Paris, France, Editions Septima, Paris, France, Vol. 1, Sub-Theme-il, 1980 pp. 1–45
25. C. Plowman, J.G. Cabrera, The hydration reaction of calcium aluminate, in *Proceedings of Cement and Concrete Symposium, Annual Meeting*, 1981, pp. 71–8 1 (Mechanism and kinetics of hydration of C3A and C4AF extracted from cement, Cem. Concr. Res. **14**, 238–248, 1984)
26. S. Diamond, D. Ravina, J. LoveIl, The occurrence of duplex films on fly ash surfaces. Cem. Concr. Res. **5**, 363–376 (1980)
27. J.D. Watt, D.J. Thorne, Composition and pozzolanic properties of pulverized fuel ashes, I: Composition of fly ashes from British power stations and properties of their component particles. J. Appl. Chem. **15**, 585–594 (1965)
28. R.E. Philleo, Recent developments in pozzolan specifica, in *Proceedings of 2nd International Conference on the Use of Fly Ash, Silica Fume, Slag, and Natural Pozzolans in Concrete*, Madrid, Spain, Apr 21–25, 1986, Supplementary Paper 27
29. A. Brizzi, M. Puccio, G.L. Valenti, Correlations between physico-chemical characteristics of fly ash and its technical properties for use in concrete, in *Proceedings of 3rd CANMET/ACI International Conference on the Use of Fly Ash, Silica Fume, Slag, and Natural Pozzolans in Concrete*, Tronctheim, Norway, June 18–23, 1989, Supplementary Paper, pp. 139–156

30. D.J. Thorne, J.D. Watt, Composition and pozzolanic properties of pulverized fuel ashes, II: Pozzolanic properties of fly ashes, as determined by crushing strength tests on lime mortars. J. Appl. Chem. **15**, 595–604 (1965)
31. J.D. Watt, D.J. Thorne, The composition and pozzolanic properties of pulverized fuel ashes. III: Pozzolanic properties of fly ashes as determined by chemical methods. J. Appl. Chem. **16**, 33–39 (1966)
32. R.C. Joshi, E.A. Rosner, Pozzolanic activity in synthetic fly ashes, I: Experimental production and characterization. Am. Ceram. Soc. Bull. **52**, 456–458 (1973)
33. ACI Committee 211.1-81, Standard practice for selecting proportions for normal, heavyweight and mass concrete, ACI Manual of Concrete Practice, No. 211, pp. 1–81, 1984
34. Electrical Power Research Institute, Investigation of high-volume fly ash concrete systems. EPRI, Palo Alto, CA, EPRI TR-103 151, 1993
35. D.G. Montgomery, D.C. Hughes, R.T.Z. Williams, Fly ash in concrete—A microstructure study. Cem. Concr. Res. **11**, 591–603 (1981)
36. R.O. Lane, J.F. Best, Properties and use of fly ash in Portland cement concrete, Concr. Int. **4**(7):81–92 (1982)
37. V. Ramakrishnan, W.V. Coyle, J. Brown, A. Tiustus, P. Venkataramanujam, Performance characteristics of concretes containing fly ash, in *Proceedings of Symposium on Fly Ash Incorporation in Hydrated Cement Systems*, ed. by S. Diamond (Materials Research Society, Boston, 1981), pp. 233–243
38. P.L. Owens, Fly ash and its usage in concrete, Concr. J. Concr. Soc. **13**: 21–26 (1979)
39. L.J. Minnick, W.C. Webster, E.J. Purdy, Predictions of the effect of fly ash in Portland cement mortar and concrete. J. Mater. **6**, 163–187 (1971)
40. K.D. Stuart, D.A. Anderson, P.D. Cady, Compressive strength studies of Portland cement mortars containing fly ash and superplasticizers. Cem. Concr. Res. **10**, 823–832 (1980)
41. R.A. Helmuth, Water-reducing properties of fly ash in cement pastes, mortars, and concretes, in *Causes and Test Methods*, ed. by V.M. Malhotra. Proceedings of 2nd CANMET/ACI International Conference on the Use of Fly Ash, Silica Fume, Slag, and Natural Pozzolans in Concrete, Madrid, Spain, Apr. 21–25, 1986 (American Concrete Institute, Detroit, Special Publication SP-91), pp. 723–737
42. J.H. Brown, The strength and workability of Concrete with PFA substitution, in *Proceedings of International Symposium on the Use of PFA in Concrete*, University of Leeds, Leeds, UK, Apr 14–16, 1982, ed. by J.G. Cabrera, A.R. Cusens, Department of Civil Engineering, University of Leeds, Leeds, UK, 1982, pp. 151–161
43. C. Ellis, Some aspects of PFA in concrete. M. Phil. Thesis, Sheffield City Polytechnic, Sheffield, UK, 1977
44. B.G.T. Copeland, PFA concrete for hydraulic tunnels and shafts, Dinorwick pumped storage scheme case history, in *Proceedings of International Symposium on the Use of PFA in Concrete*, University of Leeds, Leeds, UK, Apr 14–16, 1982, ed. by J.G. Cabrera, A.R. Cusens, Department of Civil Engineering, University of Leeds, Leeds, UK, pp. 323–343
45. B.D.G. Johnson The use of fly ash in Cape Town RMC operations, in *Proceedings of 5th International Conference on Alkali-aggregate Reaction in Concrete*, Cape Town, South Africa, Mar 30–Apr 3, 1981, Paper S252/33
46. R.E. Philleo, Fly ash in mass concrete, in *Proceedings of 1st International Symposium on Fly Ash Utilization*, Pittsburgh, PA, Mar 14–16, 1967. Bureau of Mines, Washington, DC, Information Circular IC 8348, pp. 69–79
47. R.J. Elfert, Bureau of Reclamation experiences with fly ash and other pozzolans in concrete, in *Proceedings of 3rd International Ash Utilization Symposium*, Pittsburgh, PA, Mar 13–14, 1973, Bureau of Mines, Washington, DC, Information Circular IC 8640, pp. 80–93
48. J.T. Williams, P.L. Owens, The implications of a selected grade of United Kingdom pulverized fuel ash on the engineering design and use in structural concrete, in *Proceedings of International Symposium on the Use of PFA in Concrete*, University of Leeds, Leeds, UK, Apr 14–16, 1982, ed. by J.G. Cabrera, A.R. Cusens, Department of Civil Engineering, University of Leeds, Leeds, UK, pp. 301–313

49. P.B. Bamforth, In situ measurement of the effect of partial Portland cement replacement using either fly ash or ground granulated blast furnace slag on the performance of mass concrete. Proc. Inst. Civ. Eng. **69**, 777–800 (1980)
50. R.D. Crow, E.R. Dunstan, Properties of fly ash concrete. in *Proceedings of Symposium on Fly Ash Incorporation in Hydrated Cement Systems*, ed. by S. Diamond (Materials Research Society, Boston, MA, 1981), pp. 214–225
51. Personal communication from Dr.T. Naik, University of Wisconsin, Milwaukee, WI, Dec 1985
52. R.L. Yuan, J.E. Cook, Study of a Class C fly ash concrete, in *Proceedings of 1st International Conference on the Use of Fly Ash, Silica Fume, Slag, and Other Mineral By-products in Concrete*, Montebello, PQ, July 31–Aug 5, 1983, ed. by V.M. Malhotra (American Concrete Institute, Detroit, Special Publication SP-79, 1983), pp. 307–319
53. F. Raba Jr, S.L. Smith, M. Mearing, Subbituminous fly ash utilization in concrete, in *Proceedings of Symposium on Fly Ash Incorporation in Hydrated Cement Systems*, ed. by S. Diamond (Materials Research Society, Boston, 1981), pp. 296–306
54. J.F. Lamond, Twenty-five years' experience using fly ash in concrete, in *Proceedings of 1st International Conference on the Use of Fly Ash, Silica Fume, Slag, and Other Mineral By-products in Concrete*, Montebello, PQ, July 31–Aug 5, 1983, ed. by V.M. Malhotra (American Concrete Institute, Detroit, Special Publication SP-79, 1983), pp. 47–69
55. S.H. Gebler, P. Klieger, Effect of fly ash on physical properties of concrete, in *Proceedings of 2nd CANMET/ACI International Conference on the Use of Fly Ash, Silica Fume, Slag, and Natural Pozzolan in Concrete*, Madrid, Spain, Apr 21–25, 1986, ed. by V.M. Malhotra (American Concrete Institute, Detroit, Special Publication SP-91, 1986), pp. 1–50
56. P.J. Tikaisky, P.M. Carrasquillo, R.L. Carrasquillo, Strength and durability considerations affecting mix proportioning of concrete containing fly ash. ACT Mater. J. **85**, 505–511 (1988)
57. J.G. Cabrera, C.J. Hopkins, G.R. Wooley, R.E. Lee, J. Shaw, C. Plowman, H. Fox, Evaluation of the properties of British pulverized fuel ashes and their influence on the strength of concrete, in *Proceedings of 2nd International Conference on the Use of Fly Ash, Silica Fume, Slag, and Other Mineral By-products in Concrete*, Madrid, Spain, Apr 21–25, 1986, ed. by V.M. Malhotra (American Concrete Institute, Detroit, Special Publication SP-91, 1986), pp. 115–144
58. K. Wesche, W. von Berg, Germany, in *Proceedings of Symposium on Fly Ash Incorporation in Hydrated Cement Systems*, ed. by S. Diamond (Materials Research Society, Boston, 1981), pp. 45–53
59. R.C. Joshi, Effect of coarse fraction (+ #325) of fly ash on concrete properties, in *Proceedings of 6th International Symposium on Fly Ash Utilization*, Reno, NV, Mar 1982, U.S. Department of the Environment, Washington, DC, DOEIMETC/82-52, 1982, pp. 77–85
60. D. Ravina, Production and collection of fly ash for use in concrete, in *Proceedings of Symposium on Fly Ash Incorporation in Hydrated Cement Systems*, ed. by S. Diamond (Materials Research Society, Boston, 1981), pp. 2–11
61. M. Monk, Portland—PFA cement: A comparison between intergrinding and blending. Mag. Concr. Res. **35**(124), 131–141 (1983)
62. Y. Matsufuji, H. Kohata, K. Tagaya, H. Teramoto, Y. Okawa, S. Okazawa, Study on properties of concrete with ultra-fine particles produced from fly ash, in *Proceedings of 4th CANMET/AC International Conference on the Use of Fly Ash, Silica Fume, Slag, and Natural Pozzolans in Concrete*, Istanbul, Turkey, May 3–8, 1992, ed. by V.M. Mathotra, vol 1 (American Concrete Institute, Detroit, Special Publication SP-132, 1992), pp. 351–365
63. A.M. Nevile, *Properties of concrete*, 2nd edn. (John Wiley, New York, 1973), p. 382
64. M. Kobayashi, Utilization of fly ash and its problems in use in Japan. Japan—US, Science Seminar, San Francisco, CA, Sept 10–13, 1979, pp. 61–69

65. M. Kokubu, I. Miura, S. Takano, R. Sugiki, Effect of temperature and humidity during curing on strength of concrete containing fly ash, Doboku Gakkai Robunshu (Transactions of the Japan Society of Civil Engineers), Extra Papers (4-3), 7L (Dec), pp. 1–10, 1960
66. D. Ravina, Efficient utilization of coarse and fine fly ash in precast concrete by incorporating thermal curing. J. Am. Concr. Inst. **78**, 194–200 (1981)
67. M.K. Gopalan, M.N. Haque, Effect of curing regime on the properties of fly-ash concrete. ACI Mater. J. **84**, 14–19 (1987)
68. M.N. Haque, R.L. Day, B.W. Langan, Realistic strength of air-entrained concretes with and without fly ash. ACI Mater. J. **85**, 241–247 (1988)
69. E.A. Abdun-Nur, Fly ash in concrete: An evaluation. Highway Research Board, Washington, DC, Highway Research Bulletin 284, 1961
70. R.E. Davis, Pozzolanic materials and their use in concrete, in *Proceedings of Symposium on the Use of Pozzolanic Materials in Mortars and Concretes,* Special Publication 99 (American Society for Testing and Materials, Philadelphia, 1949)
71. Bureau of Reclamation, Concrete mix investigations for Canyon Ferry Dam. Bureau of Reclamation, Denver, CO, Report C-656, 1953
72. Bureau of Reclamation, Investigation of the properties of concrete for use in the design and construction of Yellowtail Dam. Bureau of Reclamation, Concrete Laboratory, Denver, CO, Report C-705, 1953
73. Bureau of Reclamation, Laboratory and field investigations of concrete for Hungry Horse Dam. Bureau of Reclamation, Concrete, 1953
74. R.S. Ghosh, J. Timusk, Creep of fly ash concrete. J. Am. Concr. Inst. **78**(35), 1–357 (1981)
75. K.W. Nasser, H.M. Marzouk, Properties of concrete made with sulphate-resisting cement and fly ash, in *Proceedings of 1st International Conference on the Use of Fly Ash, Silica Fume, Slag, and Other Mineral By-products in Concrete*, Montebello, PQ, July 31–Aug 5, 1983, ed. by V.M. Malhotra (American Concrete Institute, Detroit, Special Publication SP-79, 1983), pp. 383–395
76. K.W. Nasser, R.N. Ojha, Properties of concrete made with Saskatchewan fly ash, *CANMET International Workshop on Fly Ash in Concrete*, Calgary, AB, Oct 3–5, 1990. CANMET, Energy, Mines and Resources Canada, Ottawa, ON, Compilation of Papers
77. R.P. Lohtia, B.D. Nautiyal, O.P. Jam, Creep of fly ash concrete. J. Am. Concr. Inst. **73**, 469–472 (1976)
78. P.M. Gifford, M.A. Ward, Results of laboratory tests on lean mass concrete utilizing PFA to a high level of cement replacement, in *Proceedings of International Symposium on the Use of PFA in Concrete*, University of Leeds, Leeds, UK, Apr 14–16, 1982, ed. by J.G. Cabrera, A.R. Cusens (Department of Civil Engineering, University of Leeds, Leeds, 1982), pp. 221–230
79. K.W. Nasser, A.A. Al-Manaseer, Creep of concrete containing fly ash and admixtures at different stress/strength ratios, *CSCE Annual Conference*, Saskatoon, SK, May 1985, vol hA (Structural), pp. 361–373
80. K.W. Nasser, A.A. Al-Manaseer, Shrinkage and creep of concrete containing 50% lignite fly ash at different stress/strength ratios, in *Proceedings of 2nd CANMET/ACI International Conference on the Use of Fly Ash, Silica Fume, Slag, and Natural Pozzolans in Concrete*, Madrid, Spain, Apr 21–25, 1986, ed. by V.M. Malhotra (American Concrete Institute, Detroit, Special Publication SP-91, 1986), pp. 433–448
81. R.L. Yuan, J.E. Cook, Time-dependent deformation of high strength fly ash concrete, in *Proceedings of International Symposium on the Use of PFA in Concrete*, University of Leeds, Leeds, UK, Apr 14–16, 1982, ed. by J.G. Cabrera, A.R. Cusens (Department of Civil Engineering, University of Leeds, Leeds, 1982), pp. 255–261
82. J.G.L. Munday, L.T. Ong, L.B. Wong, R.K. Dhir, Load-independent movements in OPC/ PFA concrete, in *Proceedings of International Symposium on the Use of PFA in Concrete*, University of Leeds, UK, Apr 14–16, 1982, ed. by J.G. Cabrera, A.R. Cusens (Department of Civil Engineering, University of Leeds, Leeds, 1982), pp. 243–261

83. R.S. Ravindrarajah, C.T. Tam, Properties of concrete containing low-calcium fly ash under hot and humid climate, in *Proceedings of 3rd CANMET/ACI International Conference on the Use of Fly Ash, Silica Fume, Slag, and Natural Pozzolans in Concrete*, Trondheim, Norway, June 18–23, 1989, vol 1, ed. by V.M. Malhotra (American Concrete Institute, Detroit, Special Publication SP-1 14, 1989), pp. 139–155
84. R.E. Davis, Pozzolanic materials-with special reference to their use in concrete pipe. American concrete pipe Association, Technical Memo, 1954
85. I.M. Kanitakis, Permeability of concrete containing pulverized fuel ash, in *Proceedings of 5th International Symposium on Concrete Technology,* Nuevo Leon, Mexico, Mar 1981, Department of Civil Engineering, University of Nuevo Leon, Monterrey, Mexico, 1982, pp. 311–322
86. D. Manmohan, P.K. Mehta, Influence of pozzolanic, slag and chemical admixtures on pore size distribution and permeability of hardened cement pastes. Cem. Concr. Aggregates **3**, 63–67 (1981)
87. R.D. Browne, Mechanisms of corrosion of steel in concrete in relation to design, inspection and repair of offshore and coastal structures. in *Performance of Concrete in Marine Environment*, ed. by V.M. Malhotra (American Concrete Institute, Detroit, Special Publication SP-65, 1980), pp. 169–204
88. R.M. Barrer, *Diffusion in and through solids* (Cambridge University Press, Cambridge, 1951)
89. N.R. Short, C.L. Page, The diffusion of chloride ions through Portland and blended cement pastes. Silic. Indus. **47**, 237–240 (1982)
90. Y. Kasai, I. Matsui, U. Fukishima, H. Kamohara, Air permeability and carbonation of blended cement mortars, in *Proceedings of 1st International Conference on the Use of Fly Ash, Silica Fume, Slag, and other mineral By-products in Concrete*, Montebello, PQ, July 31–Aug. 5, 1983, ed. by V.M. Malhotra (American Concrete Institute, Detroit, MI, Special Publication SP-79, 1983), pp. 435–451
91. M.D.A. Thomas, J.D. Matthews, C.A. Haynes, The effect of curing on the strength and permeability of PFA concrete, in *Proceedings of 3rd CANMET/ACI International Conference on the Use of Fly Ash, Silica Fume, Slag, and Natural Pozzolans in Concrete*, Trondheim, Norway, June 18–23, 1989, ed. by V.M. Malhotra (American Concrete Institute, Detroit, Special Publication SP-114, 1989), pp. 191–217
92. H. Abe, S. Nagataki, R. Tsukayama, Written discussion on Takemoto and Uchikawa (1980) (Ref. [19] of this chapter), in *Proceedings of 5th International Symposium on Chemistry of Cement*, Tokyo, Japan, Oct 7–11, vol 4, 1969, pp. 105–111
93. M. Hamada, Neutralization (carbonation) of concrete and corrosion of reinforcing steel, in *Proceedings of 5th International Symposium on the Chemistry of Cement*, Tokyo, Japan, Oct 7–11, 1968, Sub-Theme 111-3, vol 3, pp. 343–369
94. A. Meyer, Investigation on the carbonation of concrete, in *Proceedings of 5th International Symposium on the Chemistry of Cement*, Tokyo, Japan, Oct 7–11, 1968, Sub-Theme 111-52, vol 3, pp. 394–401
95. M. Kokubu, S. Nagataki, Carbonation of concrete correlating with the corrosion of reinforcement in fly ash concrete, in *Proceedings of Symposium on Durability of Concrete*, International Union of Testing and Research Laboratories, Paris, France, RILEM Final Report. Part II, 1969, pp. D71–D79
96. P. Schubert, W. von Berg, Coal ash with test mark as an additive for concrete in accordance with D 1045, Betonwerk + FertigteilTechnik, Wiesbaden, Germany, 11-692-696, 1979
97. R. Tsukayama, S. Nagataki, H. Abe, Long term experiments on the neutralization of concrete mixed with fly ash and the corrosion of reinforcement, in *Proceedings of 7th International Congress on the Chemistry of Cement*, Paris, France, 1980, Editions Septima, Paris, France, Sub-Theme VII-ISCC
98. J. Gebauer, Some observations on the carbonation of fly as concrete. Silic. Indus. **47**, 155–159 (1982)

99. D.W.S. Ho, R.K. Lewis, Carbonation of concrete incorporating fly ash or a chemical admixture, in *Proceedings of 1st International Conference on the Use of Fly Ash, Silica Fume, Slag, and Other Mineral By-products in Concrete*, Montebello, PQ, July 31–Aug. 5, 1983, ed. by V.M. Malhotra (American Concrete Institute, Detroit, Special Publication SP-79, 1983), pp. 333–346
100. F.G. Butler, M.H. Decter, G.R. Smith, Studies on the desiccation and carbonation of systems containing Portland cement arid fly ash, in *Proceedings of 1st International Conference on the Use of Fly Ash, Silica Fume, Slag, and Other Mineral By-products in Concrete*, Montebello, PQ, July 31–Aug. 5, 1983, ed. by V.M. Malhotra (American Concrete Institute, Detroit, Special Publication SP-79, 1983), pp. 71–86
101. S. Nagataki, H. Ohga, E.K. Kim, Effect of curing conditions on the carbonation of concrete with fly ash and the Corrosion of reinforcement in long term tests, in *Proceedings of 2nd CANMET/ACI International Conference on the Use of Fly Ash, Silica Fume, Slag, and Natural Pozzolans in Concrete*, Madrid, Spain, Apr 21–25, 1986, ed. by V.M. Malhotra (American Concrete Institute, Detroit, Special Publication SP-91, 1986), pp. 521–540
102. H.G. Smolczyk, Exploration of the German long time study on the rate of carbonation, in *Proceedings of RILEM International Symposium on Carbonation of Concrete*, London, UK, 1976. Theme 3, Paper 2
103. A. Meyer, Investigations on the carbonation of concrete, in *Proceedings of 5th International Symposium on the Chemistry of Cement*, Tokyo, Japan, 1968, vol 3, pp. 394–397
104. A.A. Ramezanianpour, J.G. Cabrera, Durability of OPC-trass, OPC PFA, and OPC-silica fume mortars and concretes. in *Proceedings of 14th International Conference on Our World in Concrete and Structures*, Singapore, Aug 24–25, 1989, pp. 85–89
105. A.A. Ramezanianpour, Accelerated and long term carbonation of mortars containing some natural and artificial pozzolans, in *Proceedings of 15th International Conference on Our World in Concrete and Structures*, Singapore, Aug 23–24, 1990, pp. 277–283
106. H. Ohga, S. Nagataki, Prediction of carbonation depth of concrete with fly ash, in *Proceedings of 3rd CANMET/ACI Internationa Conference on the Use of Fly Ash, Silica Fume, Slag, and Natural Pozzolans in Concrete*, Trondheim, Norway, June 18–23, 1989, ed. by V.M. Malhotra (American Concrete Institute, Detroit, Special Publication SP-114, 1989), pp. 275–294
107. D.W. Hobbs, Carbonation of concrete containing PFA. Mag. Concr. Res. **40**(143), 69–78 (1988)
108. T.D. Larson, Air entrainment and durability aspects of fly ash concrete. Proc. Am. Soc. Test. Mater. **64**, 866–886 (1964)
109. S. Gebler, P. Kheger Effect of fly ash on the air-void stability of concrete, in *Proceedings of 1st International Conference on the Use of Fly Ash, Silica Fume, Slag, and Other Mineral By-products in Concrete*, Montebello, PQ, July 31–Aug. 5, 1983, ed. by V.M. Malhotra (American Concrete Institute, Detroit, Special Publication SP-79, 1983), pp. 103–142
110. V.R. Sturrup, R.D. Hooton, T.G. Clendenning, Durability of fly ash concrete, in *Proceedings of 1st International Conference on the Use of Fly Ash, Silica Fume, Slag, and Other Mineral By-products in Concrete*, Montebello, PQ, July 31–Aug. 5, 1983, ed. by V.M. Malhotra (American Concrete Institute, Detroit, Special Publication SP-79, 1983), pp. 47–69
111. T.G. Clendenning, N.D. Dune, Properties and use of fly ash from a steam plant operating under variable load. Proc. Am. Soc. Test. Mater. **62**, 1019–1040 (1962)
112. V.R. Sturrup, T.G. Clendenning, The evaluation of concrete by outdoor exposure. Indian Roads Congress, Highway Research Board, New Dethi, India, Highway Research Board Record HRR268, 1969, pp. 48–61
113. P.W. Brown, J.R. Clifton, G. Frohnsdorff, R.L. Berger, Limitations to fly ash use in blended cements, in *Proceedings in 4th International Ash Utilization Symposium*, St. Louis, MO, Mar 24–25, 1976. Energy Research and Development Administration, Washington, DC, ERDA MERC/SP-76/4, 1976, pp. 518–529

114. C. Johnston, Effect of microsilica and Class C fly ash on resistance of concrete to rapid freezing and thawing and scaling in the presence of deicing agents, in *Proceedings of Katharine and Bryant Mather International Conference on Concrete Durability*, Atlanta, GA, Apr 27–May 1, 1987, vol 2, ed. by J.M. Scanlon (American Concrete Institute, Detroit, Special Publication SP-100, 1987), pp. 1183–1204
115. J. Virtanen, Freeze–thaw resistance of concrete containing blast-furnace slag, fly ash or condensed silica fume, in *Proceedings of 1st International Conference on the Use of Fly Ash, Silica Fume, Slag, and Other Mineral By-products in Concrete*, Montebello, PQ, July 31–Aug. 5, 1983, ed. by V.M. Mathotra (American Concrete Institute, Detroit, Special Publication SP-79, 1983) pp. 923–942
116. ACI Committee 201, *Guide to Durable Concrete*, American concrete Institute, Detroit, ACI 201.2R-77, Chap. 1, 1977
117. K.W. Nasser, P.S.H. Lai, Resistance of fly ash concrete to freezing and thawing, in *Proceedings of 4th CANMET/ACI International Conference on the Use of Fly Ash, Silica Fume, Slag, and Natural Pozzolans in Concrete*, Istanbul, Turkey, May 3–8, 1992, ed. by V.M. Malhotra, vol 1 (American Concrete Institute, Detroit, Special Publication SP-132, 1992), pp. 205–226
118. P. Klieger, S. Gebler, Fly ash and concrete durability, in *Proceedings of Katharine and Bryant Mather International Conference on Concrete Durability*, Atlanta, GA, Apr 27–May 1, 1987, ed. by J.M. Scanlon, vol 1 (American Concrete Institute, Detroit, Special Publication SP-100, 1987), pp. 1043–1069
119. A. Bilodeau, G.G. Carette, V.M. Malhotra, W.S. Langley, Influence of curing and drying on salt scaling resistance of fly ash concrete, in *Proceedings of 2nd CANMET/ACI International Conference on the Durability of Concrete*, Montréal, PQ, Aug 4–9, 1991, ed. by V.M. Malhotra, vol 1 (American Concrete Institute, Detroit, Special Publication SP-126, 1991), pp. 201–228
120. G.G. Carette, W.S. Langley, Evaluation of de-icing salt scaling of fly ash concrete, in *Proceedings of International Workshop on Alkali-aggregate Reactions in Concrete: Occurrences, Testing and Control*, Halifax, NS, 1990, CANMET, Ottawa, ON
121. K.W. Nasser, R.P. Lohtia, Mass concrete properties at high temperatures, J. Am. Concr. Inst. **68**, 180–186 (1971)
122. K.W. Nasser, R.P. Lohtia, Mass concrete properties at high temperatures. J. Am. Concr. Inst. **68**, 276–281 (1971)
123. K.W. Nasser, H.M. Marzouk, Properties of mass concrete containing fly ash at high temperatures. J. Am. Concr. Inst. **76**, 537–550 (1979)
124. G.G. Carette, K.E. Painter, V.M. Maihotra, Sustained high temperature effect on concretes made with normal Portland cement, normal Portland cement and slag, or normal Portland cement and fly ash. Concr. Int. **4**, 41–51 (1982)
125. T.C. Liu, Maintenance and preservation of concrete structures. Report 3: Abrasion-erosion resistance of concrete. U.S. Army Corps of Engineers, Waterways Experiment Station, Vicksburg, MS, Technical Report C-78-4, 1980, p. 129
126. P. Carrasquillo, Durability of concrete containing fly ash for use in highway applications, in *Proceedings of Katharine and Bryant Mather International Conference on Concrete Durability*, Atlanta, GA, Apr–27 May 1, 1987, ed. by J.M. Scanlon, vol 1 (American Concrete Institute, Detroit, Special Publication SP-100, 1987), pp. 843–861
127. T.R. Naik, S.S. Singh, M.M. Hossein, Abrasion resistance of high-volume fly ash concrete systems. Electric Power Research Institute, Palo Alto, CA, EPRI Report, 1992
128. I. Biczok, *Concrete corrosion and concrete protection* (Hungarian Academy of Sciences, Budapest, 1964)
129. R. Kovacs, Effect of the hydration products on the properties of fly ash cements. Cem. Concr. Res. **5**, 73–82 (1975)
130. F.M. Lea, *The chemistry of cement and concretes* (Edward Arnold, London, 1971)

131. J.T. Dikeou, Fly ash increases resistance of concrete to sulphate attack. Bureau of Reclamation, Denver, CO, Water Resources Technical Publication, Research Report 23, 1970
132. G.L. Kalousek, L.C. Porter, E.J. Benton, Concrete for long-term service in sulphate environment. Cem. Concr. Res. **2**, 79–89 (1972)
133. E.R. Dunstan, Performance of lignite and sub-bituminous fly ash in concrete. A progress report. Bureau of Reclamation, Denver, CO., Report REC-ERC-76-1(274) E.R. Dunstan, 1980. A possible method for identifying fly ashes that will improve the sulphate resistance of concretes. Cem. Concr. Aggregates **2**, 20–30 (1976)
134. E.R. Dunstan, A possible method for identifying fly ashes that will improve the sulphate resistance of concretes. Cem. Concr. Aggregates **2**, 20–30 (1980)
135. J.S. Pierce, Use of fly ash in combating sulphate attack in concrete, in *Proceedings of 6th International Symposium on Fly Ash Utilization*, Reno, NV, Mar. 1982. U.S. Department of the Environment, Washington, DC, DOEIMETC/82-52, 1982, pp. 208–231
136. Bureau of Reclamation, *Concrete manual,* 8th ed. (Bureau of Reclamation, Denver, 1981), p. 11
137. E.R. Dunstan, Sulphate resistance of fly ash concretes. The R value, in *Proceedings of Katharine and Bryant Mather International Conference on Concrete Durability*, Atlanta, GA, Apr 27–May 1, 1987, ed. by J.M. Scanlon, vol 2 (American Concrete Institute, Detroit, Special Publication SP-100, 1987), pp. 2027–2040
138. P.J. Tikaisky, B.F. Reed, R.L. Carrasquillo, Recommendations for the use of fly ash in sulphate-resistant concrete, *CANMET International Workshop on Fly Ash in Concrete*, Calgary, AB, Oct 3–5, 1990. CANMET, Energy, Mines and Resources Canada, Ottawa, ON
139. K. Mather, Current research in sulphate resistance at the Waterways Experiment Station, in *Proceedings of the George Verbeck Symposium on Sulphate Resistance of Concrete* (American Concrete Institute, Detroit, Special Publication SP-77, 1982), pp. 63–74
140. T.E. Stanton, Expansion of— concrete through reaction between cement and aggregate, Trans. Am. Soc. Civil Eng. **2**, 68–85 (1942)
141. A.B. Poole, Alkali-carbonate reactions in concrete, in *Proceedings of 5th International Conference on Alkali-aggregate Reaction in Concrete*, Cape Town, South Africa, Mar 30–Apr 3, 1981, Paper S252/34
142. E.G. Swenson, J.E. Gillott, *Characteristics of Kingston Carbonate Rock Reaction*, Highway Research Board, Washington, DC, Bulletin 275, 1960
143. L. Pepper, B. Mather, Effectiveness of mineral admixtures in preventing excessive expansion of concrete due to alkali-aggregate reaction. Proc. Am. Soc. Test. Mater. **59**, 1178–1202 (1959)
144. M.A.G. Duncan, E.G. Swenson, J.E. Gillott, M. Foran, Alkali-aggregate reaction in Nova Scotia, I: Summary of five year study. Cem. Concr. Res. **3**, 55–69 (1973)
145. D.W. Hobbs, Expansion due to alkali–silica reaction and the influence of pulverized fuel ash, in *Proceedings of 5th International Conference on Alkali-aggregate Reaction in Concrete*, Cape Town, South Africa, Mar 30–Apr 3, 1981, Paper S252/30
146. D.W. Hobbs, The alkali–silica reaction—A model for predicting expansion in mortar. Mag. Concr. Res. **33**(117), 208–220 (1981)
147. P.K. Mehta, Pozzolanic and cementitious by-products as mineral admixtures for concrete a critical review, in *Proceedings of 1st International Conference on the Use of Fly Ash, Silica Fume, Slag, and Other Mineral By-products in Concrete*, Montebello, PQ, July 31–Aug.5, 1983, ed. by V.M. Malhotra (American Concrete Institute, Detroit, Special Publication SP-79, 1983), pp. 1–46
148. E.R. Dunstan, The effect of fly ash on concrete alkali-aggregate reaction. Cem. Concr. Aggregates **3**, 101–104 (1981)
149. P.J. Nixon, M.E. Gaze, The use of granulated blast furnace slag to reduce expansion due to alkali-aggregate reaction, in *Proceedings of 5th International Conference on Alkali-aggregate Reaction in Concrete*, Cape Town, South Africa, Mar 30–Apr 3, 1981

150. R.E. Oberhoister, W.B. Westra, The effectiveness of mineral admixtures in reducing expansion due to alkali-aggregate reaction with Malmesbury Group aggregates, in *Proceedings of 5th International Conference on Alkali-aggregate Reaction in Concrete*, Cape Town, South Africa, Mar 30–Apr 3, 1981
151. D.W. Hobbs, Influence of pulverized-fuel ash and granulated blast furnace slag upon expansion caused by the alkali–silica reaction. Mag. Concr. Res. **34**(119), 83–93 (1982)
152. P.E. Nixon, M.E. Gaze. The effectiveness of fly ashes and granulated blast furnace slags in preventing AAR, in *Proceedings of 6th International Conference on Alkalies in Concrete*, Copenhagen, Denmark, June 22–25, 1983, Edited by G.M. Doran, S. Rostan, 1983, pp. 61–68
153. L.C. Porter, Small proportions of pozzolan may produce detrimental reactive expansion in mortar. Bureau of Reclamation, Denver, CO, Report No. C.-113, 1964, p. 22
154. D.W. Hobbs, Possible influence of small additions of PFA, GBFS and limestone flour upon expansion caused by the alkali–silica reaction. Mag. Concr. Res. **35**(122), 55 (1983)
155. J. Farbiarz, R. Carrasquillo, Alkali-aggregate reaction in concrete containing fly ash, in *Proceedings of Katharine and BryantMather International Conference on Concrete Durability*, Atlanta, GA, Apr 27–May 1, 1987, ed. by J.M. Scanlon, vol 2 (American Concrete Institute, Detroit, Special Publication SP-100, 1987), pp. 1787–1808
156. R.L. Carrasquilo, P.G. Snow, Effect of fly ash on alkali-aggregate reaction in concrete. ACI Mater. J. **84**, 299–305 (1987)
157. S. Nagataki, H. Ohga, T. Inoue, Evaluation of fly ash for controlling alkali—aggregate reaction, in *Proceedings of 2nd CANMET/ACI International Conference on the Durability of Concrete*, Montréal, PQ, Aug 4–9, 1991, ed. by V.M. Malhotra, vol 2 (American Concrete Institute, Detroit, Special Publication SP-126, 1991), pp. 955–972
158. M.D.A. Thomas, P.J. Nixon, K. Pettifer, The effect of pulverized fuel ash with a high total alkali content on alkali silica reaction in concrete containing natural UK aggregate, in *Proceedings of 2nd CANMET/ACI International Conference on the Durability of Concrete*, Montréal, PQ, Aug. 4–9, 1991, ed. by V.M. Malhotra, vol 2 (American Concrete Institute, Detroit, MI, Special Publication SP-126, 1991), pp. 919–940
159. RILEM Technical Committee 12-CRC, Corrosion of reinforcement and prestressing tendons a state of the art report, Matérial de construction **9**, 187–206 (1974)
160. Anonymous, Relationship of fly ash and corrosion (A letter to the editor), J. Am. Concr. Inst. **47**, 74 (1951)
161. J.L. Gilliland, Relationship of fly ash and corrosion (Response to the letter in reference 300), J. Am. Concr. Inst. **47**, 397 (1951)
162. J.P. Ryan, Relationship of fly ash and corrosion (Response to the letter in reference 300), J. Am. Concr. Inst. **47**, 481–484 (1951)
163. S.S. Rehsi, Studies on Indian fly ashes and their use in structural concrete, in *Proceedings of 3rd International Ash Utilization Symposium*, Pittsburgh, PA, Mar 13–14, 1973, Bureau of Mines, Washington, DC, Information Circular IC 8640, 1973, pp. 231–245
164. J. Kondo, A. Takeda, S. Hideshima, Effect of admixtures on electrolytic corrosion of steel bars in reinforced concrete. J. Jpn. Soc. Civil Eng. **43**, 1–8 (1958)
165. A. Paprocki, The inhibitory effect of fly ash with respect to the corrosion of steel in concrete, in *Proceedings of 2nd International Ash Utilization Symposium*, Pittsburgh, PA, Mar 10–11, 1970, Bureau of Mines, Denver, CO, Information Circular IC 8488, pp. 17–23
166. T.J. Larsen, G.C. Page, Fly ash for structural concrete in aggressive environments, in *Proceedings of 4th International Ash Utilization Symposium*, St. Louis, MO, Mar 24–25, 1976. Energy Research and Development Administration, Washington, DC, ERDA MERC/SP-76/4, 1976, pp. 573–587
167. T.J. Larsen, W.E. McDaniel, R.P. Brown, J.L.Sosa, Corrosion-inhibiting properties of Portland and Portland pozzolan cement concrete, Transp. Res. Rec. **613**, 21–29 (1976)
168. S. Diamond, Effects of two Danish fly ashes on alkali contents of pore solutions of cement-fly ash pastes. Cem. Concr. Res. **11**, 383–394 (1981)

169. N.S. Berke, M.J. Scali, J.C. Regan, D.F. Shen, Long-term corrosion resistance of steel in silica fume and/or fly ash containing concretes, in *Proceedings of 2nd CANMET/ACI International Conference on the Durability of Concrete*, Montréal, PQ, Aug 4–9, 1991, ed. by V.M. Malhotra, vol 1 (American Concrete Institute, Detroit, Special Publication, SP-126, 1991), pp. 393–422
170. B. Mather, Concrete in sea water. Concr. Int. **4**, 28–34 (1982)
171. P.K. Mehta, Durability of concrete in marine environment—A review, in *Performance of concrete in marine environment*, ed. by V.M. Malhotra (American Concrete Institute, Detroit, Special Publication SP-65, 1980), pp. 1–20
172. V.M. Malhotra, G.G. Carette, T.W. Bremner, Durability of concrete containing granulated blast furnace slag or fly ash or both in marine environment, Energy, Mines and Resources Canada, Ottawa, ON, CANMET Report 80-18E, 1980
173. V.M. Malhotra, G.G. Carette, T.W. Bremner, Current status of CANMET's studies on the durability of concrete containing supplementary cementing materials in marine environment, in *Proceedings of 2nd CANMET/ACI International Conference on the Performance of Concrete in Marine Environment*, St. Andrews, NB, Aug. 21–26, 1988, ed. by E.M. Malhotra (American Concrete Institute, Detroit, Special Publication SP-109, 1988), pp. 31–72
174. V.M. Malhotra, G.G. Carette, T.W. Bremner, CANMET investigations dealing with the performance of concrete containing supplementary cementing materials at Treat Island, Maine, Energy, Mines and Resources. Canada, Ottawa, ON, CANMET Report MSL 744, 1992
175. H. Ogha, S. Nagataki, Effect of fly ash on alkali-aggregate reaction in marine environment, in *Proceedings of 4th CANMET/ACI International Conference on the Use of Fly Ash, Silica Fume, Slag, and Natural Pozzolan in Concrete*, Istanbul, Turkey, May 3–8, 1992, ed. by V.M. Malhotra, vol 1 (American Concrete Institute, Detroit, Special Publication SP-132, 1992), pp. 577–590
176. T. Cheewaket, C. Jaturapitakkul, W. Chalee, Long term performance of chloride binding capacity in fly ash concrete in a marine environment, Constr. Build Mater. J. **24**, 1352–1357 (2010)
177. A.R. Boga, I.B. Topcu, Influence of fly ash on corrosion resistance and chloride ion permeability of concrete. Constr. Build. Mater. J. **31**, 258–264 (2012)
178. C.-W. Chung, C.-S. Shon, Y.-S. Kim, Chloride ion diffusivity of fly ash and silica fume concretes exposed to freeze-thaw cycles. Constr. Build. Mater. J. **24**, 1739–1745 (2010)
179. M. Thomas, A. Dunster, P. Nixon, B. Blackwell, Effect of fly ash on the expansion of concrete due to alkali-silica reaction-exposure site studies. J. Cem. Concr. Compos. **33**, 359–367 (2011)
180. P. Nath, P. Sarkar, Effect of fly ash on the durability properties of high strength concrete. Procedia Engineering **14**, 1149–1156 (2011)
181. R. Siddique, Properties of self compacting concrete containing class F fly ash, J. Mater. Des. **32**, 1501–1507 (2011)
182. R.L. Blick, C.F. Petersen, M.E. Winter, Proportioning and controlling high strength concrete (American Concrete Institute, Detroit, Special Publication SP 46-9, 1974), pp. 141–163
183. J. Wolsiefer, Ultra high strength field-placeable concrete in the range 10,000 to 18,000 psi (69–124 MPa). Paper presented at the Annual Convention of the American Concrete Institute, Atlanta, GA, Jan 19–23, 1982
184. K.L. Saucier, High-strength concrete, Past, present, future. Concr. Int. **2**, 46–50 (1980)
185. J.E. Cook, Research and application of high-strength concrete using Class C fly ash. Concr. Int. **4**(7), 72–80 (1982)
186. T.R. Naik, V.M. Patel, L.E. Brand, Performance of high strength concrete incorporating mineral by-products, in *Proceedings of CBU/CANMET International Symposium on the Use of Fly Ash, Silica Fume, Slag, and Other By-products in Concrete and Construction Materials*, University of Wisconsin, Milwaukee, WI, Nov 2–4, 1992

187. K. Ukita, M. Ishii, K. Yamamoto, K. Azuma, Properties of high strength concrete using classified fly ash, in *Proceedings of 4th CANMET/ACI International Conference on the Use of Fly Ash, Silica Fume, Slag, and Natural Pozzolans in Concretes*, Istanbul, Turkey, May 3–8, 1992, ed. by V.M. Maihotra, vol 1 (American Concrete Institute, Detroit, Special Publication SP-132, 1992), pp. 37–52
188. K. Ukita, S. Shigematsu, M. Isliii, Quality Improvement concrete utilizing classified fly ash, in *Proceedings of 3rd CANMET/ACI International Conference on the Use of Fly Ash, Silica Fume, Slag, and Natural Pozzolans in Concrete*, Trondheim, Norway, June 18–23, 1989, ed. by V.M. Malhotra, vol 1 (American Concrete Institute, Detroit, Special Publication SP-1 14, 1989), pp. 219–240
189. J.M. Raphael, The optimum gravity dam, in *Rapid Construction of Concrete Dams* (American Society of Civil Engineers, New York, 1971), pp. 221–247
190. R.W. Cannon, Concrete dam construction using earth compaction methods, in *Economical Construction of Concrete Dams* (American Society of Civil Engineering, New York, 1972), pp. 143–152
191. M.R.H. Dunstan, Development of high fly ash content in concrete, Proc. Inst. Civil Eng. Part 1 Des. Constr. **74**, 495–513 (1983)
192. K.L. Saucier, Use of fly ash in no-slump roller compacted concrete, in *Proceedings of 6th International Symposium on Ash Utilization*, Reno, NV, Mar 1982, U.S. Department of the Environment, Washington, DC, DOEIMETC/82-52, 1982, pp. 282–293
193. ACI Committee, Roller compacted concrete, J. Am. Concr. Inst. **77**(4), 215–236, Report No. ACI 207.5R-80 (1980)
194. M.R.H. Dunstan, The use of high fly ash content concrete in roads, in *Proceedings of International Symposium of the Use of PFA in Concrete*, University of Leeds, Leeds, UK, Apr 14–16, 1982, ed. by J.G. Cabrera, A.R. Cusens (Department of Civil Engineering, University of Leeds, Leeds, 1982), pp. 277–289
195. J.E. Oliverson, A.T. Richardson, Upper stiliwater dam: design and construction concepts, Concr. Int., **6**, 20–28 (1984)
196. F.A. Anderson, RCC does more. Concr. Int. **6**, 35–37 (1984)
197. R.C. Joshi, G.S. Natt, Roller compacted high fly ash concrete (Geocrete), in *Proceedings of 1st International Conference on the Use of Fly Ash, Silica Fume, Slag, and Other Mineral By-products in Concrete*, Montebello, PQ, July 31–Aug. 5, 1983, ed. by V.M. Malhotra (American Concrete Institute, Detroit, Special Publication SP-79, 1983), pp. 347–366
198. D.M. Golden, A.M. Digioia, Fly ash for highway construction and site development, in *Proceedings of Shanghai 1991 Ash Utilization Conference*, Shanghai, People's Republic of China, Sept 10–12, 1991, Electric Power Research Institute, Palo Alto, CA, vol 3, No. 95
199. T.R. Naik, B.W. Ramme, J.H. Tews, Pavement construction with high volume Class C and Class F fly ash concrete, in *Proceedings of CBU/CANMET International Symposium on the Use of Fly Ash, Silica Fume, Slag, and Other By-products in Concrete and Construction Materials*, University of Wisconsin, Milwaukee, WI, Nov 2–4, 1992
200. K. Torii, M. Kawamura, Some properties of dry lean rolled concrete containing fly ash, in *Proceedings of 3rd CANMET/ACI International Conference on the Use of Fly Ash, Silica Fume, Slag, and Natural Pozzolans in Concrete*, Trondheim, Norway, June 18–23, 1989, ed. by V.M. Malhotra (American Concrete Institute, Detroit, MI, Special Publication SP-1 14, 1989), pp. 222–234

Chapter 3
Granulated Blast Furnace Slag

Introduction

Metallurgical industry produces slag as by-products. Iron blast furnace slag is the major non-metallic product consisting of silicates and aluminosilicates of calcium. They are formed either in glassy texture used as a cementitious materials or in crystalline forms used as aggregates. Other slags such as copper slag have pozzolanic properties and react with lime. Steel slags are usually produced in crystalline form and are used as base materials for road construction or as aggregates in special concrete productions. The other utilizations of slags are in the production of slag wool for thermal isolation in the building industry and as lightweight aggregates for lightweight concretes. This part deals with the production and properties of iron blast furnace slag to be used as a binder in the cement and concrete industries.

Production

Iron oxide ore and coke are used in a blast furnace to produce pig iron at the bottom of the furnace and molten slag on top of molten iron. The molten slag usually consists of silica and alumina which combine with lime and magnesia from the added limestone into the furnace. Small amounts of manganese and sulfur are observed in the slag chemical composition.

If a molten slag with a temperature between 1400 and 1600 °C is cooled slowly in the air, a stable solid and crystalline material is obtained. They have little or no pozzolanic properties and used as aggregates in concrete or as base materials for road construction.

In the rapid cooling process, slags are formed as granular glassy materials higher in energy than the crystalline materials. Crystallization is avoided by the usage of large quantities of water (100 m^3 per t of slag) or spraying jets under 0.6 MPa pressure (Water, 3 m^3 per t of slag) for rapid cooling of molten slag.

A. A. Ramezanianpour, *Cement Replacement Materials*,
Springer Geochemistry/Mineralogy, DOI: 10.1007/978-3-642-36721-2_3,

After this process, the moisture of the slag which is usually less than 30 % is eliminated in filter basins or dryer mills.

A semi-dry process called pelletizing system was developed in Canada for granulation of molten slags [1]. As shown in Fig. 3.1, the molten slag is first cooled with water and then flung into the air by a rotary drum at a speed of 300 rpm. The consumption of water is about 1 m^3 per t of slag and the moisture content of slag is about 10 %. Different fractions are obtained at the end of cooling system. The larger pellets between 4 and 15 mm have porous structure and are partially crystalline. This part is usually used as expanded slag for lightweight aggregate concretes. The smaller fraction with particles less than 4 mm and mostly in glassy form is used as a hydraulic binder in separately ground or interground with Portland cement clinker.

Table 3.1 presents the world blast furnace slag (BFS) and granulated blast furnace slag (GBFS) production in the year 1999–2000.

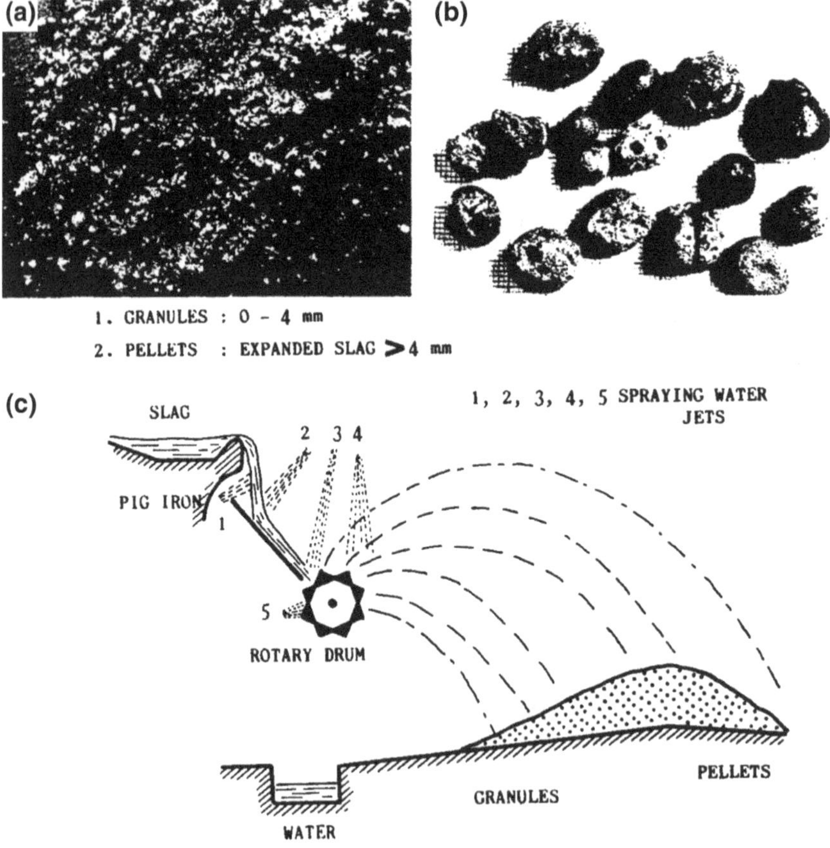

Fig. 3.1 Canadian method for slag granulation. **a** Pelletizer. **b** Granules, 0–4 mm. **c** Pellets: expanded slag > 4 mm [5]

Table 3.1 World blast furnace slag (BFS) and granulated blast furnace slag (GBFS) productions in the year 1999–2000

Region	Total BFS production (Mil.t)	GBFS production (Mil.t)	Granulation rate (%)
Asia	99.6	74.8	75
Europe	56.4	33.8	60
America	25.2	12.0	48
Africa	2.3	2.3	100
Oceania	1.7	1.0	52

Physical, Chemical, and Mineralogical Properties

Physical Properties

Fineness of granulated blast furnace slag is a major factor affecting the strength of mortars and concretes. Granulated slags can be ground to a desired particle size or surface area, depending on the degree of activation needed and economic considerations. Water granulated slag seems to be less easily ground than Portland cement. The pelletized slag, because of its porous structure requires less energy to grind than granulated slag.

It is reported that slag particles less than 10 μm contribute to early strength development up to 28 days. Particles in the range of 10–45 μm continue to hydrate beyond 28 days and contribute to later-age strength. Larger particles greater than 45 μm show little or no activity. Most standard specifications limit the proportion of particles greater than 45 μm.

The Blaine surface area of granulated blast furnace slag ranges between 4000 and 6000 cm^2/g in order to obtain satisfactory strength development in concrete.

Chemical Composition

Depending on the sources of iron ore, the limestone flux composition, and the coke composition, the chemical composition of the slag varies widely. The four major components are lime, silica, alumina and magnesia. Minor components and oxides are sulfur in the form of sulfide, ferrous and manganese oxides, titanium, sodium and potassium oxides. Chemical analysis of some metallurgical slags is given in Table 3.2.

Mineralogical Properties

Table 3.3 shows the mineral composition of crystallized slag. At a slow cooling system, a stable solid consisting of calcium, aluminium, and magnesium silicates,

Table 3.2 Chemical composition of some blast furnace slags (%)

Source	CaO	CiO_2	Al_2O_3	MgO	Fe_2O_3	MnO	S
UK	40	35	16	6	0.8	0.6	1.7
Canada	40	37	18	10	1.2	0.7	2.0
France	43	35	16	8	2.0	0.5	0.9
Germany	42	35	16	7	0.3	0.8	1.6
Japan	43	34	12	5	0.5	0.6	0.9
Russia	39	34	14	9	1.3	1.1	1.1
South Africa	34	33	16	14	1.7	0.5	1.0
USA	41	34	10	11	0.8	0.5	1.3

$Na_2O + K_2O = 0–2$ %
$TiO_2 = 0.4$ %

Table 3.3 Mineral composition of crystallized iron slag

Main components			
Melilite:	Solid solution of gehlenite and akermanite	2CaO. Al_2O_3. SiO_2 2CaO. MgO_3. $2SiO_2$	C_2AS C_2MS_2
Mervinite[a]		3CaO. MgO. $2SiO_2$	C_3MS_2
Diopside[b]		CaO. MgO. $2SiO_2$	CMS_2
Lime[c]		CaO	C
Wustite[c]		FeO	F
Ferrite[c]		Ca_2. Fe_2O_3	C_2F
Minor components			
Dicalcium silicate[a,c] (α, α', β, γ)		2CaO. $2SiO_2$	C_3S
Monticellite[a]		CaO. MgO_3. SiO_2	CMS
Rankinite		3CaO. $2SiO_2$	C_3S_2
Pseudo-wollastonite		CaO. SiO_2	CS
Oldhamite		CaS	

[a] In basic slags
[b] In acid slags
[c] In LD slags

specially melilite which is a solid solution of gehlenite (C_2 AS) and akermanite (C_3 MS_2) is obtained. Other main minerals are merwinite, diopside, lime, wustite and ferrite. Minor components such as dicalcium silicate, monticellite, rankinite, pseudo-wollastonite, and oldhamite can be observed in different slags.

Structure of Glassy Slags

The structure of glassy slag as a silicate glass is approached by considering the vitreous silica in which some Si–O–Si bonds are broken and neutralized by metal cations called structure modifiers [1–4]. As seen in Fig. 3.2, SiO_4 tetrahedra are polymerized with bridging oxygen atoms [5]. Negative charges of these anionic condensed groups are neutralized by cations such as calcium and magnesium ions in the cavities of the network. Calcium atoms are octahedrally coordinated to the oxygen atoms. The coordination of magnesium atoms is either octahedral or both tetrahedral and octahedral [1]. Some of the Si positions in a slag glass are occupied by atoms of other elements, especially Al. Each such replacement of Si^{4+} by Al^{3+} is equivalent to one of SiO_2 by AlO_2^- and hence introduces a negative charge which is balanced by further introduction of metal cations. Tetrahedral atoms are network formers while octahedral atoms are network modifiers [3].

The glassy part of a slag is characterized as a large halo in the X-ray diffraction pattern peaking at 0.3 nm. As seen in Fig. 3.3, the crystalline phases in the pattern are merwinite, melilite, calcite and quarts [6].

The crystallization of merwinite and melilite is very often dendritic (see Fig. 3.4). In this form of crystallization, the thin edges represent an initial stage from seeding

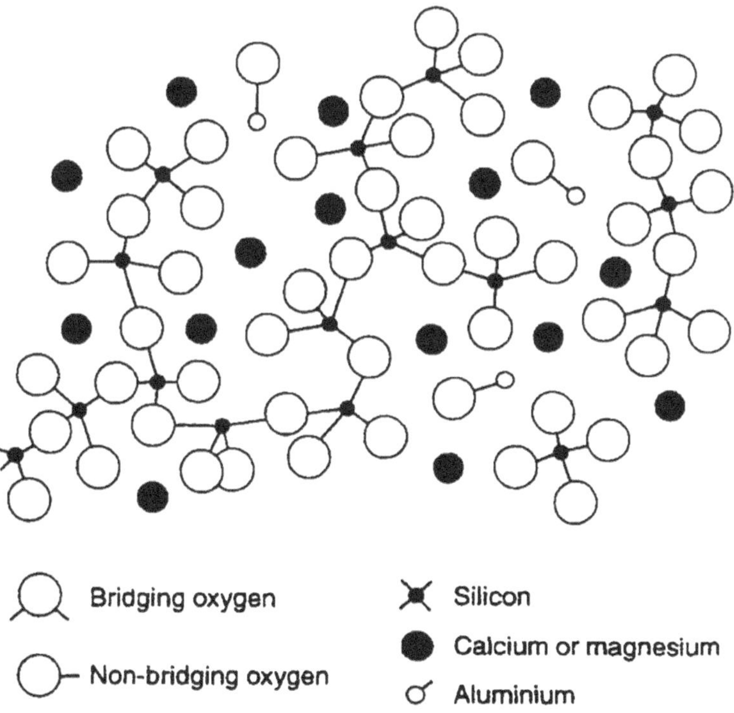

Fig. 3.2 Glass structure of slags [5]

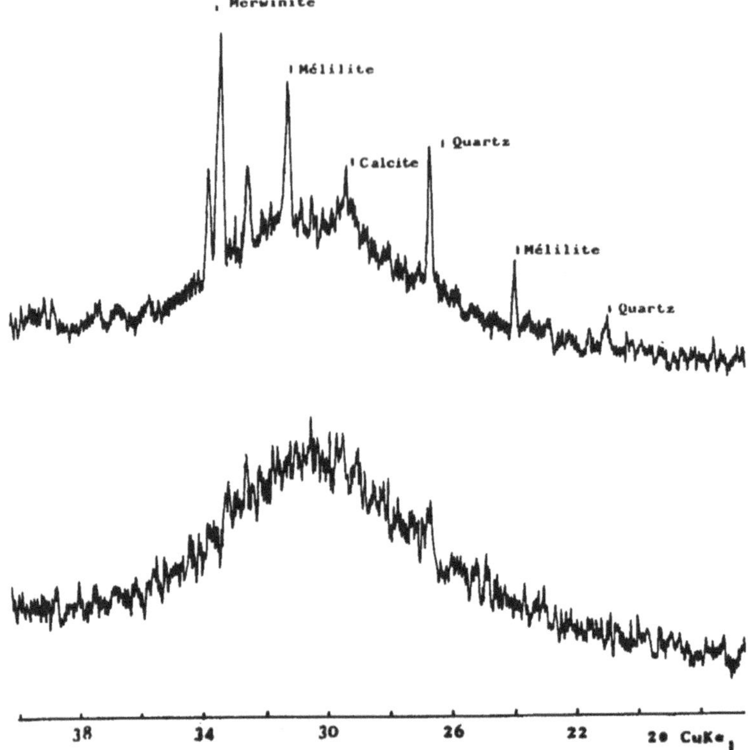

Fig. 3.3 X-ray diffraction patterns of two slags. **a** Glass:halo. **b** Glass + crystals: halo + peaks [6]

points in the melt which has been stopped by rapid quenching [6]. Inside each dendrite and between dendrites themselves there is a glassy phase which in the case of merwinite and melilite crystallization is enriched in alumina [7].

Hydraulic Properties of Slags

Hydraulic properties of slags depend on their production systems and chemical composition. Slow cooled slag has no or very little cementitious or hydraulic properties. Granulated slag has also little cementitious properties but with some activator, all except more silicious slags show cementitious properties. The activators are lime, Portland cement, alkalis such as sodium carbonate or sulfates, calcium or magnesium.

The cause of the hydraulic activity and its relation to physical state and chemical composition of a slag has not been determined yet. Basic slags show higher reactivity when activators are used. The temperature of the molten slag and

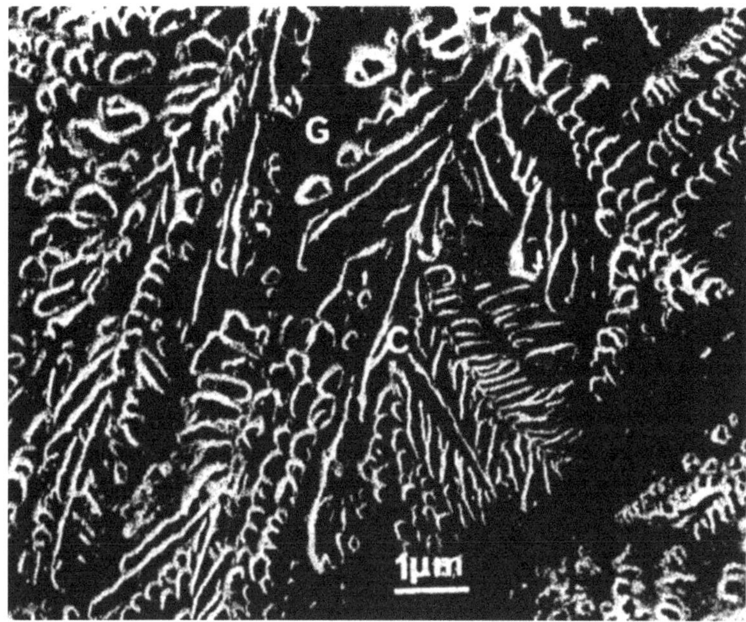

Fig. 3.4 SEM image of a polished section of slag. Crystals *C* of merwinite and glass *G* in dendritic form [6]

its glass content also affect the hydraulic activity. However the crystalline components reduce the hydraulic properties of slags though some of them contribute to the strength properties.

Several chemical hydraulic factors or hydraulic moduli have been defined as slag activity indices [2]. One of the commonly moduli used in the prediction of hydraulic reactivity of slags or compressive strength of blended cement mortars include its main oxides C, A, S, and M. Table 3.4 depicts some of the hydraulic moduli of granulated slags [4].

The most applied moduli is (C + M + A)/S. As included in the regulation, it must be ≥1 in Germany or ≥1.4 for Japan slags. Magnesium oxide has been found efficient up to 18 wt %. It has the same effect as calcium oxide in the range up to 11 wt %.

Table 3.4 Some hydraulic moduli of granulated slags derived from chemical analysis

Slag	1	2	3	4	5	6	7
C/S	1.31	1.26	1.22	0.95	0.93	1.30	1.34
$\frac{C+M+A}{S}$	1.88	1.88	1.79	1.63	1.39	1.98	1.92
$\frac{C+M+1/3A}{S+2/3A}$	1.52	1.57	1.51	1.32	1.15	1.55	1.41
A/S	0.43	0.36	0.34	0.41	0.34	0.51	0.45

Several researchers have tried to predict the mechanical properties of slag cements at various ages by using mineralogical composition, glass content and hydraulic moduli. It was found that low values of moduli correspond to low hydraulic reactivity but the correlation between the hydraulic and the mechanical strength at different ages was not good. This is observed in Fig. 3.5.

A model which was proposed by Hooton and Emery correlates the hydraulic indices at different ages to the glass content and chemical analysis of slags [8].

The hydraulic index (HI) at 7 days is given by the following formula:

$$HI(7) = 0.688(\text{glass}) + 3.03(\text{CaO}) - 7.97(\text{SiO}_2) - 14.24(\text{Mn}) \pm 177.1 \quad (3.1)$$

With coefficient of correlation r = 0.943 and standard error ±7.4.

$$HI(7) = 0.580(\text{glass}) + 153.8(C + M + A)/S - 243.0 \quad (3.2)$$

With coefficient of correlation r = 0.919 and standard error ±8.7.

There are several rapid methods for assessing granular slags. A fluorescence test is used in Germany [9]. This method is claimed to distinguish between glasses with different hydraulic value and therefore to be preferable to the microscopical estimation of glass. Extraction method has also been used to determine the rate of hydration of slag by a number of investigators [10, 11].

Similar to some methods used for pozzolanic activity index measurements, compressive strength of mortars containing slag has been compared with the identical mixes with Portland cement. The higher the hydraulic index, the higher is the activity of slag [12]. This method of determination of compressive strengths of blended cements for quality control purposes is recommended by some standards such as ASTM C989 test method.

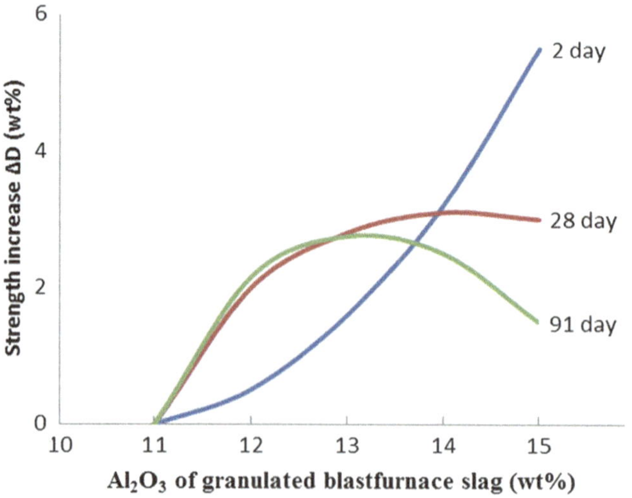

Fig. 3.5 The effect of Al_2O_3 on the development of strength for a hydraulic modulus (C + M)/S = 1.4 [3]

Hydration of Slag–Cement Mixture

The hydration of granulated slag in the presence of lime or Portland cement is different from other pozzolans due to reaction of slag with water as a cementitious material. The hydration of slag is affected by the lime liberated from the hydration of cement. In the absence of hydraulic calcium dioxide, no hydration products can be observed in the slag–water mixture. The surface of the slag particles modifies in contact with water as seen in X-ray photoelectron spectrometry analysis [13]. The hydration of slag starts with an incongruent dissolution and then precipitation of calcium silicate hydrate.

Activators can affect the hydration of slags. The most common activators in slag Portland cements are gypsum and portlandite. Other activators are classified as alkaline activators such as sodium hydroxide, sodium carbonate, sodium silicate, and sulfate activators like calcium sulfate or phosphogypsum. As in Portland cement, two kinds of activators can co-exist as slag activators. The hydration products of slag in the presence of different activators are observed in Table 3.5 [14].

The hydration of 40–50 % slag cements was studied by Feng et al. [15]. They observed the formation of a gel-like layer around slag grains with higher silica content and denser than the cement hydration products.

In general, the Portland cement clinker hydrates faster than the slag cements. Calorimetric curves in Fig. 3.6 show that both cements react at early ages and generate heat. Slag cement with 70 % slag no. 1 gives a peak at the same place as pure Portland cement. As shown from the peaks, the slag no. 2 is less reactive than slag no.1 but cement 1 is a low heat cement compared to normal Portland cement [16].

Table 3.5 Hydration products of slag with different activators [14]

	Scotland				USA		
CaO	60.90	56.83	57.34	58.15	54.97	56.32	59.36
SiO_2	23.90	24.70	24.50	24.12	26.29	25.94	24.12
Al_2O_3	8.16	9.48	8.40	8.23	8.25	7.64	6.55
Fe_2O_3	2.34	1.89	2.48	2.65	2.22	2.00	2.52
Mn_2O_3	–	–	0.44	0.25	0.33	0.40	0.10
TiO_2	0.25	0.28	0.44	0.40	–	–	–
MgO	1.50	3.22	2.55	1.85	3.76	3.69	2.31
Na_2O	0.28	0.32	0.26	0.31	0.07	0.21	0.06
K_2O	0.56	0.54	0.71	0.55	0.19	0.76	0.63
SO_3	1.32	1.36	1.41	1.97	2.39	1.76	2.07
Loss on ignition	0.73	0.82	1.62	0.15	0.48	0.94	1.97

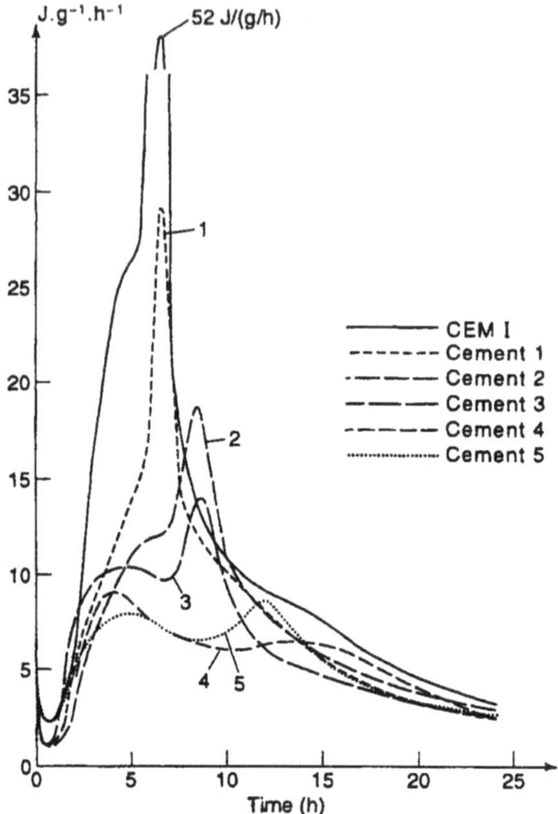

Fig. 3.6 Calorimetric curves of five blast furnace slags containing 70 % slag and 30 % normal Portland cement [16]

Effect of Thermal Treatments

Thermal treatment can accelerate the hydration of slags. The Arrhenius law can be applied to hydration reactions and to the evolution of the heat of hydration. The energy of activation (E) of chemical reactions during cement hydration can be calculated after the Arrhenius law, at two temperatures T_1 and T_2 and at the time of hydration (t), according to the following equation:

$$t_1/t_2 = \exp E/R(1/T_1 - 1/T_2). \tag{3.3}$$

The activation energy calculated for slag cements is lower than that of their corresponding Portland cement, i.e. 50 and 46 kJ/mole respectively. Therefore thermal treatment shows to be favorable to slag cements as an effective supplier of activation energy to hydration of slag which can be considered as a latent cementitious material. This is not the case for all types of slags. Those active slags rich in alumina do not need a long thermal treatment due to the rapid formation of hydrated aluminate crystals [17].

Effects of Slag on the Properties of Fresh Concrete

Workability

Granulated blast furnace slag improves the workability of concrete. This is due to the surface characteristics of the slag particles having smooth dense slip planes [18]. It is believed that unlike Portland cement the slag cement particles absorb little water during initial mixing. In another investigation the greater workability of slag cements is attributed to the increased paste content and increased cohesiveness of the paste [19].

Consolidation of concretes containing slag is much easier than that of normal Portland cement with mechanical vibration. This can be observed in Fig. 3.7. The Vebe apparatus was used for comparison of workability [19]. In all mixtures, the workability of 50 % slag cement concretes was higher than that of plain Portland cement concretes.

Research on concrete mixtures at the same water cement ratio showed that concretes containing slag cements have higher slump when compared with normal Portland cement concretes [20]. This is clearly observed in Fig. 3.8.

As the percentage of slag increases, the water binder ratio has to be reduced to maintain workability properties same as the control concrete without slag. This observation requires confirmation under field conditions with high replacement level [21].

The rate of loss of slump and air is comparable to the normal Portland cement concrete without slag. Concrete containing slag cement at 50 % replacement

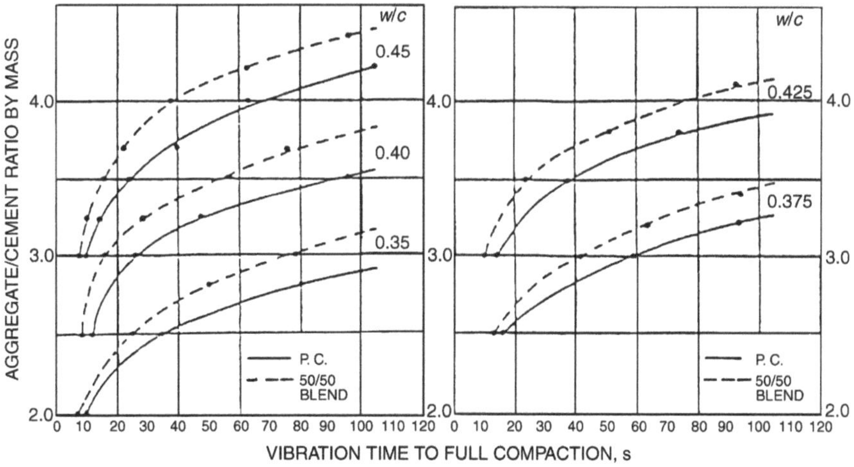

Fig. 3.7 Comparison of vibration time for full compaction of concretes with and without slag [19]

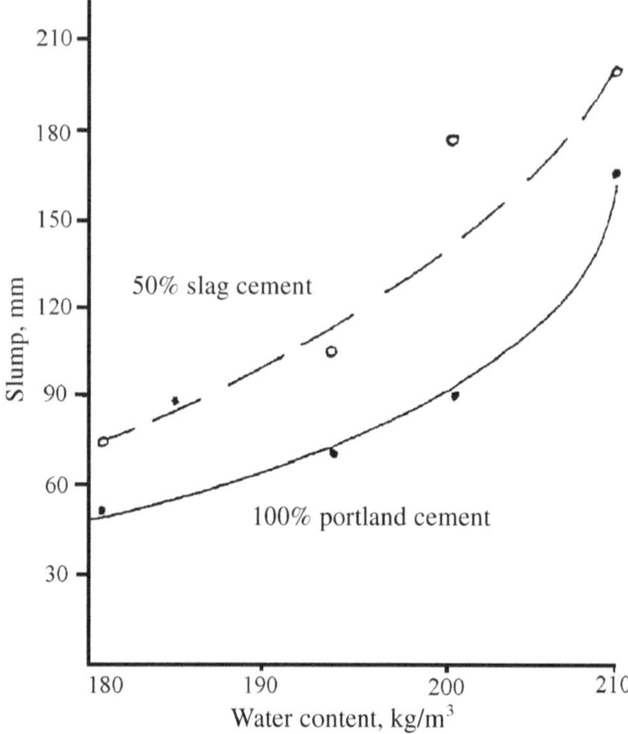

Fig. 3.8 Comparison of workability of concretes containing slag cement with normal concrete [20]

showed similar slump loss to that concrete without slag [20]. In order to maintain the same level of air content a higher dosage of admixture is required for high volume slag cement concretes, mainly because of the higher fineness of the slag.

Bleeding and Segregation

Bleeding of concrete is known to be influenced more by the size distribution, shape and surface texture of the cementitious materials. Reactivity of the cementitious materials also affects the bleeding potential of the concrete mixtures. Extended time of setting at low temperatures could also play a role in the bleeding capacity. Slow reactive slags with coarser particles may increase the bleeding of concrete. Incorporation of slags finer than Portland cement in the concrete mixture may result in lower bleeding. In most of the available data, no concrete containing slag has shown higher bleeding than normal Portland cement concretes of the same 28 day compressive strength.

Reports on the investigation of segregation of concretes containing slag are rare, but no significant difference has been observed in such concretes when compared with concretes without slag cement.

Setting Time

The partial replacement of cement with slag generally increases setting time of concrete. The final setting could be delayed significantly depending upon the ambient temperature and the amount of replacement. Lower temperatures would further extend the setting time. In an investigation results on the slag cements at 35–40 % replacement indicate that at approximately 21 °C, the setting time was increased by 1 h. Setting time was also increased by increasing the amount of slag replacement level [22]. In a 50 % slag cement replacement and at 23 °C, the initial setting time was increased 0.5–1 h while little change was obtained at temperature above 29 °C [23]. Results on time of setting of concretes incorporating a pelletized slag and two different granulated slags carried out by CANMET is shown in Table 3.6 [24].

In this investigation, there was no significant increase in the initial setting time at 25 % cement replacement, while the increase in the final setting time ranged from 16 to 101 min. At 50 % replacement level, the increase in the initial setting time of the concretes ranged from 17 to 80 min, whereas the increases in the final setting time ranged from 93 to 192 min.

Effects of Slag on the Mechanical Properties of Hardened Concrete

Strength Properties

The slow reaction of slag with water will result in lower strength gain at early ages of concrete containing various amounts of slag. Several factors including chemical composition, physical properties such as fineness, activity index, mixture design and moist curing period affect the strength of slag cement concretes. Generally, for all blast furnace slags, the strength development is slower at early ages, but between 7 and 28 days, the strength approaches that of the plain cement concrete and slightly higher than that at later ages. This trend may not always be true and at lower water binder ratio, the slag cement concrete will give a strength lower than that of control concrete even after 1 year.

Test results illustrated in Fig. 3.9 show the range applicable to mixes containing granulated blast furnace slag of different chemical compositions; as the reactivity of the slag increases or the percentage of slag decreases, so curves B and C will

Table 3.6 Bleeding and setting-time of slag concrete: comparative data on three different slags [24]

Mixture no.	Type and source of slag	Blaine fineness of slag, cm^2/g	Cement replacement by slag, %	W/(C + S)	Setting time, h:min		Total bleeding water, $cm^3/cm^2 \times 10^{-2}$
					Initial	Final	
1 C	–	–	0	0.50	4:56	6:27	4.16
2 ST	Pelletized (Canada)	4200	25	0.50	4:49	7:15	4.29
3 AT	Granulated (USA)	5400	25	0.50	4:23	6:43	3.08
4 AL	Granulated (Canada)[a]	3700	25	0.50	4:54	7:25	6.11
5 AL	Granulated (Canada)[a]	4600	25	0.50	4:49	7:20	5.67
6 AL	Granulated (Canada)[a]	6080	25	0.50	5:30	8:08	5.00
7 ST	Pelletized (Canada)	4200	50	0.50	5:54	9:32	4.13
8 AT	Granulated (USA)	5400	50	0.50	5:13	8:00	3.33
9 AL	Granulated (Canada)[a]	3700	50	0.50	6:06	9:39	9.50
10 AL	Granulated (Canada)[a]	4600	50	0.50	–	–	6.20
11 AL	Granulated (Canada)[a]	6080	50	0.50	6:16	9:37	2.78

[a] Same source but ground to different finenesses

Note Cement type: ASTM Type I

C.A, Crushed limestone 19-mm max size

F.A, Natural sand

A.E.A, Sulphonated hydrocarbon type

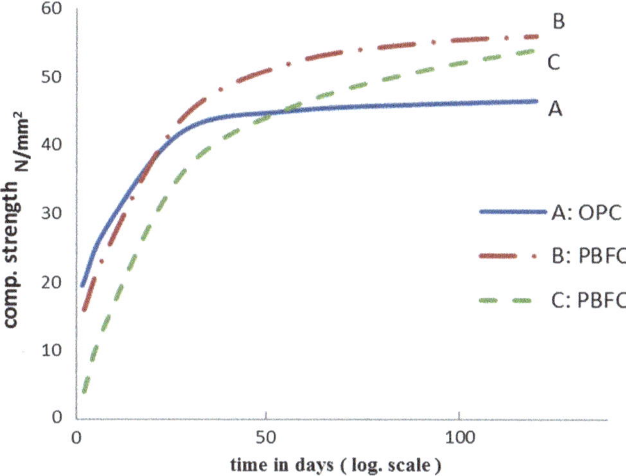

Fig. 3.9 Development of cement strength with different ggbfs [25]

tend toward curve A [25]. Figure 3.10 shows that the later age development of ggbfs concrete is faster than that of Portland cement concrete. This shows that in order to have a similar 28 day compressive strength for concretes with and without slag, more cement is needed for ggbfs concretes.

The effect of water to cementitious material ratio on the compressive strength of slag cement concretes is shown in Figs. 3.11 and 3.12. It is worth noting that the highest strength gain at 28 days was for the slag cement concrete containing 40 % slag [23].

In an investigation of high volume US slag concretes containing 65 % slag, the later age strength was found higher than that with 50 % slag and lower than the 40 % slag concretes but in all cases the strength was higher than the concretes without slag [26].

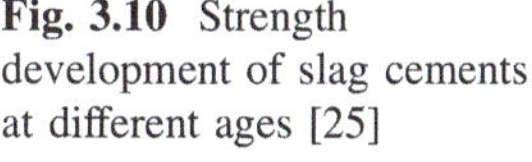

Fig. 3.10 Strength development of slag cements at different ages [25]

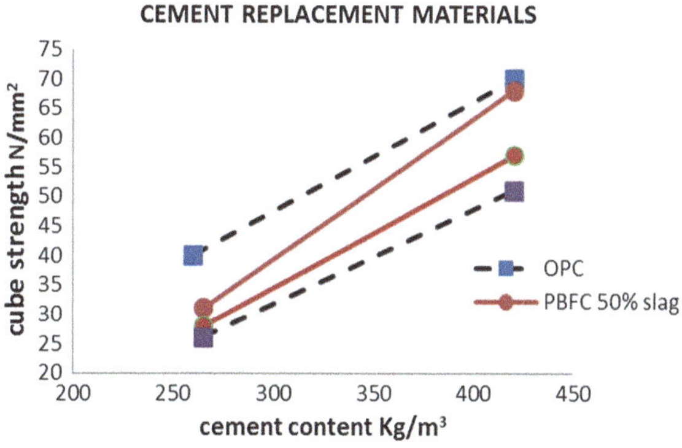

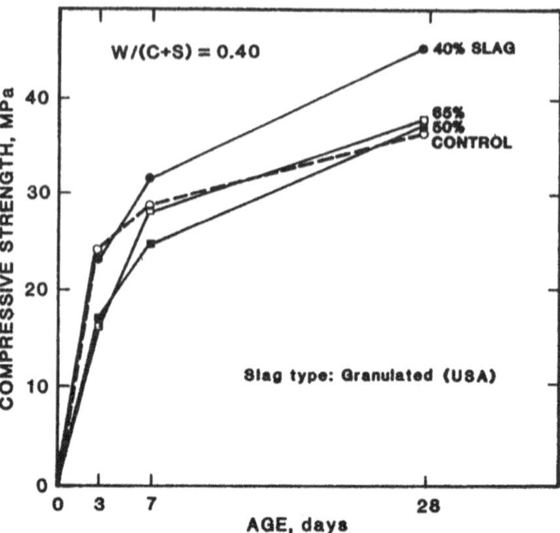

Fig. 3.11 Compressive strength at various ages for air entrained slag cement concretes: W/(C + S) = 0.4

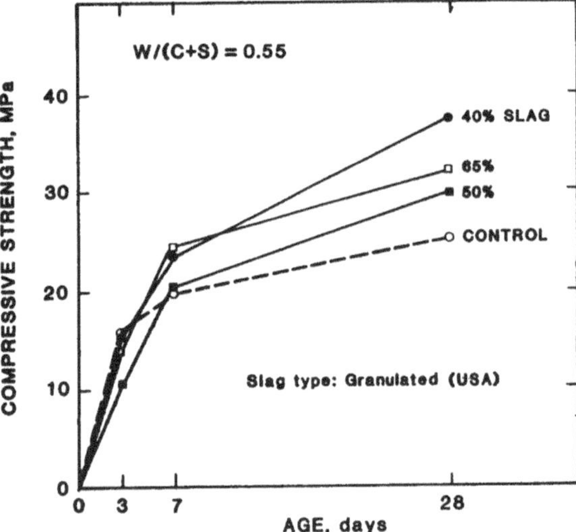

Fig. 3.12 Compressive strength at various ages for air entrained slag cement concretes [23]: W/(C + S) = 0.55

In concrete mixtures containing up to 60 % slag, the early strength is attributed to the fineness of clinker fraction while the later strengths are mainly due to the fineness of slag fraction. With cement of higher slag content the fineness of the slag is of major importance at all ages while this influence is not significant in cement containing low amount of slag.

Similar to normal Portland cement concretes, the strength of slag cement concretes is influenced by the curing temperature and relative humidity. Due to lower activation of slag than clinker, an increase in the curing temperature will be more favorable to the strength development of slag cement concretes. The influence of relative humidity is rather more important for the strength development of concretes containing slow reactive slags.

Generally the tensile and flexural strengths of concretes containing slag are comparable or greater than the corresponding strength of control concrete at ages beyond 7 days. Higher tensile strength was observed for slag concrete in comparison with no slag concrete on the basis of equal compressive strength of 70 MPa [27]. Higher flexural strengths at ages beyond 7 days have also been reported for slag cement concretes designed for greatest strength in comparison with control concretes [19, 23, 28]. This increase is probably due to the stronger bonds in the slag cement aggregate system because of the shape and surface texture of the slag particles. Table 3.7 shows the flexural strength of concretes containing three different North American slags. It is seen that at both 7 and 14 days the flexural strengths of the slag cement concretes are higher than that for control concrete [24].

Modulus of Elasticity

Modulus of elasticity of concretes containing slags varies with time similar to the compressive strength. There is no significant difference between the concretes containing slag with that without slag at the same strength level. There is some evidence to suggest that at lower values of compressive strength (below about 30 MPa) the concretes containing slag have a lower modulus of elasticity than the control concrete. As seen in Fig. 3.13 this trend is opposite at higher percentages of slag replacement [27].

Investigation on the effect of slag cement source on modulus of elasticity of slag cement concretes showed no significant differences [29]. Under normal temperatures, slag cement concretes develop modulus more slowly than the concretes without slag.

Effect on Volume Changes of Concrete

Results on the drying shrinkage of slag cement concretes are conflicting. This is due to the variations in the slag properties, cement properties, curing conditions and test methods used in investigations. Some researchers indicate that the replacement of cement by slag has little or no influence on the drying shrinkage of concretes. In an investigation a greater shrinkage was observed in concretes containing various slag cement blends when compared with no slag cement concretes [19]. Higher shrinkage for slag cements was also observed in an investigation shown in Fig. 3.14. They have attributed this increase in shrinkage to the

Table 3.7 Compressive and flexural strength test results: comparative data on three different slags [24]

Mixture no.	Type and source of slag	Blaine fineness of slag, cm^2/g	Cement replacement by slag, %	W/(C + S)	Compressive strength on 150 × 300-mm cylinders, MPa					Flexural strength on 75 × 75 × 400-mm prisms, MPa	
					3-day	7-day	28-day	91-day	365-day	7-day	14-day
1 C	–	–	0	0.50	17.8	21.1	26.9	31.6	35.0	3.8	4.8
2 ST	Pelletized (Canada)	4200	25	0.50	16.0	19.9	28.6	34.1	36.6	4.0	5.3
3 AT	Granulated (USA)	5400	25	0.50	17.6	22.9	29.8	33.9	38.1	4.6	5.1
4 AL	Granulated (Canada)[a]	3700	25	0.50	14.9	18.5	25.4	31.0	33.1	3.9	5.4
5 AL	Granulated (Canada)[a]	4600	25	0.50	13.8	18.8	26.0	30.2	32.9	4.1	4.8
6 AL	Granulated (Canada)[a]	6080	25	0.50	15.6	21.0	28.2	31.5	35.8	4.4	5.0
7 ST	Pelletized (Canada)	4200	50	0.50	9.0	14.1	25.4	29.5	33.8	3.2	5.2
8 AT	Granulated (USA)	5400	50	0.50	14.9	22.7	33.6	37.6	40.4	4.3	5.7
9 AL	Granulated (Canada)[a]	3700	50	0.50	10.4	15.6	24.7	29.1	33.0	3.7	5.0
10 AL	Granulated (Canada)[a]	4600	50	0.50	9.5	15.9	25.5	29.8	34.3	3.6	5.1
11 AL	Granulated (Canada)[a]	6080	50	0.50	12.2	20.7	30.1	32.9	37.2	4.3	5.2

[a] Same source but ground to different finenesses

Note Cement type: ASTM Type I

C.A, Crushed limestone 19-mm max size

F.A, Natural sand

A.E.A, Sulphonated hydrocarbon type

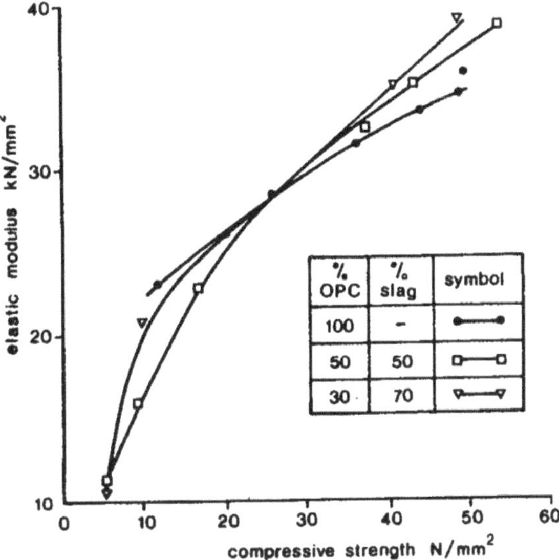

Fig. 3.13 Elastic modulus versus compressive strength of normal cured concretes with and without slag [27]

increase of volume of cement paste in concretes containing slag with lower density and on equal replacement level by weight [23].

Another research showed that for concretes containing 50 % slag and under 60 % relative humidity and at 20 °C after 28 days water storage, the drying shrinkage was about 10 % lower than that of concrete without slag [30]. Similar shrinkage has been reported in a work on time dependent properties of slag concretes containing different slag sources and varying slag cement replacement levels between 30 and 70 % by mass of total cementitious material [29].

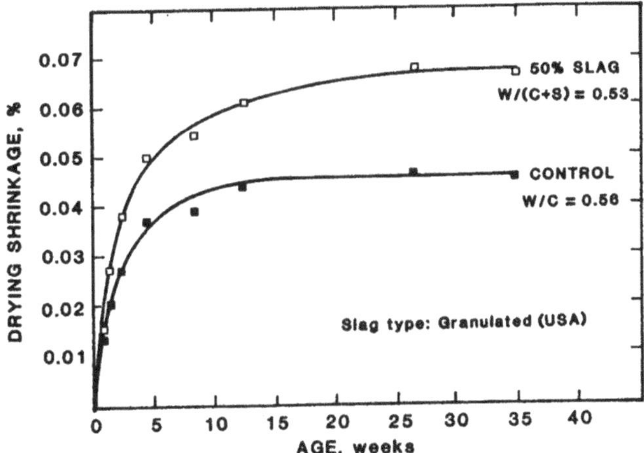

Fig. 3.14 Drying shrinkage of air entrained concrete [23]

Contradictory results have also been reported for the creep of concretes containing slag. This conflict is attributed to the properties of slag and cement, test method and stress-strength ratio, age of applied load, curing and environmental conditions. There is a general agreement among most researchers that concretes containing slag will show lower creep than identical plain concretes if tested under no moisture loss condition. As seen in Fig. 3.15 and under such conditions, the creep of slag cements up to 50 % replacement is lower than that of no slag cement concretes at equal 28 day compressive strength and equal applied load [30].

Similar concrete mixtures were tested after 28 day moist curing and maintained in a room controlled at 23 °C and 60 % relative humidity [30]. The total creep of concretes containing slag was greater than that of the control concrete (see Fig. 3.16).

In an investigation of basic creep of sealed concrete specimens subjected to a temperature cycle prior to the load application at a constant initial stress–strength ratio of 0.25, lower creep was observed for concretes containing slag cements. The reduction in basic creep was proportional to the level of replacement of slag [31]. Results on a work on time dependent properties of slag concretes containing different slag sources and varying slag cement replacement levels between 30 and 70 % by mass of total cementitious material show a lower basic creep and similar or lower total creep for concretes incorporating slag [29].

It is worth mentioning that slag concretes usually exhibit higher creep than an equivalent Portland cement concrete subjected to the same stress at early ages. This is attributed to the lower strength gain of slag cement concretes and would therefore be subjected to a higher stress-strength ratio.

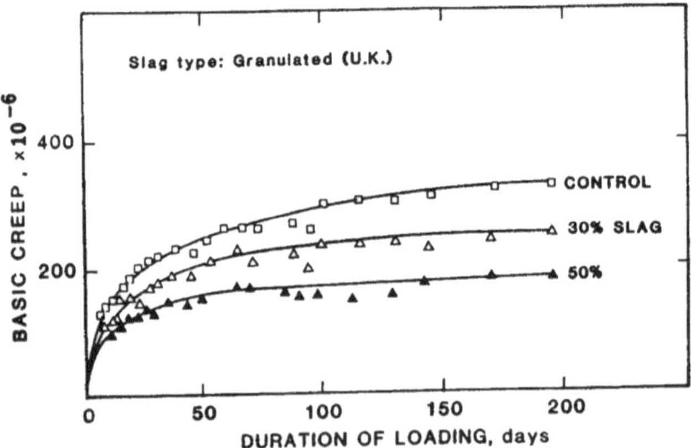

Fig. 3.15 Basic creep of slag cement concretes under 10 MPa stress and water cured at 22 °C [30]

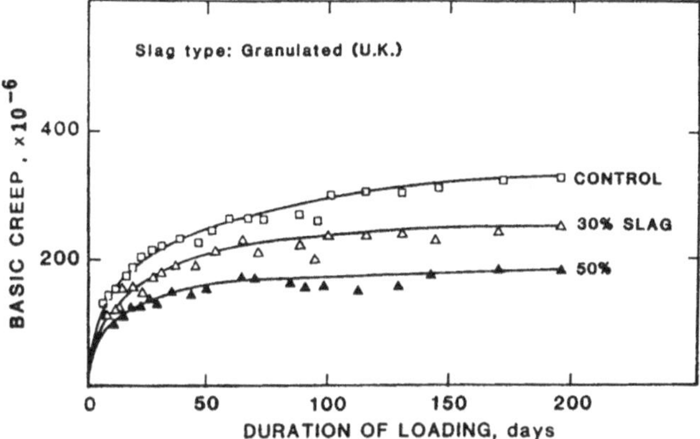

Fig. 3.16 Total creep of slag cement concretes under 10 MPa stress and maintained at 60 % RH and 21 °C [30]

Microstructure, Porosity and Permeability

The microstructure, porosity and pore size distribution of concrete containing slag is to some extend different from the Portland cement concretes. At early ages of reaction, the porosity of slag cement is similar to that of Portland cement [32]. At longer ages and after reaction of slag, the volume of very small pores in the nano scale range becomes larger. Measurement of porosity and pore size distribution of slag cement paste by mercury intrusion porosimetry shows that the pore structure in blended cements is characterized by relatively large but discontinuous and thin-walled pores [33]. It has also been reported that incorporation of granulated slags in cement paste has transformed the large pores in the paste into smaller pores. The mechanism of this pore refinement is not very clear but it is attributed to the reaction of slag with calcium hydroxide and alkalis released during the hydration of Portland cement [34, 35]. Continuous capillary pores in Portland cements pastes should be due to deficient in the interface of calcium hydroxide and calcium silicate hydrate. The amount of calcium hydroxide in the slag cement is lower and this can affect the deficiency of the interface. Pores in cement paste that normally contain calcium hydroxide are partly filled with calcium silicate hydrates resulting from the hydration of the slag cement and change the total porosity and pore size distribution [34, 35]. Figure 3.17 Shows the difference between total porosity and pore size distribution of cement pastes with and without slag by mercury porosimetry method [36].

Permeability of concrete depends mainly on the permeability of pastes which, in turn, depends on its porosity and pore size distribution [37]. Pore refinement effect and reduction in the porosity of slag cement concretes will result in lower permeability of the mix especially in well cured concretes [38]. Several other investigators have also reported the lower permeability of concretes containing slag in comparison with normal Portland cement concretes [36, 39].

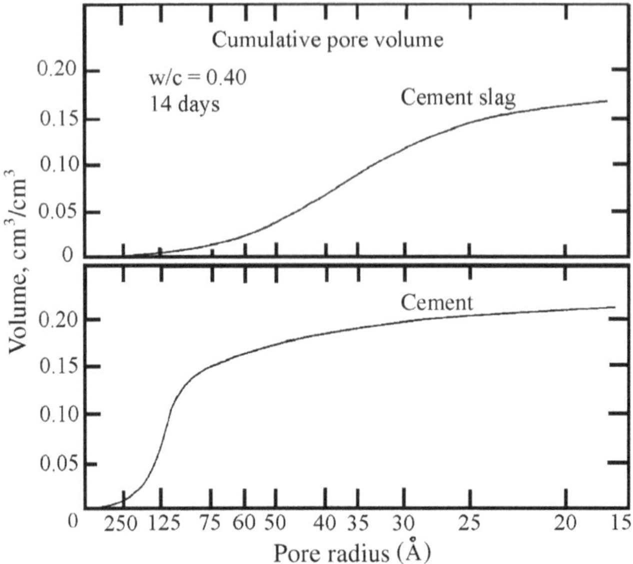

Fig. 3.17 Pore size distribution of pastes containing 40 % slag

Effect of Slag on Durability of Concrete

Durability of concretes containing slag cements in severe environments has been studied by many investigators. As mentioned earlier and due to the pore refining effect of slag cements, lower porosity and permeability has been observed in most of slag cement concretes. Low permeability of slag cement concretes controls the physical and chemical processes of degradation caused by the intrusion of water resulting in lower deterioration and higher durability.

Effect of Slag on Carbonation of Mortars and Concrete

The rate of carbonation of slag cement concrete has been reported to be higher than that of Portland cement concretes especially at high slag replacement levels [40]. In a concrete containing 50 % slag the carbonation was 1.5 times as deep as in Portland cement concrete, whereas it was twice as deep in concrete containing 70 % slag. Curing of concrete is very important in reducing the carbonation depth especially in slag cement concretes. Figure 3.18 shows the effect of water curing up to 1 year on the carbonation depth of concrete containing 50 % slag in comparison with normal cement concrete [41].

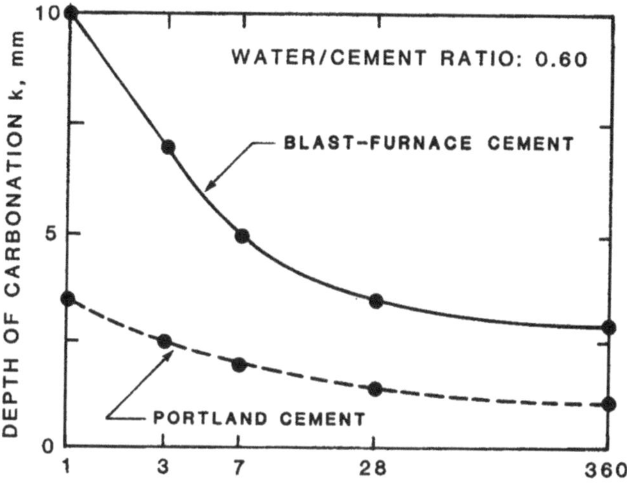

Fig. 3.18 Effect of moist curing on the depth of carbonation stored in air for 1 year after water curing and protected from rain [41]

In an investigation on the carbonation of concrete cores after 20 years exposure, the depth of carbonation was higher in concretes containing 65 % slag when compared with ordinary Portland cement concrete at same age [42].

They concluded that carbonation increases the permeability of blast furnace slag concrete because the small pores originally present become larger.

Higher carbonation depth was also observed in concrete containing 70 % slag than the concrete containing no slag or less than 50 % slag [43]. This finding is related to the high content of micropores formed at early ages in concretes incorporating slag. It has also been attributed to the porous silica gel as the product of calcium silicate hydrates in highly carbonated samples [44]. The increase in larger pores has resulted in an increase in permeability and hence higher carbonation depth. Similar to normal Portland cement concretes, carbonation of concretes made with slag can be reduced by better compaction of concrete and sufficient moist curing.

Effect of Chloride Ions on Durability of Concretes Containing Slag

The incorporation of granulated blast furnace slag in concrete increases its resistance to chloride ion diffusion especially at later ages [34, 45]. It also improves the chloride binding capacity of the concrete. This is very advantageous for protecting reinforcement steel from corrosion due to chloride penetration.

Figure 3.19 shows the chloride diffusion in concretes containing slag cements. Concrete mixtures containing higher amounts of slag (up to 70 %) have shown lower capillary pores and permeability and therefore lower chloride diffusion [4].

The reason for reduction in chloride penetration is not simply the lower permeability of concretes containing slag; even badly cured slag cement concrete has shown a better resistance to chloride ion intrusion. This suggests that chemical resistance of concrete containing slag might be a reason for that. Chemical reaction of chloride ions with hydration products of slag cement increases the chloride binding capacity of the slag cement pastes. Further improvement can be achieved by replacing more slag with cement and prolonging the moist curing period. Some concerns were expressed regarding potentially specified in standards is adequate for most practical applications. Therefore resistance to corrosion of reinforcement is not an issue even with high slag content in a good quality concrete.

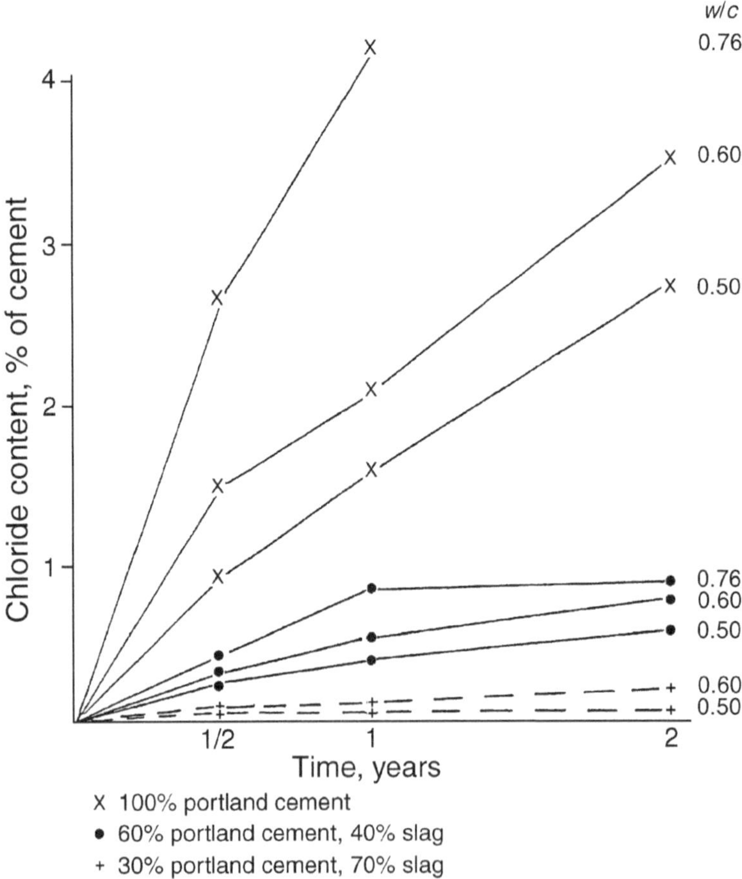

Fig. 3.19 Diffusion of chloride ions in concrete containing slag [4]

Sulfate Resistance of Mortars and Concretes Containing Slag

It is generally accepted that concrete containing slag improves its resistance to sulfate attack or sea water when compared to no slag concretes. The level of improvement depends upon the amount of slag replacement, the alumina content of slag, the C_3A content of the cement and permeability of concrete.

Several investigations report that slag cement concretes will have higher sulfate resistance irrespective of the slag composition and Portland cement components if they contain more than 65 % slag. Others believe that the alumina content of the slag is also an important factor and increasing the amount of slag with high alumina content may result in lower resistance to sulfate attack.

The major mechanisms for the sulfate resistance of slag cement concretes are the reduction in the formation of ettringite and lower free calcium hydroxide in the mixtures. Slag cements usually have less C_3A content and produce less ettringite to cause expansion in sulfate environments. Free calcium content of mixtures containing slag cement also reduces the formation of gypsum in reaction with sulfate ions. Formation of calcium silicate hydrate in the pores also reduces the permeability of concretes with slag cement and hence reduces the intrusion of sulfate ions and enhances the durability of concrete.

In an investigation of mortar containing slag cement in sodium sulfate solution, cement mortars containing more than 65 % slag showed higher sulfate resistance than plain Portland cement mortars. In mortars containing 20–50 % slag, the use of high alumina slag and up to 17.7 % decreased the sulfate resistance, while low alumina slag and about 11 % increased the sulfate resistance independently of the alumina content of the slag in both cases [46]. Similar results have been observed for mortars maintained in magnesium sulfate for sulfate resistance evaluation [47].

In another investigation with type II Portland cement, mixtures containing more than 50 % slag showed higher sulfate resistance [23]. Mixtures containing 50 % slag (with 7 % Al_2O_3) with type I Portland cement having up to 12 % C_3A showed similar sulfate resistance to that of type V Portland cement [48]. Results of the study on the effect of three levels of cement replacement by slag indicate that as the amount of slag increases, the resistance to sulfate improves. Figure 3.20 shows that replacement up to 65 % has markedly reduced the expansion of mortar bars [23].

Comparison of sulfate resistance of mortars containing various amounts of slag with mortars made with type V Portland cement showed that increasing the amount of slag increases the sulfate resistance and 70 % slag replacement performed better than that of type V Portland cement. Mortars containing less than 30 % slag showed higher sulfate resistance than normal Portland cement mortars but lower than that made with type V Portland cement [48].

It has been reported that a minimum amount of slag is required to provide a high sulfate resistance mixture [39]. This is seen in Fig. 3.21 that the minimum content would be about 50 % or greater. This is valid for type I Portland cement having a C_3A content up to 12 % and the slag with alumina content lower than 11 %.

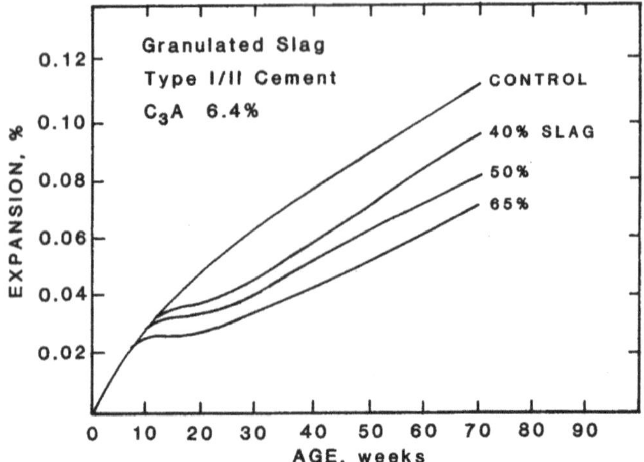

Fig. 3.20 Effect of slag content on sulfate resistance of mortar bars [23]

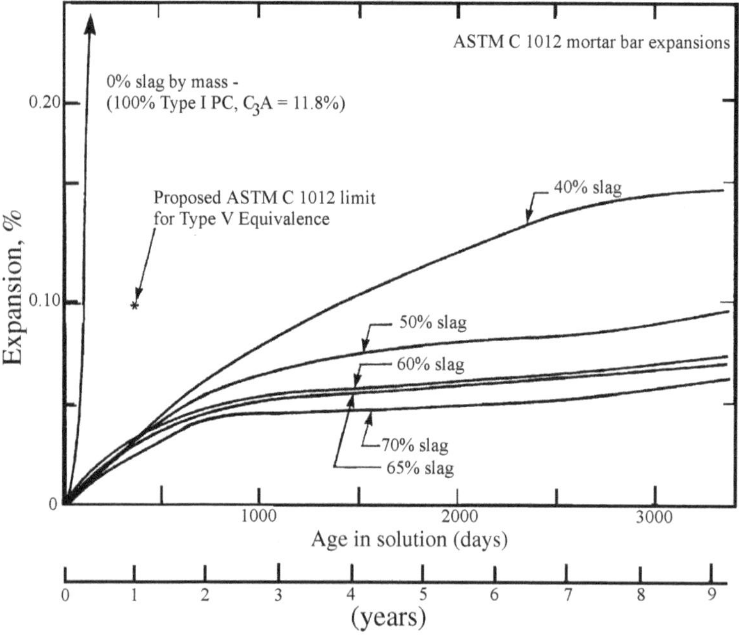

Fig. 3.21 Expansion of mortars containing various amounts of slag in sodium sulfate solution (ASTM C1012, Slag C_3A = 8.4 %) [39]

Resistance to sea water of concretes containing slag cement was also better than that of normal Portland cement concretes. A well cured low water to cementing materials ratio concrete containing a high replacement level of slag performs very well. However, some concretes with slag cements have shown deterioration in sea water due to the insufficient moist curing before exposure to marine environment and high permeability of mixtures with high water cemetitious material ratio. In an investigation on the long term performance of concretes containing slag in marine environment, mixtures containing 65 % slag with water to cementitious materials ratio of 0.4 performed very well. Those mixtures with 80 % slag have also a satisfactory performance. However, mixtures containing slag but at water cementitious ratio of 0.6 started to show severe deterioration after 7 years [49]. Salt crystallization occurs to concrete sections above the sea water level due to capillary action of salt water and subsequent evaporation. This will result in deterioration of concretes exposed to sea water. A low permeable concrete containing slag with sufficient curing would prevent such penetration of salt water and hence enhances the durability of concrete.

Effect of Slag on Suppressing the Alkali Aggregate Reaction

The use of slag in concrete generally reduces expansion due to any potential reactivity of aggregates. The improved resistance to alkali silica and alkali carbonate reactions is attributed to the changes in chemistry of the solution and to the microstructure and permeability of paste. The use of slag will change the alkali-silica ratio, dissolution and consumption of the alkali species, reduction in calcium hydroxide to support the reaction, and direct reduction of available alkali in the system. Further reaction of slag reduces the permeability and hence reduces the alkali aggregate reaction of concrete.

The first report showing the effect of slag on suppressing alkali silica reaction was published in 1950 [50]. Several other researchers showed that the incorporation of slag can reduce the ASR in concrete [51–53]. In a review of the effect of slag on controlling the AAR in concrete, it was concluded that the efficiency of slag cement depends on the reactivity of aggregate, quality and quantity of slag, and the alkali content of the Portland cement [54]. In most cases the expansion of mixtures were controlled by incorporation of 50 % slag even in concretes having highly reactive aggregates and high alkali cement content.

Figure 3.22 shows that increasing the amount of slag can reduce the expansion of Pyrex glass as a very reactive aggregate. The expansion was ten times less with 50 % slag cement than the normal Portland cement [55].

In another investigation, the effect of slag cement content on five reactive aggregates from Canada was examined using ASTM C1293 concrete prism test at 2 years. As seen in Fig. 3.23, between 35 and 50 % slag cement was required to control the expansion of reactive aggregates concretes [56]. In this work the low

SLAGS AND SLAG CEMENTS

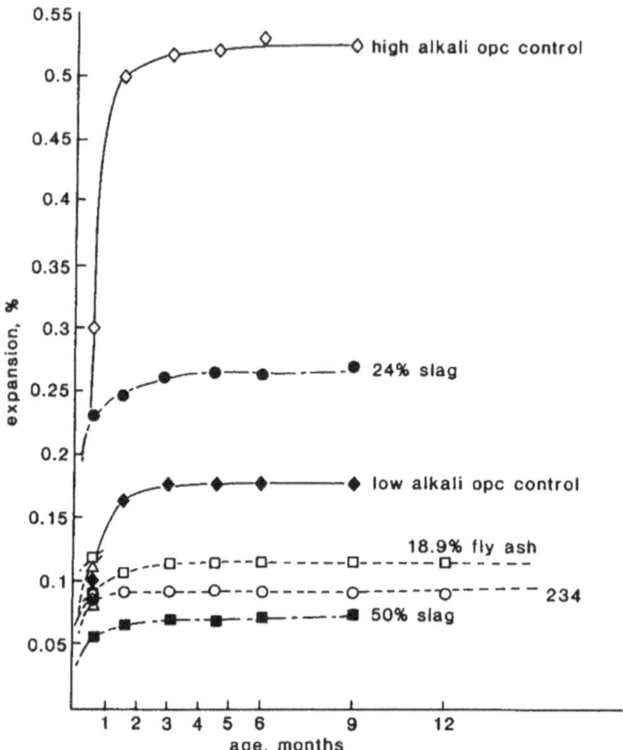

Fig. 3.22 Effect of slag content on ASR reduction of mortars [55]

alkali cement having 0.5 % Na_2O equivalent was found ineffective in controlling the expansion of such reactive aggregate.

The German regulation regarding the permissible value of alkalis in cement when the aggregates used in concrete are potentially reactive increases the limit of Na_2O equivalent from 0.6 to 2 % from Portland cement blast furnace slag cement with an addition of 70 % slag. Table 3.8 shows the relation between the amount of alkalis and slag content for controlling the expansion of reactive aggregate.

Comparison of the expansion reduction of low alkali cement and slag cement in reactive aggregate concretes was reported after 2 years. The slag cement mixtures were found effective in reducing the expansion but the reduction was less than that with low alkali cement. With high alkali cement the amount of slag needed to reduce the potential reaction of ASR was about 50 % [57].

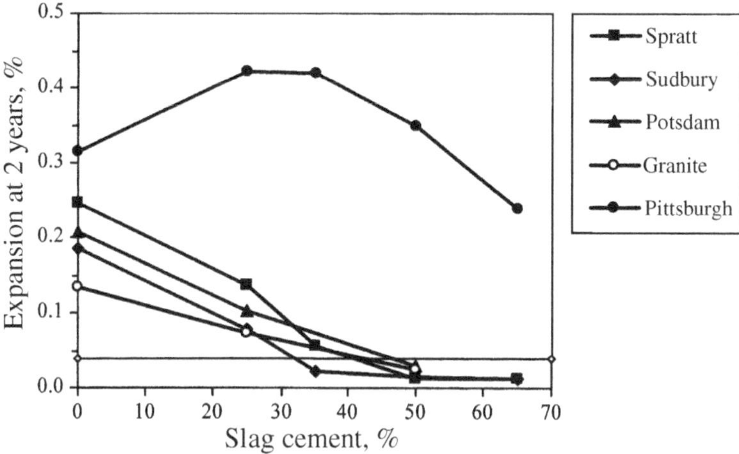

Fig. 3.23 The effect of slag cement on controlling the expansion of various reactive aggregates [56]

Table 3.8 German regulation for the amount of alkalis in cement in the presence of potentially reactive aggregates

Cement	Na_2O_{eq} (wt %)	Blast furnace slag (wt %)
Portland cement	≤0.60	–
Slag cement	≤1.10	≥50
Slag cement	≤2.00	≥65

Freezing and Thawing of Concretes Containing Slag

As with all Portland cement concretes, proper air content and bubble spacing is necessary for adequate protection against freezing and thawing. Slag cement concretes also require sufficient air content to resist frost action. Several investigations on the effect of slag cement on resistance to freezing and thawing in water show similar results with normal Portland cement concretes provided both are air entrained properly [19, 58]. Amount of slag also affects the freezing thawing resistance of concretes. In comparison with type II Portland cement mixtures, concretes containing 50 % slag with sufficient air entraining agents showed a good resistance to freezing and thawing cycles [23]. An increase in critical factor spacing with higher slag content has been reported which resulted in deterioration in frost action [59]. It seems that higher dosage of air entraining admixture is needed for higher slag concretes to achieve similar frost resistance as type I and type II Portland cement concretes.

An investigation on the frost resistance of air entrained concretes containing 50 % slag in accordance with Procedure A, ASTM C666 test method was carried out up to about 300 cycles. Although a measurable difference was found in the durability factors of the concretes with and without slag, both mixtures were rated as frost resistant [23]. In another research work concrete mixtures containing 25–65 % slag were tested in accordance with ASTM C666, Procedure B test method [28]. Specimens cured for 14 days and then subjected to freeze–thaw cycles. Visual inspection, weight loss, resonant frequency and ultrasonic pulse velocity methods were used for concrete evaluation. Table 3.9 shows the results after 700 freeze–thaw cycles. The test results indicate that regardless of the water cementitious materials ratio and whether the concretes were air entrained or air entrained and superplasticized, all specimens showed a very good performance with high durability factors greater than 90 %.

Salt scaling resistant of concretes containing slag also depends on the air content, slag content and water to cementitious materials ratio of the mixtures. A properly air entrained with low water to cementing materials ratio concrete shows a good resistance to salt scaling. A little difference was observed on the deicing salt frost resistance of concretes containing 35 % and up to 50 % slag when compared with control concrete [23, 60]. ACI 301 and ACI 318 allow up to 50 % slag cement in severe exposure condition environments applicable to salt scaling resistance. In Canadian standard (CSA), the air content requirements for freeze–thaw resistance and deicing salt scaling are given. Also this standard does not give any limit on slag replacement for concrete exposed to deicing salts. Slag Cement Association suggests limiting it between 25 and 50 % with water to cementing materials ratio of 0.45.

Application of Slag in Concrete

Granulated blast furnace slag blended Portland cement was first produced in Germany in 1892. Due to considerable suspicion on the mechanical properties and durability of slag cements, the replacement level was only 20 %. Slag cement was used as a raw material for the manufacture of cement or as cementitious material in combination with Portland cement, gypsum anhydrite and hydrated lime. It is now used as blended cement or added to the mixer during concrete production in most of European countries, North America and Japan.

Long term good performance of slag cements in various environmental conditions has resulted in their applications in many mortar and concrete works. It has been used as mortar and grout for masonry works and injection purposes. Slag cements are now used for reinforced concrete structures such as buildings, dams and hydraulic structures, bridges, and precast products. The replacement level was about 50 % in ready mix concrete producers for warm weather applications. Due to slow development of strength of slag cement concretes in cold conditions, the level of replacement has been reduced to 20 % in such environments. In precast concrete elements where high early strength is required, the amount of slag in the blended cement is also reduced in order to achieve required strength.

Table 3.9 Frost resistance of concretes containing various amounts of slag [28]

Mix series	W/C + S	Type of mix	Summary of freeze–thaw tests results								Durability factor, %	Relative durability factor, %
			At zero cycles				At completion of 700 cycles					
			Weight, kg	Length, mm	Longitudinal resonant frequency, HZ	Pulse velocity, m/s	Weight, kg	Length, mm	Longitudinal resonant frequency, HZ	Pulse velocity, m/s		
B	0.38	Control + AEA	8.703	2.89	5150	4717	8.693	2.90	5200	4747	102	100
		Control + AEA + SP	8.499	2.70	5150	4684	8.486	2.72	5138	4661	99	97
		25 % slag + AEA	8.697	3.00	5300	4788	8.673	3.05	5225	4788	97	95
		25 % slag + AEA + SP	8.540	2.96	5125	4684	8.517	3.01	5100	4656	99	97
		65 % slag + AEA	8.622	2.74	5140	4684	8.626	2.91	4950	4568	93	91
		65 % slag + AEA + SP	8.302	1.59	5025	4589	8.302	1.68	4875	4531	94	92
D	0.56	Control + AEA	8.331	2.56	5000	4568	8.299	2.56	5010	4600	100	–
		Control + AEA + SP	8.443	2.76	4980	4568	8.394	2.76	4980	4504	100	100
		25 % slag + AEA	8.451	2.85	5000	4573	8.416	2.88	5000	4606	100	100
		25 % slag + AEA + SP	8.544	2.83	5040	4639	8.483	2.91	5050	4622	100	100
		65 % slag + AEA	8.465	2.61	4950	4546	***	2.88	****	****	–	59
		65 % slag + AEA + SP	8.471	2.52	4930	4563	***	2.75	****	****	70	

Fig. 3.24 Rajai port complex in the south of Iran

Granulated blast furnace slag cements have been used in many countries all over the world. Some of the most important projects are:

Foundations, superstructure and tower legs of Humber Bridge in UK, Reactor top cap concrete of Heysham and Hartlepool Nuclear Power Stations in UK, Conventional Bridge in Canada, Brighton Marina Marine Structure in UK, concrete pavements in several motorway standard roads in UK and USA, 7 World Trade center and Georgia Aquarium in USA, Serre-Poncon dam and Rance tidal power station in France, Rajai Port Complex and Bushehr Nuclear Power station in Iran (Fig. 3.24) and the most recent one Burj Dubai in U.A. Emirates.

References

1. M. Regourd, Slags and slag cements, in *Concrete Technology and Design*, vol 3, *Cement Replacement Materials,* ed. by R.N. Swamy (Surrey University Press, Guildford, 1986), pp. 73–99
2. F. Schröder, Blast furnace slag and slag cements, in *Proceedings of the 5th International Congress on the Chemistry of Cement*, Tokyo, 1968; vol IV, pp. 149–207
3. V.I. Satarin, Slag Portland cement, in *Proceedings of the 6th International Congress on the Chemistry of Cement*, Moscow, 1974. Principal paper, p. 51
4. H.G. Smolczyk, Slag structure and identification of slag, in *Proceedings of the 7th International Congress on the Chemistry of Cement*, Paris, 1980. Principal Report, vol I, Theme III, pp. 1–17
5. M. Regourd, Characteristics and activation of additive products, in *Proceedings of the 8th International Congress on the Chemistry of Cement*, Rio de Janeiro, 1980. Principal Report, vol I, pp. 199–229
6. K. Chopra, C.A. Teneja, Coordination state of aluminium, magnesium and manganese ions in synthetic slag glasses, in *Proceedings of the 5th International Symposium on the Chemistry of Cement*, Tokyo, 1968. vol IV, Theme, pp. 228–236

7. J.P.H. Frearson, J.M. Uren, Investigation of a granulated blast furnace slag containing merwinite crystallization, in *Proceedings of the 2nd International Conference on the Use of Fly Ash, Silica Fume, Slag and Natural Pozzolans in Concrete*, Madrid, 1986. vol II, pp. 1401–1421
8. R.D. Hooton, J.J. Emery, Glass content determination and strength development predictions for vitrified blast furnace slag, in *Proceedings of the International Conference on the Use of Fly Ash, Silica Fume, Slag in Concrete*, Montebello, Canada, 1983. American Concrete Institute Special Publication 79, vol 2, pp. 943–962
9. W. Kramer, Blast furnace slags and cements, in *Proceedings of the 4th International Symposium on the Chemistry of Cement*, Washington, 1960. National Bureau of Standards, Monograph 43, vol II, Paper VIII-2, pp 957–981, 1962
10. R. Kondo, S. Oshawa, Studies on a method to determine the amount of granulated blast furnace slag and the rate of hydration of slag in cement, in *Proceedings of the 5th International Symposium on the Chemistry of Cement*, Tokyo, 1968; vol IV, pp. 225–262
11. S.A. Mironov, I.I. Kurbatova, S.A. Vysotskii, Extraction method to determine the blast furnace slag content in cements, Tsement, no. 12, 1976, pp. 11–12
12. F. Keil, Slag cements, in *Proceedings of the 3rd International Symposium on the Chemistry of Cement*, London 1952, *Cement and Concrete Association*, vol Paper 15, pp. 530–580, 1954
13. M. Regourd, J.H. Thomassin, P. Baillif, J.C. Touray, Blast furnace slag hydration. Surface analysis. Cem. Concr. Res. **13**(4), 549–556 (1983)
14. F.P. Glasser, Chemical, "mineralogical and microstructural changes occurring in hydrated slag-cement blends", in *Materials Science of Concrete II*, ed. by J. Skalny, S. Mindess (The American Ceramic Society, Westerville, 1991), pp. 41–82
15. Q.L. Feng, E.E. Lachowski, F.P. Glasser, Densification and migration of ions in blast furnace slag Portland cement pastes. Mater. Res. Soc. Symp. Proc. **136**, 263–272 (1989)
16. M. Regourd, B. Mortureux, E. Gautier, H. Hornain, J. Volant, Caracterisation et activation thermique des ciments au laitier, in *Proceedings of the 7th International Congress on the Chemistry of Cement*, Paris, 1980. vol II, Theme III, p. 111
17. D.M. Roy, G.M. Idorn, Developments of structure and properties of blast furnace slag cements. J. Am. Concr. Inst. **97**, 444–457 (1982)
18. K. Wood, Twenty years of experience with slag cement, in *Symposium on Slag Cement*, University of Alabama, Birmingham, 1981
19. F.S. Fulton, *The properties of Portland cement containing milled granulated blast furnace slag* (Monograph Portland Cement Institute, Johannesburg, 1974), pp. 4–46
20. J.W. Meusel, J.H. Rose, Production of granulated blast furnace slag at sparrows point, and the workability and strength potential of concrete incorporating the slag, in *Fly Ash, Silica Fume, Slag and Other Mineral By-Products in Concrete*, SP-79, vol 1, ed. by V.M. Malhotra (American Concrete Institute Farmington Hills, Michigan, 1983), pp. 867–890
21. G.J. Osborne, Carbonation and permeability of blast-furnace slag cement concretes from field structures, in *Fly Ash, Slag and Natural Pozzolans in Concrete*, SP-114, ed. by V.M. Malhotra. Proceedings of 3rd International Conference, Trondheim, Norway (American Concrete Institute Farmington Hills, Michigan, 1989), pp. 1209–1237
22. M.D. Luther, W.J. Mikols, A.J. DeMaio, J.E. Whitliger, Scaling resistance of ground granulated blast furnace slag (GGBFS) Concretes, in *Durability of Concrete*, SP-145, ed. by V.M. Malhotra. Proceedings of the 3rd International Conference (American Concrete Institute, Farmington Hills, Michigan, 1994), pp. 47–64
23. F.J. Hogan, J.W. Meusel, Evaluation for durability and strength development of a ground granulated blast-furnace slag, Cem. Concr. Aggregates **3**(1), 40–52 (1981)
24. V.M. Malhotra, Comparative evaluation of three different granulated slags in concrete. Report MSL 187-50 (op), CANMET, Canada, Ottawa, 1986
25. C.M. Reeres, The use of granulated blast furnace slag to produce durable concrete, *Durability of Concrete Conference*, Institute of Civil Engineering, May 1985, London, UK
26. American Concrete Institute (ACI), *Report of Committee ACI 233, Ground granulated blast furnace slag cementitious constituent in concrete*, ACI, Publications ACI 233R-03, 2003

27. P.J. Wainwringt, J.J.A. Tolloczko, The early and later age properties of temperatue cycled OPC concretes, *2nd International Conference on the Use of Fly Ash, Silica Fume, Slag and Natural Pozzolans in Concrete*, CANMET, Otlawa, 1986, Madrid
28. V.M. Malhotra, Strength and durability characteristics of concrete incorporating a pelletized blast furnace slag, in *Fly Ash, Silica Fume, Slag and Other Mineral By-Products in Concrete*, SP-79, vol 2, ed. by V.M. Malhotra (American Concrete Institute, Farmington Hills, Michigan, 1983), pp. 891–922
29. J.J. Brooks, P.J. Wainwright, M. Boukendakji, Influence of slag type and replacement level on strength, elasticity, shrinkage and creep of concrete, in *Fly Ash, Silica Fume, Slag and Natural Pozzolans in Concrete*, SP-132, vol 2. Proceedings of the 4th International Conference (American Concrete Institute, Farmington Hills, Michigan, 1992), pp. 1325–1341
30. A.M. Neville, J.J. Brooks, Time-dependent behaviour of Cemsave concrete, Concr. J. **9**(3), 36–39 (1975)
31. P.B. Bamforth, In situ measurement of the effect of partial Portland cement replacement using either fly ash or ground granulated blast furnace slag on the performance of mass concrete, Part 2, vol 69, *Proceedings* (Institution of Civil Engineers, London, 1980), pp. 777–800
32. H. Uchikawa, Blended cements. Effect of blending components on hydration and structure formation, in *Proceedings of the 8th International Congress on the Chemistry of Cement*, Rio de Janeiro, 1986. vol I, pp. 250–280
33. R.F. Feldman, Pore structure, permeability and diffusivity as related to durability, in *Proceedings of the 8th International Congress on the Chemistry of Cement*, Rio de Janeiro, 1986. vol I, pp. 336–356
34. R.F.M. Bakker, On the cause of increased resistance of concrete made from blast-furnace cement to alkali reaction and to sulfate corrosion. Thesis RWTH-Aachen, 1980, p. 118
35. D.M. Roy, G.M. Idorn, Hydration, structure, and properties of blast-furnace slag cements, mortars, and concrete, ACI J. Proc. **79**(6), 445–457 (1982)
36. D.M. Roy, K.M. Parker, Microstructures and properties of granulated slag-Portland cement blends at normal and elevated temperatures, in *Fly Ash, Silica Fume, Slag and Other Mineral By-Products in Concrete*, SP-79, vol 1, ed. by V.M. Malhotra (American Concrete Institute, Farmington Hills, Michigan, 1983), pp. 397–414
37. P.K. Mehta, Durability of concrete in marine environment a review", in *Performance of Concrete in Marine Environment*, (American Concrete Institute, Farmington Hills, Michigan, 1980), pp. 1–20
38. P.K. Mehta, Pozzolanic and cementitious byproducts as mineral admixtures for concrete—A critical review, in *Fly Ash, Silica Fume, Slag and Other Mineral Byproducts in Concrete*, vol 1 (American Concrete Institute, Farmington Hills, Michigan, 1983), pp. 1–46
39. R.D. Hooton, J.J. Emery, Sulfate resistance of a Canadian slag, ACI Mater. J. **87**(6), 547–555 (1990)
40. V.M. Malhotra, *Supplementary Cementing Materials for Concrete*, SP 86-8E (CANMET Publication, Canada, 1987)
41. A. Meyer, Investigation on the carbonation of concrete, in *Proceedings of 5th International Symposium on Chemistry of Cement*, Tokyo, 1968
42. G.G. Lirvan, A. Mayer, Carbonation of granulated blast furnace slag cement concrete during twenty years of field exposure, in *Proceedings of the 2nd International Congress on Fly Ash, Silica Fume, Slag and Natural Pozzolans in Concrete*, Madrid, 1986
43. G.J. Osborne, Carbonation and permeability of blast furnace slag cement concretes from field structures, in *Proceedings of the 3rd International Congress on the Use of Fly Ash, Silica Fume, Slag and Natural Pozzolans in Concrete*, Trondheim, 1989
44. T.A. Bier, J. Kropp, H.K. Hilsdorf, The formation of silica gel during carbonation of cementitious systems containing slag cements, in *Proceedings of the 3rd International Conference on the Use of Fly Ash, Silica Fume, Slag, and Natural Pozzolans in Concrete*, Trondheim, 1989. American Concrete Institute Special Publication 114, vol 2, pp. 1413–1428

45. R.F.M. Bakker, Permeability of blended cement concrete, in *Fly Ash, Silica Fume, Slag and Other Mineral By Products in Concrete*, SP-79, vol 1 (American Concrete Institute, Farmington Hills, Michigan, 1983), pp. 589–605
46. F.W. Locher, Zement-Kalk-Gips **19**, 395 (1966)
47. J.H.P. van Aardt, S. Vissor, Research Bulletin of the National Building Research Institute of the Union of South Africa, p. 47 (1967)
48. B. Chojnacki, Sulphate resistance of blended (slag) cement. Report No. EM-52, Ministry of Transportation and Communications, Downsview, Ontario, Canada, 1981
49. V.M. Malhotra, G.G. Caratte, T.W. Bremmer, Durability of concrete containing supplementary cementing materials in marine environment, in *Concrete Durability*, SP-100, ed. by J.M. Scanlon. Katharine and Bryant Mather International Conference (American Concrete Institute, Farmington Hills, Michigan, 1987), pp. 1227–1258
50. H.P. Cox, R.B. Coleman, L. Wite, Effect of blast furnace slag cement on alkali aggregate reaction in concrete. Pit Quarry **45**(5), 95–96 (1950)
51. H.G. Smolczyk, Slag cements and alkali-reactive aggregates, in *Proceedings of the 6th International Congress on the Chemistry of Cement*, Moscow, Section III, 1974
52. R.F.M. Bakker, About the cause of the resistance of blast furnace cement concrete to the alkali-silica reaction, in *Proceedings of the 5th International Congress on Alkali-Aggregate Reaction in Concrete*, Cape Town, South Africa. National Building Research Institute, S252/29, 1981
53. R.E. Oberholster, N.B. Westra, The effectiveness of mineral admixtures in reducing expansion due to alkali-aggregate reaction with Malmesbury aggregates, in *Proceedings of the 5th International Congress on Alkali-Aggregate Reaction in Concrete*, Cape Town, South Africa. National Building Research Institute, S252/32, 1981
54. M.D.A. Thomas, Review of the effect of fly ash and slag on alkali-aggregate reaction in concrete. Building Research Establishment Report BR 314, Construction Research Communication Ltd., Watford, UK, 1996, p. 117
55. P.J. Nixon, M.E. Gaze, The use of fly ash and granulated slag to reduce expansion due to alkali-aggregate reaction, in *Proceedings of the 5th International Conference on Alkali-Aggregate Reaction in Concrete*, Cape Town, South Africa. National Building Research Institute, S252/32, 1981
56. M.D.A. Tomas, F.A. Innis, Effect of slag on expansion due to alkali-aggregate reaction in concrete. ACI Mater. J. **95**(6), 716–724 (1998)
57. J.A. Soles, V.M. Malhotra, H. Chen, CANMET Investigations of supplementary cementing materials for reducing alkali-aggregate reactions: Part I-granulated/pelletized blast furnace slags, in *Fly Ash, Silica Fume, Slag and Natural Pozzolans in Concrete*, SP-114, vol 2, ed. by V.M. Malhotra. Proceedings of 3rd International Conference, Trondheim, Norway (American Concrete Institute, Farmington Hills, Michigan, 1989), pp. 1637–1656
58. B. Mather, Laboratory tests of Portland blast-furnace slag cements, ACI J. Proc. **54**(3), 205–232 (1957)
59. M. Pigeo, M. Regourd, Freeze and thawing durability of three cement with various granulated blast furnace concretes, *International Conference on Fly Ash, Silica Fume, and Slag in Concrete*, vol 2 (ACI Special Publication SP-79, Canada, 1983)
60. P. Klieger, A.W. Isberner, Laboratory studies of blended cement-portland blast furnace slag cements, PCA Res. Devel. Dept. Lab. **9**(3), 2–22 (1967)

Chapter 4
Silica Fume

Introduction

Silica fume or microsilica is very fine non-crystalline silica produced in electric arc furnaces as a by-product of the production of elemental silicon or alloys containing silicon. Micro-silica was first tested in concrete in Norway in the early 1950s. Higher strength was obtained for concretes containing silica fume. Performance of silica fume concretes in sulfate environment was also better than normal Portland cement concretes. After twenty years and due to the environmental restrictions for the smelting industry in Norway, a large amount of silica fume was produced in filtering process. Intensive research efforts were carried out to find various applications for silica dust. Durability enhancement of concretes incorporating silica fume increased the usage of silica fume in the concrete industry. It has been used in other industries but it is known as a super pozzolan for the improvement of concrete properties.

Production

The high–purity quarts is usually heated to 2000 °C in an electric arc furnace with coal, coke and wood chips added to remove the oxygen from Silicon dioxide. The alloy is collected at the bottom of the electric furnace. Silicon dioxide vapor which is released from the quarts oxidizes and condenses into microspheres at the upper parts of the furnace. Then these fumes are drawn from the furnace through a cyclone by powerful fans. In this process the large coarse particles of unburned wood or carbon are removed and the fumes are blown into a series of special filter bags in the bag house. Figure 4.1 shows the schematic diagram of the silica fume production.

At higher temperatures and more than 1500 °C in the furnace, silicon dioxide reacts with carbon to produce SiC.

$$SiO_2 + 3C = SiC + 2CO$$

A. A. Ramezanianpour, *Cement Replacement Materials*,
Springer Geochemistry/Mineralogy, DOI: 10.1007/978-3-642-36721-2_4,

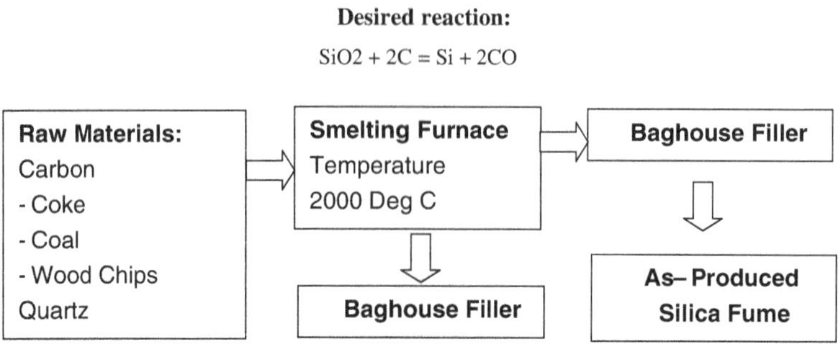

Desired reaction:

SiO2 + 2C = Si + 2CO

Raw Materials:
Carbon
- Coke
- Coal
- Wood Chips
Quartz

Smelting Furnace
Temperature
2000 Deg C

Baghouse Filler

Baghouse Filler

As– Produced Silica Fume

Fig. 4.1 Schematic diagram of silica fume production

At temperature higher than 1800 °C SiC reacts with silicon dioxide to produce Si.

$$3SiO_2 + 2SiC = Si + 4SiO + 2CO$$

Finally the unstable gas at the top of the furnace reacts with Oxygen to produce silicon dioxide or silica fume.

$$4SiO + 2O_2 = 4SiO_2$$

The small particles of silicon dioxide condense as the temperature drops in the smokestack. The average diameter of silica fume is about 0.2–0.3 μ. Larger particles more than 0.3 μ may be observed when small particles agglomerate.

Silica fume is produced and delivered in different forms. It can be used in undensified, densified, pelletized, and slurry forms. The bulk density of undensified silica fume is about 200–300 kg/m^3 and is difficult to use it in concrete industry. Densified silica fume has greater bulk density (More than 500 kg/m^3) which makes it easier to handle and cheaper to transport than the undensified silica fume. Pelletizing process involves forming the silica fume into pellets of about 0.7–1 mm in diameter on a pelletizing table. The bulk density of pelletized silica fume is more than 600 kg/m^3. It is not suitable to use silica fume in pelletized form for concrete making. The slurry form silica fume has a specific gravity of 1400 kg/m^3 and is produced by mixing the undensified silica fume with equal water content by weight. Better dispersion of small particles of silica fume in concrete mix is obtained with slurry form resulting in more homogenous mixture.

Physical, Chemical and Mineralogical Properties

Physical Properties

Silica fume particles are very small, with more than 95 %of the particles being less than 1 μ. Most particles are in spherical shape with a mean diameter of 0.2 μ.

Fig. 4.2 Photomicrograph of Portland cement grains (*left*) and silica fume particles (*right*) at the same magnification

Figure 4.2 shows the difference between the particle sizes of cement grains and silica fume particles.

Typical grain size distribution of silica fume is also shown in Fig. 4.3.

This figure represents the size of the primary agglomerates. Because the particles of silica fume are very small, the surface area is very large. The specific surface area of silica fume particles varies between 15000 and 30000 m^2/Kg. The high surface area of silica fume particles is an important factor affecting the reactivity of particles.

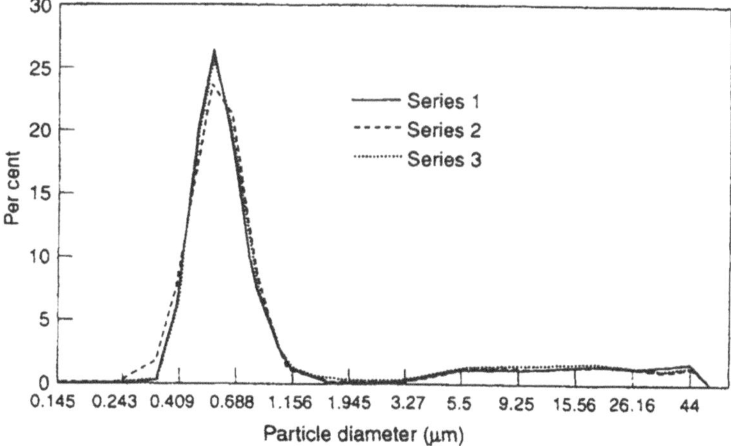

Fig. 4.3 Typical particle size distribution of various silica fumes

Silica fume has a specific gravity of about 2.2, which is lower than Portland cement. Thus the addition of silica fume to a concrete mixture will not increase its density.

Chemical and Mineralogical Properties

The chemical composition of silica fume is related to the composition of raw materials used in the furnace. The composition is also depends on the furnace design which may produce lower or higher ignition loss corresponds to lower and higher carbon content in the final production. Typical chemical compositions of silica fume obtained from seven different operations are shown in Table 4.1 [1].

The silica content of silica fume is generally higher than 80 % except for the fume obtained from other alloys such as Ca–Si and Si–Mn which are not suitable materials to be used as pozzolans. Other oxides such as Ca0, Al2O$_3$, Fe_2O_3 and alkali contents of silica fume are relatively low. The magnesium oxide is also low in most of the raw materials but slightly high in Fe–Cr–Si fume which may not be deleterious to concrete. The carbon content varies between 0.5 and 1.5 % and usually is less than 2 %.

The chemical composition of silica fume from a single source does not vary from one day to another. This is mainly attributed to the pure raw materials usually used for production of silicon metal or ferrosilicon alloys.

Mineralogical composition of silica fume is different from natural pozzolans. It consists essentially of an amorphous silica structure with very little crystalline particles. Figure 4.4 depicts the mineral composition of different silica fumes by X-ray diffraction analysis. A wide scattering peak centered at about 22° is observed for different types of silica fumes.

Table 4.1 Typical chemical composition of different silica fumes

Component	Content of fume (wt %)						
	Si	FeSi 75 %	FeSi 75 % (heat recovery)	FeSi 50 %	FeCrSi	CaSi	SiMn
SiO_2	94	89	90	83	83	53.7	25
Fe_2O_3	0.03	0.6	2.9	2.5	1.0	0.7	1.8
Al_2O_3	0.06	0.4	1.0	2.5	2.5	0.9	2.5
CaO	0.5	0.2	0.1	0.8	0.8	23.2	4.0
MgO	1.1	1.7	0.2	3.0	7.0	3.3	2.7
Na_2O	0.04	0.2	0.9	0.3	1.0	0.6	2.0
K_2O	0.05	1.2	1.3	2.0	1.8	2.4	8.5
C	1.0	1.4	0.6	1.8	1.6	3.4	2.5
S	0.2		0.1				2.5
MnO		0.06		0.2	0.2		36.0
LOI	2.5	2.7		3.6	2.2	7.9	10.0

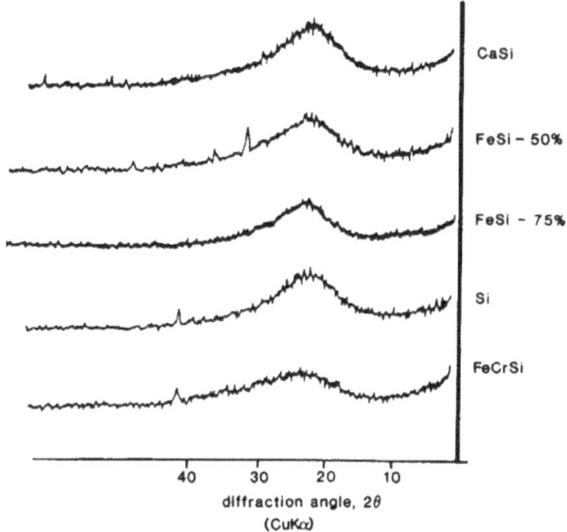

Fig. 4.4 Typical X-ray diffraction analysis of different types of silica fumes [1]

Hydration Reactions and Pozzolanic Activity

Hydration of Silica Fume with Lime

The reaction of silica fume with lime is very rapid and causes a phase to precipitate on the silicon dioxide particles as a high silica hydrated layer. The formed layer is unstable and rapidly turns into calcium silicate hydrate gel [2]. In normal consistency pastes, due to high reactivity of silica fume particles, free lime disappears between 7 and 28 days and even sooner [2, 3]. This form of CSH gel as the reaction product is much more crystalline than the CSH gel formed in Portland cement hydration [4]. In the 1:1 mixes of $Ca(OH)_2$ and SiO_2, the C/S ratio of CSH increases in the first days up to 1.30 and then decreases, reaching 1.10 after 70 days of reaction [3].

Sixteen percent silica fume was found sufficient to combine all the free lime contained in the pastes with lower water/cement ratio (0.2–0.4) cured for 550 days [5]. This difference from the previous results obtained for silica fume and free lime reaction is related to the low water cement ratio and incomplete hydration of cement paste. In cement pastes having higher water/cement ratio (0.6), the calcium hydroxide content does not change after 90 days of hydration while with lower water/cement ratio, it decreases after 90 days [6]. This means that the progress of pozzolanic reaction strongly depends on the water/cement ratio similar to the progress of hydration of Portland cement.

Pozzolanic Activity

As mentioned earlier the pozzolanic activity can be evaluated by the consumption of lime in a pozzolan-lime solution. The overall amount of combined lime depends on the active phases content in the pozzolan, SiO_2 content in the pozzolan, the fineness of the pozzolan, lime/pozzolan ratio, curing regime and temperature and the water to solid ratio of the mixture.

Silica fume contains very active reactive silica phase, very rich in amorphous silica content, and also contains very fine particles. Therefore in comparison with other natural and artificial pozzolans, silica fume is the most active pozzolan.

Silica fume containing about 85 % of silica combines most of the available lime within 28 days while other natural pozzolans and fly ashes containing about 50–65 % silica are capable combining about 30–50 % of the lime in the lime-pozzolan mixture [2, 3].

Hydration of Silica Fume with Clinker

Silica fume accelerates the hydration of C_3S in the cement-silica mixture depending on the C_3S/S ratio. It causes an early decrease in the Ca^{2+} and OH^- concentrations in the mixing water [7, 8]. The decrease occurs in water dispersions as well as in the paste.

Very fine amorphous silica having a BET surface area of about 200 m^2/g (Aerosil) reduces the length of the dormant and increases the intensity of the second peak in the heat evolution curve [7]. If the C_3S/S ratio decreases beyond a certain level (0.67), the dormant period and the second peak will disappear, the Ca^{2+} concentration in solution will increase and the curve of heat evolution will show only an initial peak whose height increases as the ratio decreases [8]. When mixed with alite, amorphous silica having the same specific surface area of about 20 m^2/g behaves like Aerosil [9]. The slight differences observed can depend on the composition of the tricalcium silicate and the different C_3S/S ratios used.

The reduction in the induction period has been observed in blends with silica fume having a specific surface area in the range 50–380 m^2/g, but not when silica fume had had a lower surface area at about 19 m^2/g [9, 10]. Figure 4.5 shows that the height of the second peak initially increases as fineness increases, but then decreases.

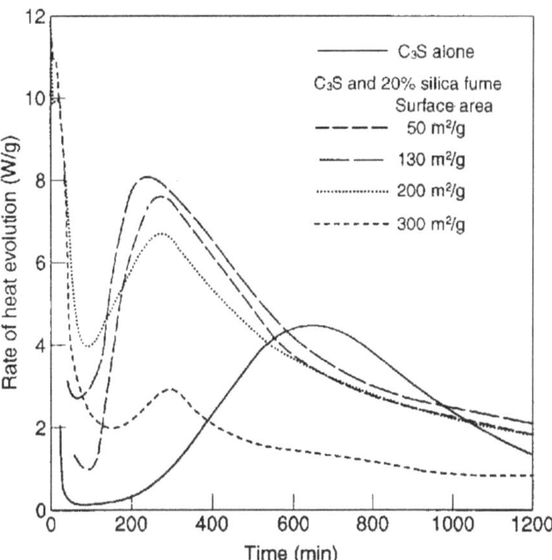

Fig. 4.5 Calorimetric *curves* from the hydration of C_3S with 20 % by weight of amorphous silica of various surface areas [9]

Effect of Silica Fume on the Properties of Fresh Concrete

Workability

Addition of very fine particles of silica fume to the concrete mixtures increases the cohesiveness of the mixture and makes it slightly stiffer. Result is the lower slump in concrete mixtures containing silica fume. In order to maintain the workability, it is recommended to use water reducing admixtures in silica fume concrete mixtures.

It is worth to note that slump test does not seem to be an appropriate method for the measurement of the workability of silica fume concrete mixtures. Due to the cohesiveness of silica fume concrete mixture, it will show lower slump than the similar normal concrete mixture. When energy such as vibration is applied to the fresh concrete, the spherical shape silica fume particles behave as lubricant and give higher mobility to the silica fume concrete than that of the normal concrete at similar slump. The difference between the two concretes can be measured by two point test method.

The water demand for concretes containing silica fume for constant workability varies by the type and fineness of the silica fume particles. The higher water demand for silica fume concrete mixtures can be reduced by decreasing the other fines in the mixture.

Segregation and Bleeding

The addition of silica fume to the fresh concrete has a stabilizing effect on its rheology. This is attributed to the very fine particles of silica fume which increase the cohesiveness and tend to reduce segregation and bleeding of the concrete mixture. Low segregation of concrete containing silica fume makes the material highly attractive and useful for pumping and shotcreting operations. The use of silica fume in concrete allows greater thickness of concrete layer and significant reduction in rebound especially for overhead shotcreting.

Very fine particles of silica fume in concrete mixtures also reduce the size of flow channels and causes segmentation in the bleed water routes. That is the reason why silica fume concretes usually have little or no bleeding. This concrete needs better moist curing to avoid plastic shrinkage particularly in hot environments. Final finishing of this concrete should be started earlier than the normal concrete due to this low bleeding.

Setting Time

Investigations show that the addition of silica fume in small amount and up to 10 % in normal concrete mixtures either has no effect or only slightly alters the setting time. Pistilli [11] reported that the use of 24 kg/m^3 silica fume in a concrete mixture containing 237 kg/m^3 ordinary Portland cement increased the initial and the final times of set by 26 and 29 min respectively.

Effects of Silica Fume on the Mechanical Properties of Hardened Concrete

Strength of Mortar and Concrete

In many investigations on the compressive strength of mortars and concretes containing silica fume, there has been a significant increase when compared with plain mortars and concretes. Several factors such as water cement ratio, cement and aggregate properties and contents, mixture design, the use of admixture and curing affect the amount of strength improvement in silica fume mixtures. Silica fume concretes follow the same behavior as normal concretes in strength properties by varying water cement ratio. The difference is the higher strength by addition of silica fume as seen in Fig. 4.6.

The compressive strength of mortars with different cements and incorporating 10 % silica fume was about 30–50 % higher than that of plain cements at 28 days [12]. The strength of silica fume mortars depends on water to binder ratio of the

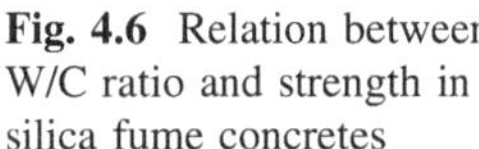

Fig. 4.6 Relation between W/C ratio and strength in silica fume concretes

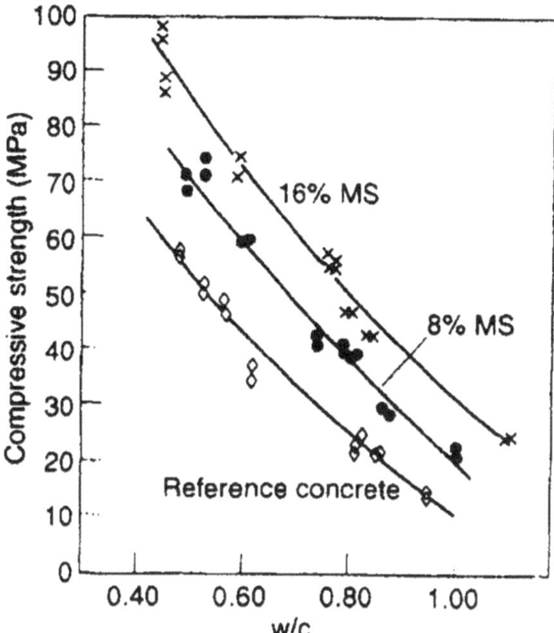

mixture. At a water binder ratio of 0.48 the strength increased at up to 15 % replacement and then decreased slowly up to 25 % replacement [13]. The compressive strength of mortars containing 10 % silica fume decreases with decreasing fineness of the parent Portland cement, but it is always higher than the strength of the plain Portland cement up to 28 days [12].

In concrete mixtures containing silica fume, the early strength (up to 3 days) may be reduced by various amounts depending on the proportion of silica fume used, especially when water to binder ratio is higher than 0.5. In an investigation three series of silica fume concretes with and without air entraining agent were used. The first contained 284 kg/m^3 of normal Portland cement, replacement of 0, 5, 10 and 15 % cement by weight with silica fume and 0.6 water to binder ratio. The second contained 340 kg/m^3 portland cement, similar replacement levels of silica fume and a water to binder ratio of 0.5. In the third series, cement content increased to 431 kg/m^3 and water binder ratio was decreased to 0.4. Concrete mixtures with air entraining agents contained 5–7 % air. All mixtures were designed to have a constant slump of 75–100 mm by adding superplasticizer at various amounts. Figure 4.7 shows the results obtained in this investigation [14].

The following conclusions were drawn from this investigation:

- Concretes containing silica fume and with water/binder ratio of 0.5 and 0.6 did not show any significant change in compressive strength at 3 days; however concretes with 0.4 water binder ratio showed an increase in compressive strength with increasing the amount of silica fume from 5 to 15 % cement replacement.

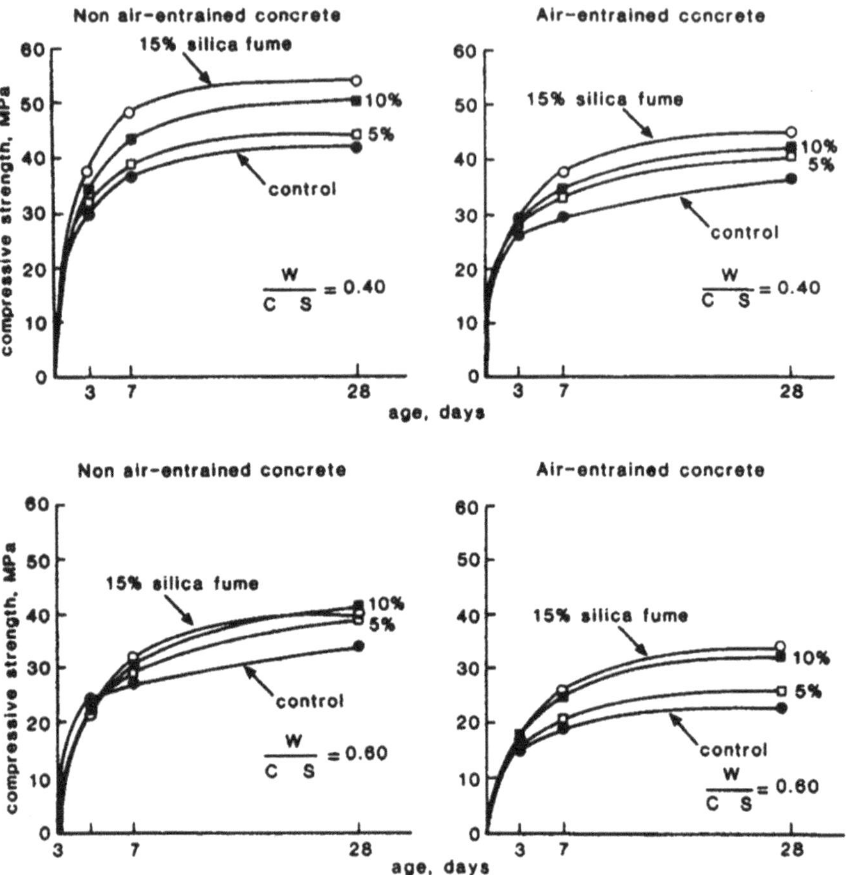

Fig. 4.7 Typical compressive strength of silica fume concretes with and without air entraining agent [14]

- Regardless of water/binder ratio in the silica fume concrete mixtures, compressive strength at 7 and 28 days were increased, and the increase was proportional to the level of replacement of silica fume with cement.
- Reduction in the compressive strength of concretes with and without silica fume was observed in all air entrained concrete mixtures. The strength loss was about 5 % for each percentage of air entrained.

In another investigation, Loland and Hustad [15] found that as a substitute for normal Portland cement in concrete, the 28 and 90 days compressive strength ability of silica fume is 2–3 times that of an equivalent mass of the Portland cement replaced, provided the addition of silica fume is combined with the use of plasticizing admixture in the mixture. In order to quantify the effect of silica fume on the compressive strength of concrete, a cement equivalent factor (K factor) is proposed, where K is equal to the mass of Portland cement necessary to be added to a concrete mixture to produce the same strength increase as one unit mass of silica fume. The relationship between water/binder ratio and the compressive strength of concretes containing silica fume was similar to that of plain concretes provided the amount of silica fume is multiplied by the cement equivalent factor.

Due to the very fine particles of silica fume, its addition to the concrete mixtures increases the water demand and hence decreases the compressive strength. This problem can be overcome by the addition of water reducing admixture to achieve silica fume concretes with even lower water to binder ratio than concrete mixtures [16].

It is worth to note that the silica fume concrete is very sensitive to the curing regime. Moist curing period is very important for silica fume concretes same as other pozzolanic materials. Early drying in the silica fume concrete may result in significant reduction in its final strength.

Investigations on the effect of curing on the strength properties of silica fume concretes showed a gradual loss of compressive strength between 90 days and 2 years in 10 % silica fume-containing concrete cured in air, while no strength loss observed in moist cured concrete specimens [17, 18]. It seems that air curing adversely affects the long-term compressive strength of both control and silica fume concretes, but it is more significant in the silica fume concrete. This phenomenon has been attributed to self stresses and probable micro cracks caused by drying of concrete [18, 19].

Very high strength concrete can be produced by proper mix design of silica fume. In recent years mixtures with compressive strength up to 120 MPa have been used in the construction of tall buildings in the world.

The flexural and tensile strengths of concretes have also increased by incorporating silica fume in the mixtures. The relationships between flexural, tensile and compressive strengths in silica fume concrete are similar to that of normal Portland cement concrete.

Modulus of Elasticity

In many investigations on the modulus of elasticity of mortars and concretes containing silica fume, there has been a significant increase when compared with plain mortars and concretes.

Ramezanianpour et al. investigated the modulus of elasticity of concretes containing silica fume [20]. The results of secant modulus of elasticity of concrete specimens containing different levels of silica fume (6–15 %), which are measured in creep tests, are shown in Table 4.2. In fact, the cylindrical specimens of 80 by 270 mm were loaded at the ages of 7 and 28 days. Because secant modulus is related to the level of applied stress and also loading rate, all the specimens of this research carried a stress of 10 MPa and the time taken to apply it was about 10 min.

Table 4.2 Secant modulus of elasticity of concretes containing silica fume (Ramezanianpour et al.) [20]

Concrete Id and age		Compressive strength (MPa)	Measured modulus (GPa)	Predicted modulus by the ACI (GPa)
OPC	7 days	46	28.81	30.24
	28 days	58	34.35	33.96
SF6	7 days	50.5	31	31.69
	28 days	65	35.5	35.95
SF8	7 days	52	31.24	32.15
	28 days	68	37.25	36.77
SF10	7 days	52	31.1	32.15
	28 days	67.5	37	36.63
SF15	7 days	53	31.5	32.46
	28 days	70	38.11	37.31

The stress/strength ratio of them was between 0.14 and 0.22. As shown in Table 4.2, increasing the silica fume replacement level increases the secant modulus of concrete. It is worth noting that sealing the specimens had no influence on their elastic modulus.

Effect on Volume Changes of Concrete

Shrinkage

Shrinkage of concrete mixtures incorporating various amounts of silica fume have been investigated by a number of researchers [21]. ACI 209R-92 [22] and also CEB-FIP 1990 [23] committees have also presented prediction methods for shrinkage. Ramezanianpour et al. investigated the shrinkage of silica fume concretes [21]. Figures 4.8 and 4.9 present the results of drying shrinkage of concretes containing 6–10 % of silica fume and compare it with the two codes. In fact, Figs. 4.8 and 4.9 show that at early ages both of those committees underestimated the shrinkage of drying specimens; nevertheless, at later ages The CEB and ACI underestimated and overestimated total shrinkage respectively.

Also this figure shows that silica fume did not have considerable influence on drying specimens (total shrinkage). The average amount of total shrinkage after 587 days of drying for the 80 by 270 mm specimens were 524 microstrain. It is worth noting that silica fume considerably affect the shrinkage of sealed specimens. It is clear that the general effect of silica fume inclusion is to increase autogenous shrinkage. This is in agreement with the results of other researchers [11–15]. In other words, as the silica fume replacement level increased the autogenous shrinkage of concrete increased.

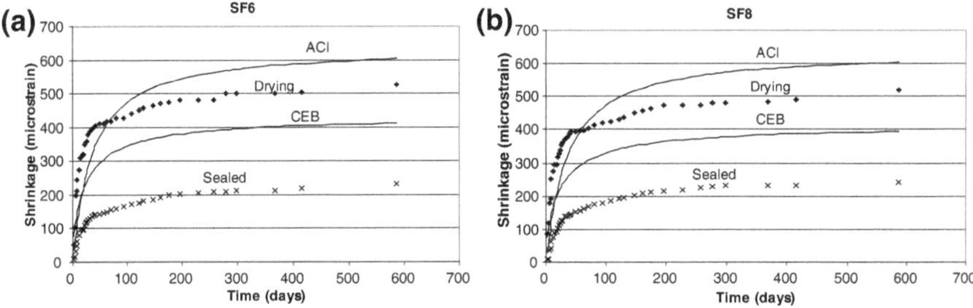

Fig. 4.8 **a** Shrinkage Of 80 by 270 mm, **b** Shrinkage Of 80 by 270 mm

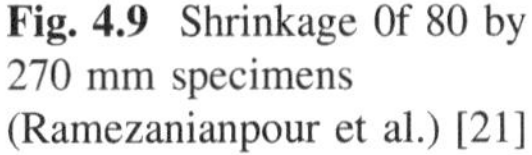

Fig. 4.9 Shrinkage Of 80 by 270 mm specimens (Ramezanianpour et al.) [21]

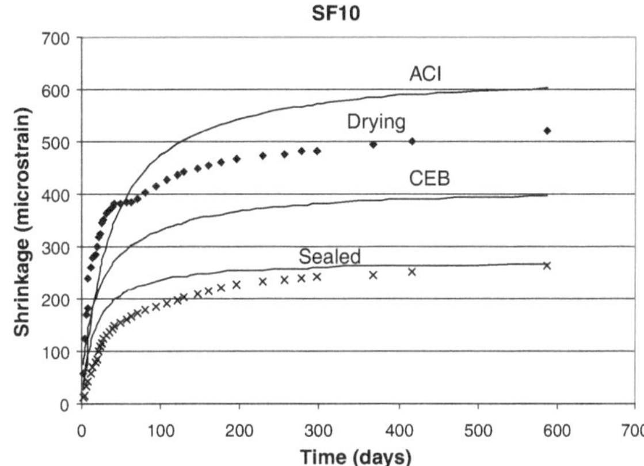

Table 4.3 shows total shrinkage, autogenous shrinkage and also the difference between them, which is called drying shrinkage, for the specimens of 80 by 270 mm after 587 days of drying. As it can be seen, there is significant increase in autogenous shrinkage at high levels of silica fume. In fact, inclusion of 10 and 15 % silica fume increases the autogenous shrinkage of concrete by 33 and 50 %, respectively. The effect of silica fume on autogenous shrinkage could be explained by its influence on the pore structure and pore size distribution of concrete as well as its pozzolanic reaction. According to Sellevold [16] the inclusion of silica fume at high replacement levels significantly increases the autogenous shrinkage of concrete due to the refinement of pore size distribution that leads to a further

Table 4.3 Values of total, autogenous and drying shrinkage of 80 by 270 mm specimens (microstrain)

Kind of shrinkage	Concrete mixes				
	OPC	SF6	SF8	SF10	SF15
Total	532	528	523	523	512
Autogenous	198	231	242	264	297
Drying	334	297	281	259	215

increase in capillary tension. Previous experimental results [17] on the pore structure of mortars using Mercury Porosimetry technique showed that as silica fume content increased, the pore size distribution was shifted toward a finer distribution, the average pore size reduced and the porosity decreased. It was found that the addition of silica fume and also the dosage of silica fume greatly influence the self-desiccation and autogenous shrinkage of cement paste. In addition, the pozzolanic reaction of silica fume, which was found to be less sensitive to self-desiccation, also leads to an increase in autogenous shrinkage.

The high amount of autogenous shrinkage of the high-strength concrete mixes investigated here is very important; because autogenous shrinkage of concrete occurs as a result of chemical reactions during the hydration of cementitious materials and it does not have any relationships with moisture movement from concrete to the atmosphere.

It should be mentioned that Loukili et al. believe this shrinkage in very high-strength concrete stops after 10 days.

This investigation shows one of the ways to lower the autogenous shrinkage and also the cracking probability of high-strength concrete is to add not more than 10 % silica fume to the mix.

Creep

Investigations on the creep of concretes containing silica fume show that at similar conditions and load applied, The creep of silica fume concrete was lower than that for normal concrete [21, 23].

Figures 4.10 and 4.11 show the results of specific creep, i.e. creep per unit stress, for cylindrical specimens of 80 by 270 mm loaded at the age 28 days. Because specimens were subjected to a sustained compressive stress of 10 MPa during all the creep tests, the real amounts of creep measured in the laboratory were 10 times more than the amounts shown in these graphs. It is worth mentioning that the above stress even in the weakest concrete of this research was lower than 30 % of its compressive strength. It means that the relationship between creep and load in all the specimens was linear. The creep of sealed specimens, which had coated by aluminum waterproofing tape, can be compared to the creep of drying specimens. Because there is no moisture movement between sealed specimens and the atmosphere, their creep is called basic creep. As it can be observed, the difference between this creep and the creep of unsealed specimens or total creep, which is named drying creep, is not considerable in control concrete and also it is almost zero in specimens containing silica fume. It means there is no interaction between creep and shrinkage and also factors affecting the rate of drying, which are specimen size and the relative humidity of atmosphere, had no influence on the creep of high-strength concrete specimens investigated here.

ACI 209R-92 [22] and also CEB-FIP 1990 [23] committees have also presented prediction methods for creep. Their predictions have been compared with the

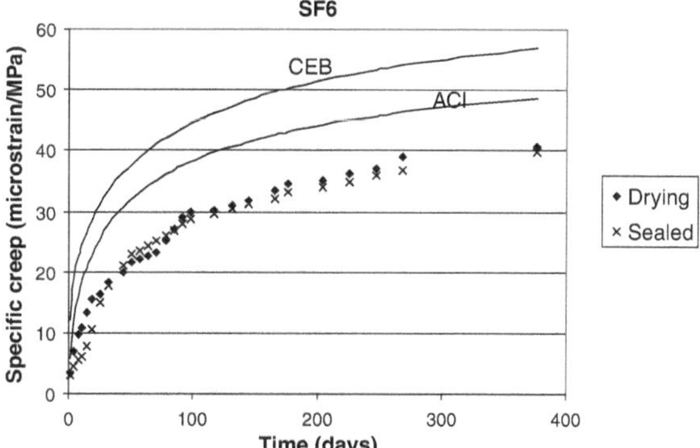

Fig. 4.10 **a** Specific creep of specimens loaded at the age of 28 days (Ramezanianpour et al.) [21]

results of the investigation carried out by Ramezanianpour et al. [21] in Figs. 4.10 and 4.11. In fact, Figs. 4.10 and 4.11 show that both of those committees overestimate the creep of investigated specimens.

It can also be seen that silica fume had considerable influence on the total creep measured in the laboratory. These values for specimens loaded at the ages of 7 and 28 days after 400 days under load are shown in Table 4.4. It is clear that as the proportion of silica fume increased the total creep of concrete decreased. This finding is in agreement with the results of other investigators. The reason for the above decreasing may be because of higher compressive strength of concrete mixtures containing higher levels of silica fume at the age of loading; because some researchers believe that higher compressive strength leads to lower creep.

Microstructure, Porosity and Permeability

Microstructure of concrete will be affected by the incorporation of silica fume in the mixture. The major influence is the refinement of the pore structure of cement paste. The total porosity may not alter by the addition of silica fume but the large

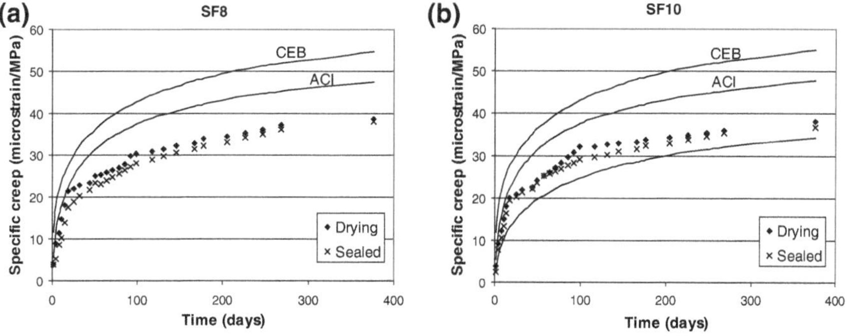

Fig. 4.11 (**a**, **b**) Specific creep of specimens loaded at the age of 28 days [21]

Table 4.4 Creep of 80 by 270 mm specimens (microstrain)

Age of loading-days	Concrete mixes				
	OPC	SF6	SF8	SF10	SF15
7	595	510	447	459	417
28	413	407	386	381	328

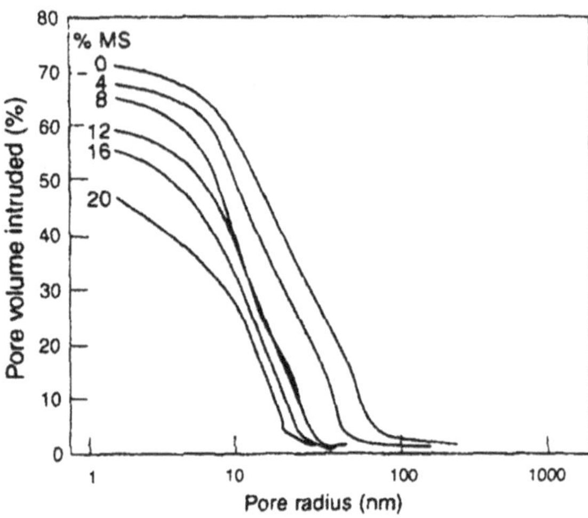

Fig. 4.12 Effect of silica fume on total porosity [24]

pores are divided into smaller pores and hence changing the microstructure of the cement paste [24]. The effect of various amounts of silica fume on total porosity is shown in Fig. 4.12.

Silica fume addition to the concrete mixture also improves the interfacial transition (ITZ) zone between cement paste and aggregates. The content of calcium hydroxide is reduced in the ITZ due to the pozzolanic reaction of silica fume. This will reduce the porosity and permeability of concretes at the interface region and enhances the bonding of paste and aggregates [25, 26].

Permeability of concrete is also affected by the incorporation of silica fume in concrete mixtures. It is even more important than the effect on compressive strength. In a research work, a concrete mixture containing 100 kg/m^3 portland cement, 20 % silica fume, and a superplasticizer showed approximately similar permeability to the plain concrete containing 250 kg/m^{3}portland cement.

The effect of silica fume on water penetration of concretes is clearly seen in Fig. 4.13. The water penetration depth in silica fume concrete is about 6 times lower than that of normal concrete [27].

Water permeability of concretes containing silica fume is also seen in Fig. 4.14 on core samples [28].

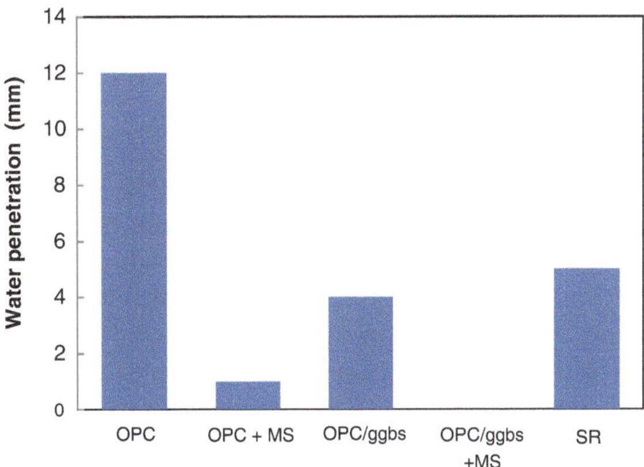

Fig. 4.13 Water penetration of concrete (DIN 1048) containing silica fume [27]

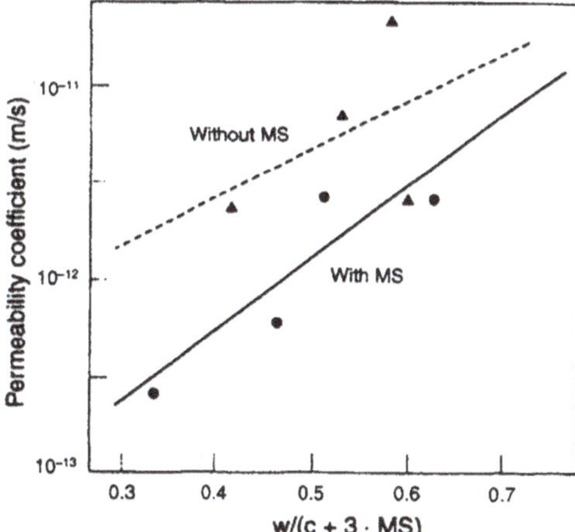

Fig. 4.14 Water permeability of concretes containing silica fume with approximately equal compressive strength

Effect of Silica Fume on Durability of Concrete

Durability of concrete is usually enhanced by the addition of silica fume. Silica fume can improve the chemical and physical properties of concrete. The reduction of sodium, potassium, and calcium hydroxides in concretes containing silica fume is observed. A refined pore structure in the hardened concrete is also attributed to the physical effect of silica fume when added to the concrete mixture.

Effect of Silica Fume on Carbonation of Mortars and Concretes

Similar to other pozzolanic materials, silica fume consumes the calcium hydroxide of the cement paste. This may increase the risk of carbonation in the silica fume mortar and concrete. On the other hand silica fume addition will reduce the permeability of concrete which may result in lower carbonation. These two contradictory effects of silica fume on concrete carbonation have been reported by many researches. However, the results of these two effects depends on other factors such as silica fume content, water to binder ratio and curing regime of concretes which may vary the calcium hydroxide content and permeability of concrete. In an investigation of the effect of silica fume on the carbonation, the use of 20 % silica fume resulted in the slight reduction of carbonation when compared with control mixture [29]. The rate of carbonation can be related to the water cement ratio of the concrete rather than the addition of silica fume. This is shown in Fig. 4.15 where silica fume was used up to 15 % in concretes with different water to cement ratio [30].

Some researchers tried to relate the depth of carbonation of silica fume concretes to their compressive strengths. They concluded that for a given strength of concrete below 40 MPa, carbonation of concrete is higher in silica fume concrete than in normal concretes, corresponding to the increase in water cement ratio. Concretes having the strength above 40 MPa, corresponding to a low water cement ratio, show little or no change in the carbonation rate [29, 31].

In an accelerated carbonation tests of different concrete mixtures, no significant change was observed in concretes with and without silica fume if the mixture designs were similar [32]. Concretes containing silica fume are usually designed with low water cement ratio to enhance durability. If such concretes are cured in a proper way, there should be no concern of the carbonation during the service life.

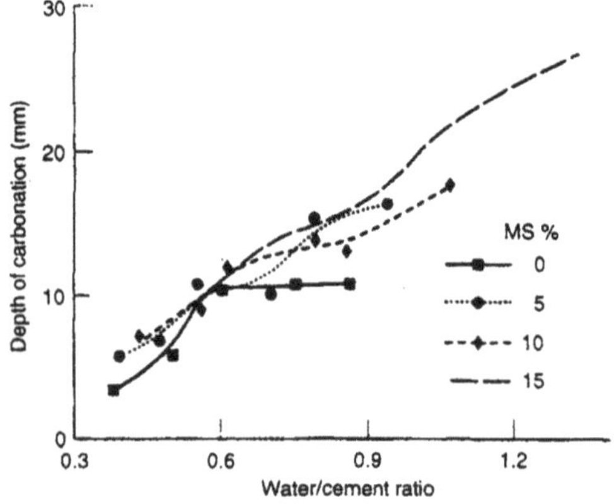

Fig. 4.15 Depth of carbonation for different water to cement ratio [30]

Effect of Chloride Ions on Durability of Concretes Containing Silica Fume

Chloride ions in the aggregate or mixing water and also from de-icing salts and sea water destroy the passive layer protecting the reinforcement from corrosion in reinforced concrete elements. The alkalinity of concrete in the pore solution is an important factor controlling the passivity of the reinforcement. Permeability of concrete is also very important in controlling the penetration of chloride ions. Many researchers have studied the combined effect of permeability and the reduction in the alkalinity of pore water on the corrosion of reinforced concrete elements.

The use of silica fume in concrete slightly reduces the alkalinity of pore water and will cause a reduction in the threshold value of chloride necessary for the initiation of corrosion of steel bars. On the contrary, the use of silica fume reduces the permeability and chloride diffusion of concrete [33].

The effect of silica fume on the chloride diffusion of cement pastes at different water cement ratios is shown in Fig. 4.16. The use of silica fume has significantly reduced the chloride ions diffusion of the cement paste [34].

The use of different silica fumes at various percentages of replacements with Portland cement show that the depth of chloride penetration decreases by increasing the amount of silica fume. Figure 4.17 shows that penetration of silica fume concretes are always lower than that for normal Portland cement concretes [35].

Several investigations show the reduction of chloride penetration of silica fume mortars and concretes in rapid chloride penetration test (RCPT). Figure 4.18 present the RCPT values at 28 days for concretes containing 7 % silica fume

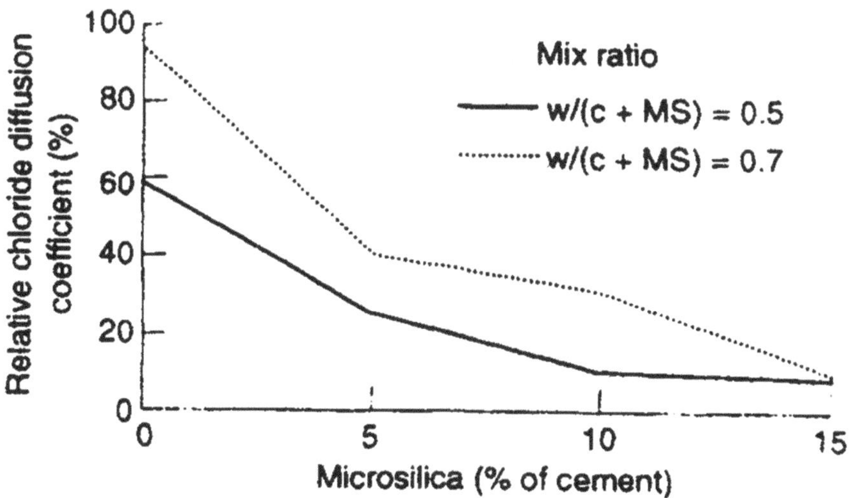

Fig. 4.16 Effect of silica fume on chloride diffusion [34]

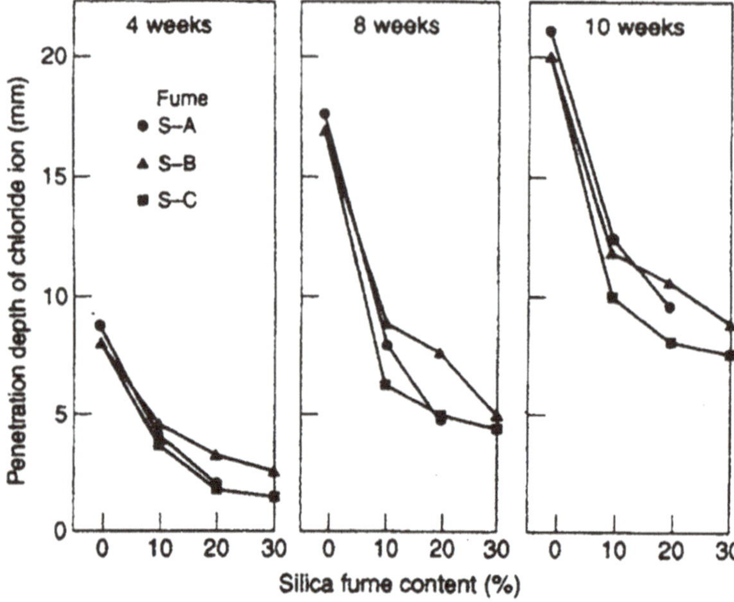

Fig. 4.17 Penetration depth of chloride ions in silica fume concretes [35]

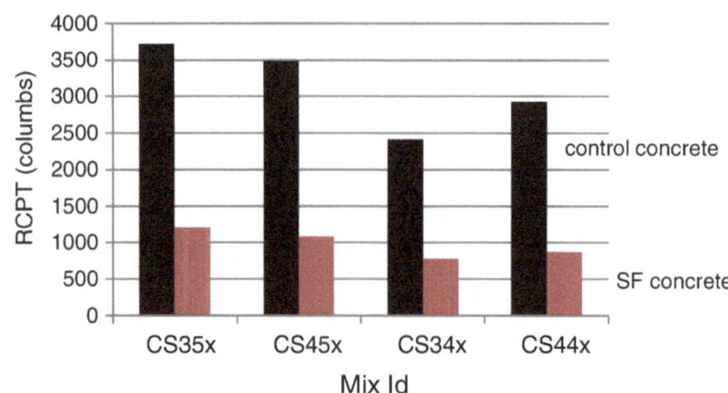

Fig. 4.18 Influence of silica fume on rapid chloride penetration test results (Ramezanianpour [36])

replacement with cement and various cement contents. The charge passed through specimens has reduced from 3500 columbs in control concretes to about 1000 columbs for silica fume concretes [36].

Sulfate Resistance of Mortars and Concretes Containing Silica Fume

The performance of silica fume mortars and concretes in sulfate solutions was better than the normal ones. This resistance depends on the replacement level of silica fume by Portland cement and the type of sulfate. It has been reported that silica fume increases resistance of Portland cement to sodium sulfate attacks especially at higher level of replacements [37]. The replacement of silica fume up to 7 % was not satisfactory but 15 % replacement for normal Portland cement and sulfate resisting Portland cement improved the resistance of mortars even after 4 years of immersion in sulfate solution [38]. In an investigation in the field exposure for more than 20 years in Oslo, concrete made from ordinary Portland cement and with 15 % cement replacement by silica fume had as much as resistance to attack as concrete made from a sulfate-resisting Portland cement when exposed to alum shale water (5 g/l SO_3, and 2.8 pH) [39]. In another investigation of the effect of 5 % sodium sulfate solution on concrete containing 10 % silica fume, expansion and sulfate consumption of concrete were significantly reduced [40]. Better performance was observed in 10 % sodium sulfate solution when replacement of silica fume reached to 20 % of the cement weight [35].

Performance of silica fume mortars and concretes in magnesium sulfate is rather different from the sulfate solution. In a solution of 3 % magnesium sulfate ($MgSO_4$), the resistance of mortars containing 10 % silica fume as cement replacement was improved up to 3 years of testing [41]. However, in a 4.2 % magnesium sulfate solution for two years, the strength of mortars containing 15 % replacement of silica fume for both normal Portland cement and sulfate-resisting Portland cement was significantly reduced [38, 42]. Similar poor performance of silica fume additions have also been observed in 6 mm thick paste beams exposed to 7.67 % magnesium sulfate solution. As seen in Fig. 4.19 there was not much difference between the extent of damage in Type I Portland cement with 10.52 % C_3A and type v Portland cement having 2.26 % C_3A but, in both cases, the replacement of 15 % silica fume strongly decreased strength and increased weight loss [43].

In a mortars containing silica fume and exposed to a magnesium solution, the surface layer showed a lower secondary mineralization than that of plain cement mortar. This is attributed to the formation of a thinner surface double layer of brucite and gypsum and less precipitated gypsum. Despite the lower secondary mineralization, the overall degree of surface deterioration in mortars containing silica fume was higher than that of plain Portland cement mortars [44].

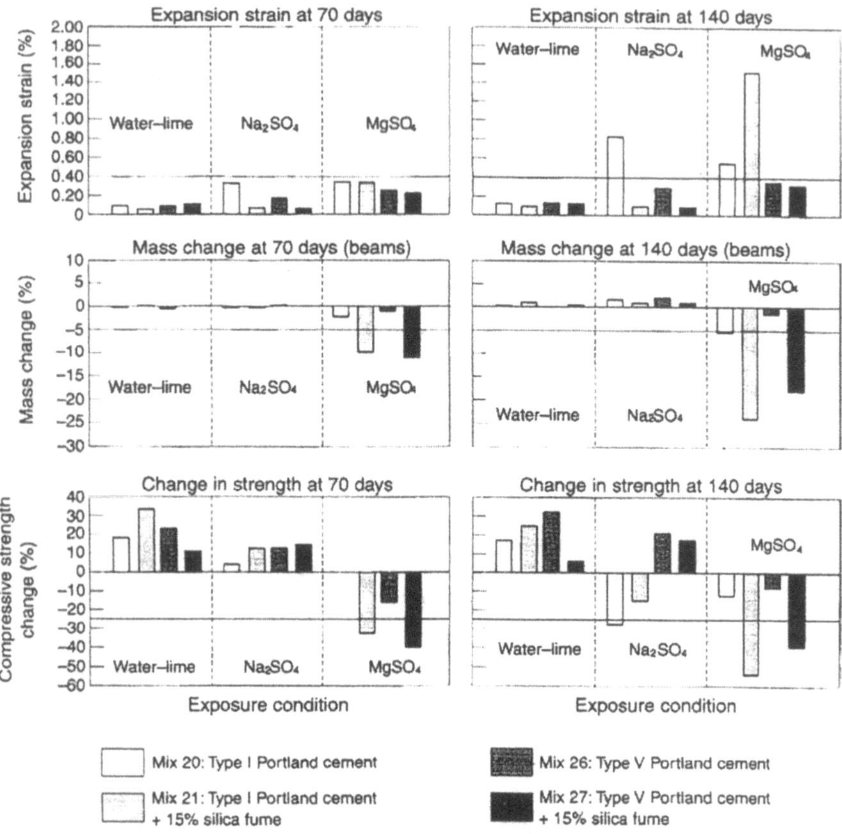

Fig. 4.19 Variations in length, weight and compressive strength of mortars exposed in 10 % Na_2SO_4 and 7.7 % $MgSO_4$ solutions [43]

Effect of Silica Fume on Suppressing the Alkali Aggregate Reaction

Very fine particles of silica fume react with the alkalis in cement paste to form alkali silicates. This reduces the available alkalis in the pore solution and would prevent attack on reactive siliceous aggregates. Mixtures containing silica fume are also less permeable and prevent the penetration of water necessary for the alkali aggregate reaction. Therefore it is expected to have less expansion in concrete mixtures due to alkali aggregate reaction when silica fume is used. The amount of silica fume necessary for prevention of alkali aggregate reaction is usually lower than that required by other pozzolanic materials due to its high reactivity.

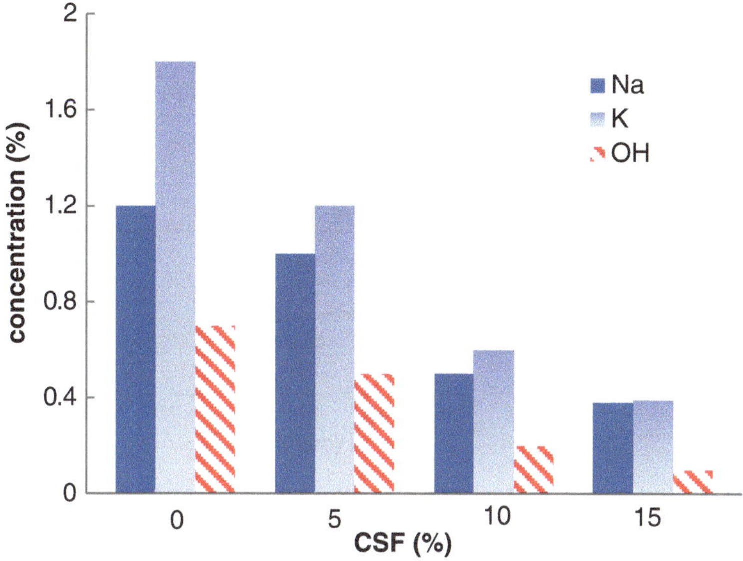

Fig. 4.20 The effect of 0, 5, 10 and 15 % silica fume on concentration of ions in pore solution at 6 months [45]

Reduction in alkalinity of cement paste containing silica fume, which is mainly given by sodium and potassium hydroxide, is exhibited in Fig. 4.20 [45]. Hydroxyl ion (OH) is also reduced by addition of silica fume. In another investigation the OH^- concentration, which was 0.5 N in plain paste solution after 76 days of hydration, reduced to 0.3, 0.15 and 0.075 when 5, 10 and 20 % silica fume was replaced with cement [46]. When silica fume is used in the cement paste, the concentration of Na^+ and K^+ reaches a maximum after about 4 days and then decreases by time. This is clearly seen in Figs. 4.20, 4.21 [45, 46].

In several research works silica fume has been used to mitigate the alkali aggregate reactions. In almost all ASR accelerated tests expansion of mortar bars was reduced when silica fume was used. The higher the silica fume replacement the lower expansion was observed. It is worth mentioning that silica fume should be dispersed thoroughly in the mixture otherwise it may even cause the ASR problem. However in long term tests the expansion reduction was not significant as observed in accelerated tests. It is concluded that for every reactive aggregates different percentages of silica fume should be tested at longer time for possible mitigation of alkali aggregate reaction.

The effect of silica fume on suppressing the ASR can be clearly seen in Fig. 4.22 [47]. One of the earliest usage of silica fume in controlling the alkali silica reaction was in Iceland. In most concretes when potentially reactive aggregates were used, 7 % silica fume was recommended [48].

Ramezanianpour et al. examined the effect of various contents of silica fume on suppressing the alkali aggregate reaction of four reactive aggregates. Results are

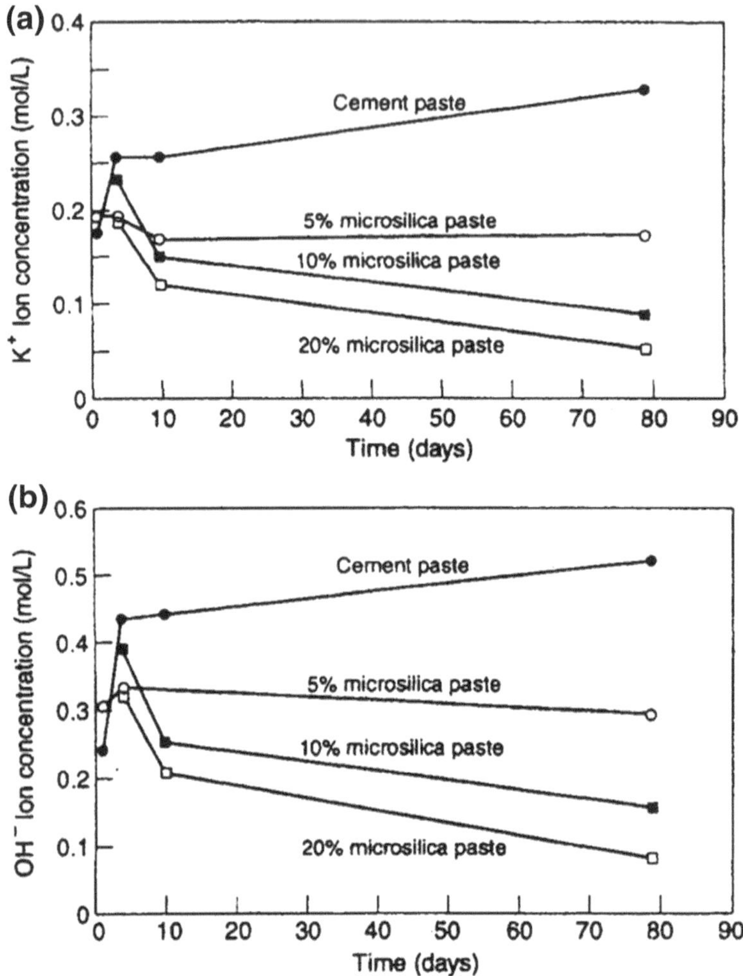

Fig. 4.21 K^+ and OH^- concentrations in pore solution at different times for cement paste containing silica fume (W/C = 0.5)

shown in Table 4.5 The addition of 5 % silica fume seems to not sufficient in controlling alkali aggregate reaction of some of the aggregates at 14 day accelerated mortar bar test. The replacement level above 7 % of silica fume has significantly reduced the expansion of concrete prisms after one year and was recommend for the controlling of alkali aggregate reaction of reactive aggregates [49].

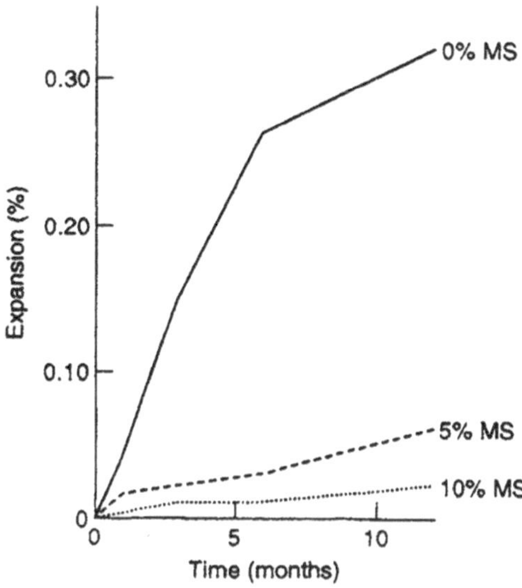

Fig. 4.22 Influence of silica fume on alkali silica reaction mitigation [47]

Table 4.5 Expansion of mortar bars and concrete prisms at 14 days and one-year (Ramezanianpour et al.) [49]

SF replacement	Expansion (%)									
	0 %		5 %		7 %		10 %		15 %	
Aggregate	14 days	One year	14 days	One year	14 days	One year	14 days	One year	14 days	One year
GW	0.232	0.050	0.085	–	0.045	0.019	0.034	0.020	0.048	–
AZ	0.265	0.024	0.125	–	0.069	0.021	0.057	0.025	0.082	–
TA	0.29	0.044	0.082	–	0.063	0.017	0.02	0.021	0.02	–
ME	0.6	0.026	0.518	–	0.231	0.027	0.07	0.018	0.038	–

Freezing and Thawing of Concretes Containing Silica Fume

The published data on the freezing and thawing of mortars and concretes containing silica fume are contradictory [50–53]. In a research work mortar prisms containing up to 25 % silica fume at different water cement ratios with and without plasticizing admixture and air entraining agent were investigated [54]. After 50 freeze–Thaw cycles, they concluded that the use of air entraining and plasticizing admixtures did not seem to give any significant additional improvement in frost resistance of mortars containing silica fume. A good resistance to frost action observed in silica fume mixtures with and without admixtures. This is attributed to the pore refinement of silica fume concrete mixtures and considerable increase in the volume of small pores.

Some authors reported that silica fume reduces the frost resistance of non air entrained concretes but improves that of air entrained one [34]. They concluded that silica fume cements do not worsen frost resistance, provided that concretes have a suitable amount of air entraining agent and similar strengths. It seems that the combination of silica fume and air entraining admixture is a good option. Silica fume lowers the permeability and maintains a better stability of the air in fresh concrete with a uniform bubble spacing of the air providing the best frost resistance. Salt scaling resistance of a concrete containing silica fume and air entrainment in a long term test was similar to the normal concrete mixture without silica fume [55]. It is worth noticing that replacement of silica fume for Portland cement at higher percentages (20–30 %) may cause a lower frost resistance even in the air entrained mixtures [56–58]. This may be attributed to the variations in the specific surface area and the changes in the air voids spacing of the concrete mixtures.

Application of Silica Fume in Mortars and Concretes

Application of silica fume in mortars and concretes has been started in the late 1970s. It was first used for the production of high strength concrete. A considerable amount of silica fume was successfully used in structural concretes by ready mix concrete industry. High strength concretes in the range of 70–80 MPa can be easily achieved by incorporating silica fume. Ultra high strength concretes in the range of 120–180 MPa have been produced with a special mix design using special aggregates, silica fume and sufficient high range water reducing agent. Construction of New Tjorn Bridge in Sweden in 1981 is believed to be the first industrial use of silica fume to achieve a compressive strength of about 50 Mpa. Construction of a 65 storey high rise building in 311 South Waker Drive in Chicago in 1990 is another example of the incorporation of silica fume in concrete for obtaining strength up to 83 MPa. In the Petronas Tower in kuala Lumpur, Malaysia, the tallest concrete building at the time of completion in 1997, silica fume concrete with compressive strength of 80 MPa was used for the columns at lowest levels.

Another application of silica fume is in the projects where mortar or concrete is shotcreted. Silica fume increases the cohesiveness of fresh concrete and improves its bonding. Therefore it is a very useful material for both dry and wet shotcreting to reduce rebound. The high cohesiveness and reactivity of the silica fume shotcrete also reduces the amount of accelerators or eliminate it for the early setting of concrete.

As mentioned before, concretes containing silica fume have low permeability and are durable in the severe environments. Due to this advantage of silica fume in concrete mixtures, it is usually recommended to be applied in concretes where enhanced durability is required. This is usually referred to high performance

Fig. 4.23 Bushehr port complex under construction with silica fume concrete

concrete (HPC). The first known project on the use of silica fume in concrete where durability to erosion was important was for the repair of the stilling basin at the kinzu Dam in the USA [59]. The number of projects in severe marine environments which high performance concrete containing silica fume was designed and used is still increasing. Tsing Ma Bridge in Hong Kong, Stolma Bridge in Norway, and Storebelt Bridge in Denmark are just some of the few examples of the large projects.

In the Middle East and Persian Gulf Region, the use of silica fume is one of the recommended options for preventing the sulfate and chloride attack of concrete structures. High performance concrete containing silica fume has been used in the construction of Baynunah Tower in Abu Dhabi, the offshore Chicago Beach Hotel, Dubai Airport extension and several projects in Saudi Arabia. Silica fume has been recommended in the construction of concrete structures in the National Code of practice for concrete durability in the Persian Gulf and Omman Sea of Iran [60]. Silica fume concrete has been used in several projects in that region. Concrete structures in Assaloyeh gas refinery plant, Bandar Abbas port complex, Bandar Imam port complex, Bushehr port complex are among them. It has also been used in Noushahr Port complex in the north and a number of dams including Kharkheh and Sattarkhan Dams. Figure 4.23 shows a picture of a large project in Bushehr Port under construction with silica fume concrete.

References

1. P.C. Aitcin, P. Pinsonneault, D.M. Roy, Physical and chemical characterization of Condenced silica fume. Bull. Am. Cer. Soc. **63**, 1487–1491 (1984)
2. M.W. Grutzeck, S. Atkinson, D.M. Roy (1983) *Mechanism of Hydration of Condensed Silica Fume in Calcium Hydroxide Solutions*, in *Proceedings of the 1st International Conference on the Use of Fly Ash, Silica Fume, Slag and Other Mineral By-Products in Concrete*, Canada, 1983, ed. by V.M. Malhotra, vol. II (American concrete Institute Special Publication 79) pp. 643–664
3. American Society for Testing and Materials. *Electrical Indication of Concrete's Ability to Resist Chloride Ion Penetration* (ASTM, W. Conshohoken, 1996) C 1202
4. A.D. Buck, J.P. Burkes, *Characterization and Reactivity of Silica Fume*, in *Proceedings of the 3rd International Congress on Cement Microscopy*, Houston, 16–19 March 1981, pp. 279–285
5. M.-H. Zhang, E. Gjorv, Effect of silica fume on cement hydration in low porosity cement pastes. Cem. Concr. Res. **21**, 800–808 (1991)
6. M. Atkins, E.E. Lachowski, F.P. Glasser, Investigation of solid and aqueous chemistry of 10-year-old Portland cement pastes; with and without silica modifier. Adv. Cem. Res. **5**(19), 97–102 (1993)
7. H.N. Stein, J.M. Stevels, Influence of silica on the hydration of $3CaO.SoO_2$. J. Appl. Chem. **14**, 338–346 (1964)
8. W. Kurdowski, W. Nocun-Wczelik, The tricalcium silica hydration in the presence of active silica. Cem. Concr. Res. **13**, 341–348 (1983)
9. S.S. Beedle, G.W. Groves, S.A. Rodger, The effect of fine pozzolanic and other particles on hydration of C_3S. Adv. Cem. Res. **2**(5), 3–8 (1989)
10. C.M. Dobson, D.G.C. Goberdhan, J.D.F. Ramsay, S.A. Rodger, ^{29}Si MAS NMR study of the hydration of tricalcium silicate in the presence of finely divided silica. J. Mater. Sci. **23**,4108–4114 (1988)
11. M.F. Pistilli, R. Winteresteen, R. Cechner, The uniformity and influence of silica fume source on the properties of Portland cement concrete. Cem. Concr. Aggreg. **6**(2), 120–124 (1984)
12. A. Kumar, D.M. Roy, A study of silica fume modified cements of varied fineness. J. Am. Ceram. Soc. **67**, 61–64 (1984)
13. V. Yogendran, B.W. Langan, M.A. Ward, *Utilization of Silica Fume in High Strength Concrete*, in *Proceedings of the Symposium on Utilization of High Strength Concrete*, Stavanger, Norway, 1987, pp. 85–94
14. V.M. Malhotra, Mechanical properties and freezing and thawing resistance of non-air-entrained and air-entrained condensed silica fume concrete using ASTM test C666. Cem. Concr. Aggreg. (1986)
15. K.E. Loland, T. Hustad, Silica in Concrete-Mechanical Properties. Cement and Concrete Inst. At the Norwegian Institute of Technology, Trondheim, Norway, Report No, STF 65 A 81031, June 1981
16. E.J. Sellevold, F.F. Radjy, *Condensed Silica Fume (microsilica) in Concrete: Water Demand and Strength Development*, in *Proceeding of the 1st Mineral By-Products in Concrete*, Montebello, Canada, 1983, ed, by V. M. Malhotra, vol. II (American Concrete institute Special publication 79) pp. 677–694
17. G.G.Carette, V.M. Malhotra, P.C. Aitcin, *Preliminary Data on Long Term Development of Condensed Silica Fume Concrete*, in *Proceedings of the 3rd International Conference on Fly ash, Silica Fume, Slag, and Natural Pozzolans in Concrete*, Trondheim, 1989, ed. by V.M. Malhotra (American Concrete Institute Special Publication SP 114; Supplementary Papers) pp. 597–617
18. F. De Larrard, J.L. Bostivironnois, On the long-term strength losses of silica-fume high-strength concretes. Mag. Concr. Res. **43**(155), 109–119 (1991)

19. G.G. Carette, V.M. Malhotra, *Long-Term Strength Development of Silica Fume Concrete*, in *Proceedings of the 4th International Conference on Fly ash, Silica Fume, Slag, and Natural Pozzolans in Concrete*, Istanbul, 1992, ed. by, V.M. Malhotra vol. II (American Concrete Institute Special Publication 132) pp. 1017–1044
20. M. Mazloom, A.A. Ramezanianpour, J.J. Brooks, *Effect of silica fume on mechanical properties of high-strength concrete*, *Cement and Concrete Composites*, (Elsevier Science ltd, March 2004), ISI, ISSN 0958-9465
21. M. Mazloom, A.A. Ramezanianpour, Time dependent behaviour of concrete columns containing silica fume. Int. J. Civil Eng. **2**(1) (2004)
22. ACI Committee 209, Prediction of creep, shrinkage and temperature effects in concrete structures, ACI Manual of concrete practice, Part 1, 1997, 209R 1-92
23. CEB-FIP Model Code for Concrete Structures 1990, Evaluation of the time dependent behavior of concrete, Bulletin d'Information No. 199,Comite European
24. V. Yogendran, B.W. Langan, *Utilization of Silica Fume in High Strength Concrete* in *Proceedings of Utilization of High Strength Concrete*. Stavanger, 1987 (Tapir Publishers, Trondheim)
25. A. Goldman, A. Bentur, Bond effects in high strength silica fume concretes. ACI186- M39. ACI Mater. J. **86**(5) 1989
26. P. Monteiro, P.K. Mehta, *Improvement of the Aggregate Cement Paste Transition Zone by Grain Reinforcement of Hydration Products*, in *Proceedings of the 8th International Congress on the Chemistry of Cement*, Rio de Janeiro, 1986, vol. 3
27. P. Fidjestel, J. Frearson, *High-Performance Concrete Using Blended and Triple Blended Binders*, in *Proceedings of the ACI International Conference on High Performance Concrete*. Singapore, 1994, ed. by, V.M. Malhotra (American Concrete Institute, SP- 149)
28. M. Maage, Effect of Micro Silica on the Durability of Concrete Structures, FCB/SINTEF, Trondheim, Report STF65 A84019 (1984)
29. O. Vennesland, O.E. Gjorv, *Silica Concrete-Protection Against Corrosion of Embedded Steel*, 1983, vol. 11 (American Concrete Institute, Publication SP-79) pp. 719–729
30. R. Johansen, Silica in Concrete. Report section no. 6, Long Term Effects. SINTEF, Trondheim Report no. STF65 A81031, pp. 24 (1981)
31. O. Vennesland, Silica in Concrete, Report3. Corrosion Properties. FCB/SINTEF, Trondheim, SINTEF Report no. STF65 A81031 (1981)
32. A.A. Ramezanianpour, A. Tarighat, *Carbonation Models for Concrete Containing Silica Fume, Second International Conference on Engineering Materials*, August 2001, San Jose, USA
33. O. Gautefall, *Effect of Condensed Silica Fume on the Diffusion of Chloride Trough Hardened Cement Paste*, in *Proceedings of the 2nd CANMET/ACI International Conference on the Use of Fly Ash, Silica Fume, Slag and Natural Pozzolans in Concrete*. Madrid, 1986 (Paper SP) pp. 91–48
34. O. Geutefall, Vennesland. "Modifisert Portland cement", Delrapport 5. Elektrik Motstand of pH-nivà., SINTEF/FCB, Trondheim, Report no. STF65 A85042 (1985)
35. T. Yamato, M. Soeda, Y. Emoto, *Chemical Resistance of Concrete Containing Condensed Silica Fume*, in *Proceedings of the 3rd International Conference on Fly Ash, Silica Fume, Slag, and Natural Pozzolans in Concrete*. Trondheim, 1989, vol. 114, ACI, SP, pp. 897–913
36. A.A. Ramezanianpour et al., Practical evaluation of relationship between concrete resistivity, water penetration, rapid chloride penetration and compressive strength. Constr. Build. Mater. **25**(3), 2472–2479 (2011)
37. R.D. Hooton, Influence of silica fume replacement of cement on physical properties and resistance to sulfate attack, freezing and thawing, and alkali-silica reactivity. ACI Mater. J. **90**, 143–151 (1993)
38. P.W. Brown, An evaluation of the sulfate resistance of cements in a controlled environment. Cem. Concr. Res. **11**, 719–727 (1981)
39. O.M. Fiskaa, Betong i. "Alunskifer (in Norwegian)", Norges Festekniske Inst., Oslo, Report No. 101 (1983)

40. K.W. Nasser, S. Ghosh, *Durability Properties of High Strength Concrete Containing Silica Fume and Lignite fly Ash*, in *Proceedings of the 3rd International Conference on Durability of Concrete*, Nice, 1994, ed. by, V.M. Malhotra, vol. 145 (American Concrete Institute Special Publication) pp. 191–214
41. C.D. Lawrence, The influence of binder type on sulfate resistance. Cem. Concr. Res. **22**, 1047–1058 (1992)
42. J. Madej, *Corrosion Resistance of Normal and Silica Fume Modified Mortars Made From Different Type of Cements*, in *Proceedings of the 4th International Conference on Fly Ash, Silica Fume, Slag, and Natural Pozzolans in Concrete*. Istanbul, 1992, ed. by, V.M. Malhotra vol.II (American concrete institute special publication 132) pp. 2042–2058
43. M.D. Cohen, A. Bentur, Durability of Portland cement-silica fume pastes in magnesium sulfate and sodium sulfate solutions. Am. Concr. Inst. Mater. J. **85**(3), 148–157 (1988)
44. D. Bonen, A microstructural study of the effect product by magnesium sulphate on plain and silica fume-bearing Portland cements mortars. Cem. Concr. Res. **23**, 541–553 (1993)
45. B. Durand, J. Berard, R. Roux, J.A. Soles, Alkali-silica reaction: the relation between pore solution characteristics and expansion test results. Cem. Concr. Res. **20**, 419–428 (1990)
46. E. Koelliker, Some observations on the carbonation of solutions of $Ca(OH)_2$ and liquid phase of cement paste and the mechanism of carbonation of concrete. RILEM International symposium of carbonation of concrete, Fulmer, 1976
47. H. Asgeirsson, Silica fume in cement and silence for counteracting of alkali-silica reactions in Iceland. Cem. Concr. Res. **16**(3) (1986)
48. H. Asgeirsson, G. Gudmundsson, Pozzolanic activity of silica dust. Cem. Concr. Res. **9**, 249–252 (1979)
49. A.A. Ramezanianpour, A.M. Raiss Ghasemi, T. Parhizkar, *Influence of Silica Fume on Alkali-Silica Reaction Mitigation*, in *Proceedings of the 12th International Conference on Alkali-Aggregate Reaction in Concrete*, October 2004, Beijing, China
50. G.G. Carette, V.M. Malhotra, Mechanical properties, durability and drying shrinkage of Portland cement concrete incorporating silica fume. Cem. Concr. Aggreg. **5**(1), 3–13 (1983)
51. C-Y. Huang, R.F. Feldman, Dependence of frost resistance on the pore structure of mortar containing silica fume. ACI J. **82**(5) (1985)
52. R.F. Feldman, C-Y. Huang, Microstructural properties of blended cement mortars and their relation to durability. RILEM seminar on durability of concrete structures under normal outdoor exposure, Hanover, 1984
53. V.M. Malhotra et al., Mechanical properties and freezing and thawing resistance of high strength concrete incorporating silica fume. International workshop on condensed silica fume in concrete, Montreal, 1987
54. A. Traettberg Frost Action in Mortar of Blended Cement with Silica Dust, in Proceedings of Conference on Durability of Building Materials and Components, 1980, vol. 691 (ASTM STP) pp. 536–548
55. P. Fidjestøl, *Salt-Scaling Resistance of Silica Fume Concrete*, in *Proceedings of the International Symposium on Utilization of By-products in Concrete*. Milwaukee, 1992, ed. by, Naik
56. V.R. Sturrup, R.D. Hootonm, T.G. Clendenning, Durability of Fly Ash Concrete, in Proceeding of the 1st International Conference on the Use of Fly Ash, Silica Fume, Slag and Other Mineral By-Products in Concrete, Montebello, Canada, 1983, ed. by, V.M. Malhotra, vol. I (American concrete institute special publication 79) pp. 71–78
57. T. Yamato, Y. Emoto, M. Soeda, Strength and Freezing-and-Thawing Resistance of Concrete Incorporating Condensed Silica Fume, in Proceeding of the 2nd International Conference on the Use of Fly Ash, Silica Fume, Slag and Natural Pozzolans in Concrete, Madrid, 1986, ed. by, V.M. Malhotra, vol.I (American concrete institute special publication 91) pp. 1095–1019
58. D. Galeota, M.M. Giammatteo, R. Marino, V. Volta, *Freezing and Thawing Resistance of Non Air-Entrained and Air-Entrained Concretes Containing a High Percentage of Condensed Silica Fume*, in *Proceedings of the 2nd International Conference on Durability*

of Concrete. Montreal, 1991, ed. by, V.M. Malhotra, vol.I (American concrete institute special publication 126) pp. 249–616
59. T.C. Holland, Abrasion-Erosion Evaluation of Concrete Mixtures for Kinzu Dam Stilling Basin Repairs, US Army Engineer Waterways Experiment Station, Misc. Paper SL-83-16, September 1983
60. A.A. Ramezanianpour et al., National Code of Practice for Concrete Durability in the Persian Gulf and Omman sea. Building and Housing research Center, PNS428, Iran

Chapter 5
Metakaolin

Introduction

Metakaolin (MK), commercially available since the mid-1990s, is one of the recently developed supplementary cementing materials (SCM) that conforms to ASTM C 618, Class N pozzolan Specifications. Metakaolin differs from other supplementary cementitious materials (SCMs), like fly ash, silica fume, and slag, in that it is not a by-product of an industrial process; it is manufactured for a specific purpose under carefully controlled conditions [1, 2]. This allows manufacturing process of metakaolin to be optimized, ensuring the production of a consistent pozzolanic material. Metakaolin is produced by heating kaolin, one of the most abundant natural clay minerals, to temperatures of 650–900 °C. The *Meta* prefix in the term is used to denote change. The scientific use of the prefix is used for a combining form denoting the least hydrated of a series. In the case of metakaolin, the change that is taking place is dehydroxylization, brought on by the application of heat over a defined period of time. This heat treatment, or calcinations, serves to break down the structure of kaolin. Bound hydroxyl ions are removed and resulting disorder among alumina and silica layers yields a highly reactive, amorphous material with pozzolanic and latent hydraulic reactivity, suitable for use in cementing applications [3, 4]. The first documented use of MK was in 1962, when it was incorporated in the concrete used in the Jupia Dam in Brazil [5].

Production

The raw material input in the manufacture of metakaolin is kaolin clay. Kaolin, also called china clay, is one of the more important industrial clay minerals. It is thought that the term kaolin is derived from the Chinese Kaolin, which translates loosely to white hill and has been related to the name of a mountain in China that yielded the first kaolins that were sent to Europe. Kaolin is a phyllosilicate, consisting of

A. A. Ramezanianpour, *Cement Replacement Materials*,
Springer Geochemistry/Mineralogy, DOI: 10.1007/978-3-642-36721-2_5,

alternate layers of silica and alumina in tetrahedral and octahedral coordination, respectively (Figs. 5.1 and 5.2). This electrically neutral crystalline layer structure leads to a fine particle size and platelike morphology and allows the particles to move readily over one another, giving rise to physical properties such as softness, soapy feel, and easy cleavage [6]. Kaolin is comprised predominantly of the mineral kaolinite, hydrated aluminum Silicate ($Al_2Si_2O_5\ (OH)_4$). Other kaolin minerals are dickite, nacrite and halloysite.

The calcining temperature plays a central role in the reactivity of the resulting MK product. Many researchers studied the effects of calcining temperature on the pozzolanic reactivity of metakaolin [8–11].

$$\underset{\text{Kaolin}}{Al_2Si_2O_5(OH_4)} \xrightarrow{700\text{–}800^\circ C} \underset{\text{Metakaolin}}{Al_2O_3.2SiO_2} + 2H_2O$$

At about 100–200 °C, clay minerals lose most of their adsorbed water. The temperature at which kaolinite loses water by dehydroxilization is in the range of 500–800 °C. This thermal activation of a mineral is also referred to as calcining. Beyond the temperature of dehydroxylization, kaolinite retains two-dimensional order in the crystal structure and the product is termed metakaolin. The key in producing metakaolin for use as a supplementary cementing material, or pozzolan is to achieve as near to complete dehydroxilization as possible without over-heating. Successful processing results in a disordered, amorphous state and is highly pozzolanic. Thermal exposure beyond a defined point will result in sintering and the formation of mullite, which is dead burnt and not reactive. In other words, kaolinite, to be optimally altered to a metakaolin state, requires that it is thoroughly roasted but never burnt (Fig. 5.3).

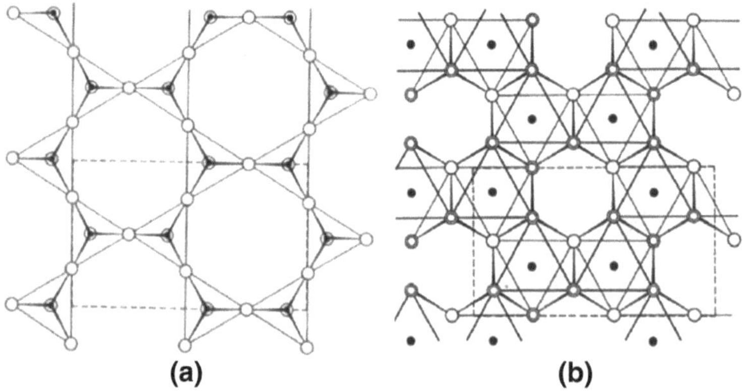

Fig. 5.1 Atomic arrangements of **a** Si_2O_5 and **b** $AlO(OH)_2$ layers: Si (a) or Al (b); Oxygen; Hydroxyl [7]

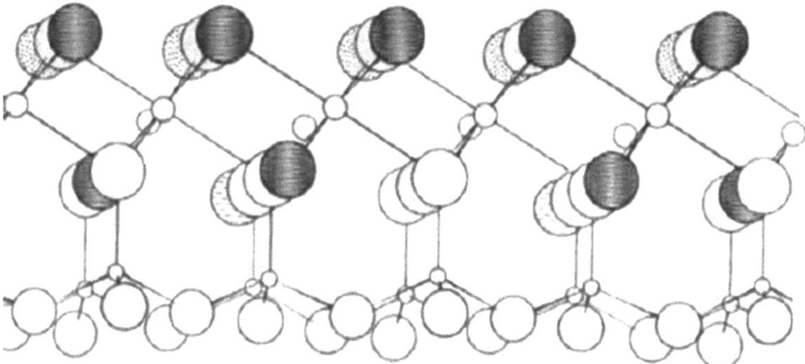

Fig. 5.2 Perspective drawing of kaolinite with Si–O tetrahedrons on the Bottom half and Al–O, OH octahedrons on the top half of the layer [7]

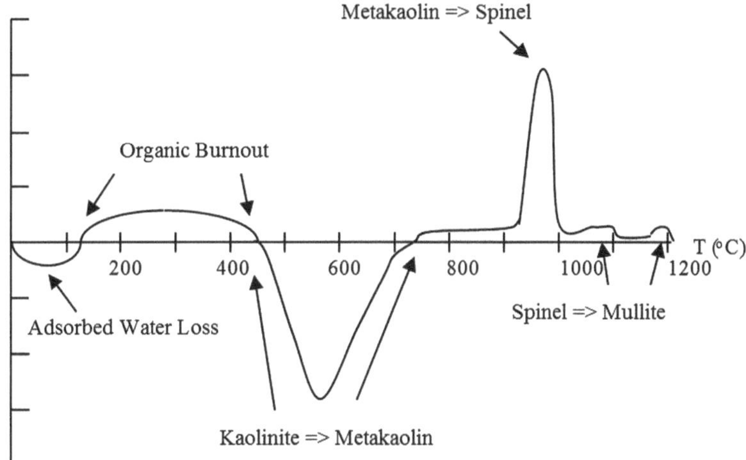

Fig. 5.3 DTA thermogram of kaolin [12]

Kaolinites are usually calcined in rotary kilns or using fluidised bed processes which allows the reduction of calcining time from hours to minutes. While prolonged soak-calcining at lower temperatures is generally quite effective at removing hydroxyl groups, some researchers used flash-calcination to reduce the calcining time to seconds. The process consists of rapid heating, calcining and cooling. They showed that different qualities of MK are obtained depending on the temperature (500–1000 °C) and time of flash-calcination (0.5–12 s) and that more active MK can be produced by this method, than by soaking [11, 13].

Ramezanianpour et al. [10] investigated the optimal temperature and burning time to produce metakaolin from three different kaolins. They found that for the investigated kaolin types, the optimal temperature for thermal activation of kaolin between 750 and 850 °C for 1 h was sufficient to obtain the product of desired performance. Increasing the heating duration from 1 to 2 h causes an increase in the reactivity at temperatures lower than 750 °C. Within the optimal activation temperatures(from 750 to 850 °C), by increasing the heating duration from 1 to 2 h, no significant improvement was observed in the compressive strength and dehydroxilation of kaolins. In addition, reactivity decreased with an increase in the heating duration at these temperatures.

Physical, Chemical, and Mineralogical Properties

Physical Properties

The particle size of MK is generally less than 2 mm, which is significantly smaller than that of cement particles, though not as fine as SF. MK is white in color (whereas silica fume is typically dark grey or black), making it particularly attractive in color matching and other architectural applications. Due to the controlled nature of the processing, MK powders are very consistent in appearance and performance [2]. Some Physical characteristics of metakaolin compare to other pozzolans are shown in Table 5.1.

The shape of particles of metakaolin is usually platy. As an example, the typical scanning electron micrograph (SEM) of the metakaolin is shown in Fig. 5.4.

Chemical and Mineralogical Properties

The chemical composition of the materials was measured by wavelength-dispersive x-ray fluorescence spectrometry (XRF). MK typically contains 50–55 % SiO_2 and 40–45 % Al_2O_3. Other oxides present in small amounts include Fe_2O_3, TiO_2, CaO, and MgO. The typical chemical composition is given in Table 5.2. The alkali content of metakaolin is low and varies between 0.3 and 1 %. In additions; Loss on ignition (LOI) is generally low but may reach to 4 % in some metakaolin. The mineralogical composition of metakaolin also varies at different sources of kaolin. Kaolinite and Quarts are usually high in metakaolin. The next mineral compositions are feldspar, mica and calcite.

Table 5.1 Physical characteristics of metakaolin and other pozzolans

Material	Mean size (10^{-6} m)	Surface area (m^2/g)	Particle shape	Specific gravity
Portland cement	10–15	<1	Angular, irregular	3.2
Fly ash	10–15	1–2	Mostly spherical	2.2–2.4
Silica fume	0.1–0.3	15–25	Spherical	2.2
Metakaolin	1–2	15	Platey	2.4

Fig. 5.4 Typical scanning electron micrograph of the metakaolin

Table 5.2 Typical chemical composition of metakaolin

	Percentage of by mass
Sio_2	51.52
Al_2o_3	40.18
Fe_2o_3	1.23
Cao	2.0
Mgo	0.12
K_2o	0.53
So_3	0.0
Tio_2	2.27
Na_2o	0.08
LOI	2.01

Pozzolanic Activity

There are several parameters influencing the pozzolanic activity of metakaolin. The sources of kaolin and nature of active phases and their contents in the pozzolan, surface area and fineness of the particles, the lime pozzolan mixture and the amount of mixing water, curing system and temperature are the most important factors affecting the reactivity. The hydration reaction of metakaolin depends of Portland cement type, the AS_2/CH ratio, the quantity of free water and the temperature that the reaction will take place [14].

The principal reaction is between AS_2 and CH derived from cement hydration, in the presence of water. This reaction forms additional cementitious C–S–H gel, together with crystalline products, which include calcium aluminate hydrates and alumosilicate hydrates (C_2ASH_8, C_4AH_{13} and C_3AH_6) [15]. In general, supplementary cementitious materials (SCM) with higher alumina contents, such as MK, tend to have higher pozzolanic capacities because formation of C–A–H has a high CH demand [12].

$$\mathrm{MK[Al_2Si_2O_7] + CH + H \rightarrow C-S-H, C_4AH_{13}, C_3AH_6, C_2ASH_8.}$$

Table 5.3 Pozzolanic activity of pozzolans [17]

Pozzolan	SF	FA	MK
Reactivity (mg) Ca $(OH)_2$/g pozzolan	427	875	1050

In addition if carbonate is freely available carbo-aluminates may also be produced. Hydration reaction depends upon the level of reactivity of MK, which in turn depends upon the processing conditions and purity of feed clay. Reactivity level of MK can be determined by the Chapelle test and is expressed as consumption rate of CH per gram of pozzolans. The amount of CH in hardened concrete can be determined by thermogravimetric analysis (TG) and differential thermal analysis (DTA) [16]. Table 5.3 presents the comparison of reactivity level of MK with silica fume (SF) and fly ash (FA).

The reactivity of pozzolans can also be assessed by chemical determination of unreacted pozzolan in hydrated PC-pozzolan pastes. Kostuch et al. reported that 20 % replacement of cement by MK was required to fully remove all the CH in concrete at 28 days [4]. Oriol and Pera reported that between 30 and 40 % MK is required to remove all the CH in MK-PC paste at a water binder ratio of 0.5 when cured in lime-saturated water for 28 days [18]. Frías Rojas and Cabrera compared the hydration of 1:1 hydrated lime to meatakolin mixtures at 23 and 60 °C. They found C_2ASH_8 (stratlingite) and C_4AH_{13} appear, and finally the C_3ASH_6 (hydrogarnet) [19].

The degree of pozzolanic reaction completed is defined as the percentage of unreacted pozzolan remaining relative to the initial amount of pozzolan present in the cement paste. The degree of pozzolanic reaction of MK was higher at a 5 % replacement level than 10 and 20 %. The higher rate of pozzolanic reaction in cement pastes with a lower replacement level can be attributed to the higher concentration of CH [20]. Because MK reacts with and consumes free CH, another method for determining the extent of pozzolanic reaction completed is to measure the remaining CH content in a paste, mortar, or concrete sample. As part of the same study, Poon et al. also determined total CH content of paste samples, both based on the ignited weight and the weight of cement, using differential scanning calorimetry. Cements blended with 20 % MK showed the least total CH at all ages. MK mixtures showed steadily decreasing CH contents up to 90 days, as illustrated in Fig. 5.5.

Effects of Metakaolin on the Properties of Fresh Concrete

The properties of fresh concrete are important because they affect the choice of equipment needed for handling and consolidation and because they may affect the properties of hardened concrete [21]. Slump is by far the oldest and most widely used test of workability. Metakaolin has been shown to produce smaller slumps than control mixtures. Many researchers reported that MK offered much better

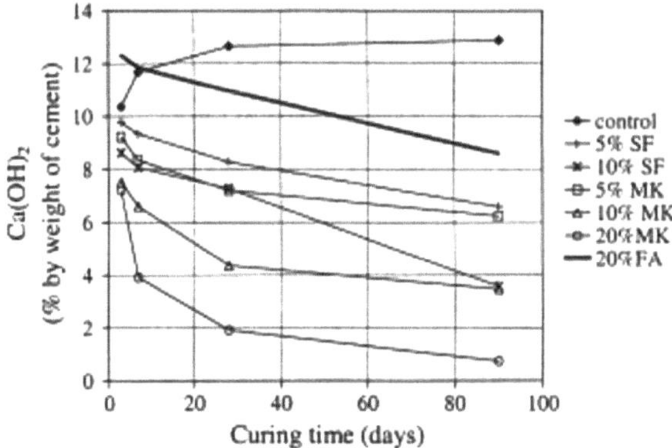

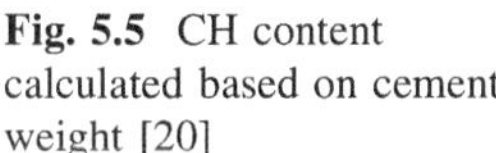
Fig. 5.5 CH content calculated based on cement weight [20]

workability than did silica fume for the same mixture proportions. Additionally, they found that at 5 and 10 % replacement, MK mixtures had a slightly higher slump than the control mixture. At 15 % replacement with MK, slump decreased approximately 10 % from the control value [2]. The results, also indicated by other researchers meant that the concrete mixtures modified by MK required less high-range water-reducing admixture than SF mixtures to achieve similar workability at the same water/binder ratio [22–24]. This reduction in high-range water-reducing admixture demand may result in fewer tendencies for surface tearing during finishing operations and lead to an overall better finish ability. In addition, the MK-modified mixtures may be more economical because of a lower dosage of high-range water-reducing admixture [2]. Workability of concrete containing MK was reported by Wild et al. (1996) is shown in Table 5.4 [25].

Due to the fineness of the metakaolin grains, consistency of mixes with metakaolin greatly decreases in comparison with the cement mortar reference. As seen in Fig. 5.6 the use of 5 % metakaolin induces a decrease of 14.6 % in the consistency and the addition of 20 % metakaolin induces a decrease of 34 % of consistency [26].

The cohesiveness of metakaolin concrete is greater than that of control concrete. Metakaolin concrete is easier to pump and place generally and it bleeds less than plain concrete. The greater volume of cementitious fines results in the production of sharp arise and high quality surface finish on cast vertical surface.

Table 5.4 Workability of metakaolin concretes [25]

Metakaolin (MK) (%)	Superplasticizer (%)	Slump (mm)	Compacting factor	Vebe time (s)
0	0	5	0.81	26
5	0.6	10	0.84	15
10	1.2	15	0.88	10
15	1.8	25	0.89	9
20	2.4	75	0.89	7
25	3	75	0.89	4
30	3.6	90	0.9	5

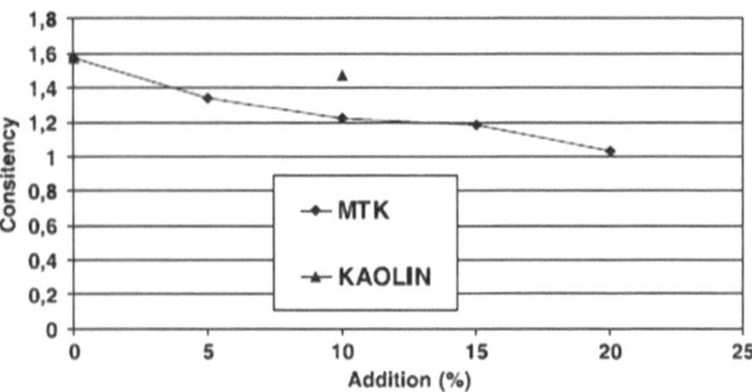

Fig. 5.6 Consistency for mortars with cement CEM I 42.5, metakaolin and kaolin [26]

Many factors influence setting time, including the w/cm, casting and curing temperature, admixture type, source, and dosage, and cement content, fineness, and composition.

Previous researchers have found MK incorporation to have varying effects on the setting behavior of mortars and pastes. Brooks et al. examined the effect of silica fume, MK, FA, and slag on setting time of high strength concrete via ASTM C 403. They found that all SCMs tended to retard setting time, and that increasing the levels of silica fume, FA, and slag resulted in greater retardation of the set [27]. For HSC containing MK, there was a progressive increase in the retarding effect up to 10 % replacement, but a reduction at higher replacement levels. Similar results were reported by Batis et al., found all MK mixtures to have significantly longer setting times than control pastes [28]. Conversely, other researchers found MK to shorten setting time, as compared to control samples [22, 29]. Also in an investigation shown in Fig. 5.7, the initial setting time decreased as much as 20 % with 15 % replacement MK whereas for the final setting time, no significant changes were observed in different concrete samples [30].

Effects of Metakaolin on the Mechanical Properties of Hardened Concrete

Strength of Mortars and Concretes

Metakaolin reacts with portlandite (CH) to form calcium-silicate-hydrate (C–S–H) supplementary to that produced by Portland cement hydration. This reaction becomes important within the interfacial transition zone (ITZ) located between aggregate and paste fractions. This region typically contains a high concentration of large, aligned CH crystals, which can lead to localized areas of increased porosity and lower strength [31, 32]. This refinement in the ITZ can result in increased strength in metakaolin concrete.

Three elementary factors, which influence the contribution that MK makes to concrete strength. These are the filler effect, which is immediate, the acceleration

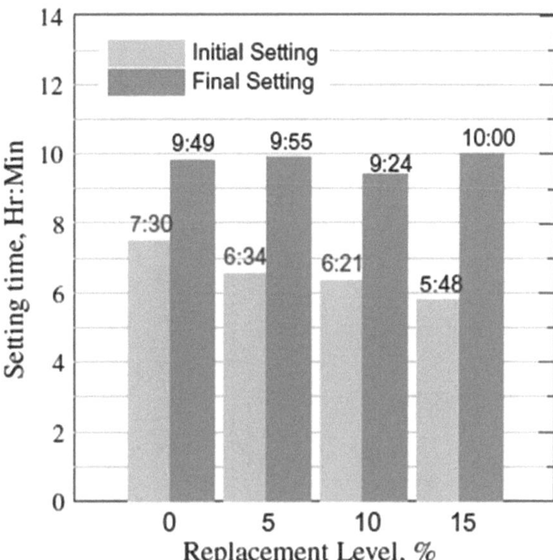

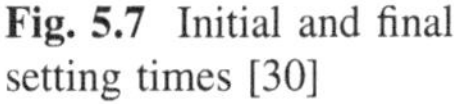
Fig. 5.7 Initial and final setting times [30]

of PC hydration, which occurs within the first 24 h, and the pozzolanic reaction, which has its maximum effect within the first 7–14 days for all MK levels between 5 and 30 %. Influences of MK on the strength of concrete have been reported by Many researches [2, 10, 25, 31]. Figure 5.8 shows the effect of a metakaolin on the compressive strength of concrete up to 120 days.

Metakaolin is a very active natural pozzolan and can produce higher early strength concrete. This is clearly seen in Fig. 5.8. The very early strength enhancement is due to a combination of the filler effect and accelerated cement hydration [34]. Subsequently, these effects are enhanced by the pozzolanic reaction between MK and the CH produced by the hydration of the cement.

The strength of metakaolin concretes depends on many factors such as the quality of cement composition and metakaolin, its particle size and specific surface area, fineness, temperature and time of thermal activation of kaolin and approach of mix design.

Moodi et al. show that reaction ability and dehydroxilation of kaolinite increase with an increase in temperature. However, as temperatures approach 900 °C, recrystallization begins and consequently the reactivity decreases. In addition, by way of comparison of diagrams in Fig. 5.9, it can be deduced that increase of heating duration from 1 to 2 h does not have much effect on the improvement of compressive strengths except at low temperatures. At low temperatures, an increase of heating duration leads to an increase of dehydroxilation and this in turn causes a further increase in the reactivity and compressive strength of mortar samples. It was seen that peak of compressive strength has shifted towards lower temperatures as the heating duration increases [10].

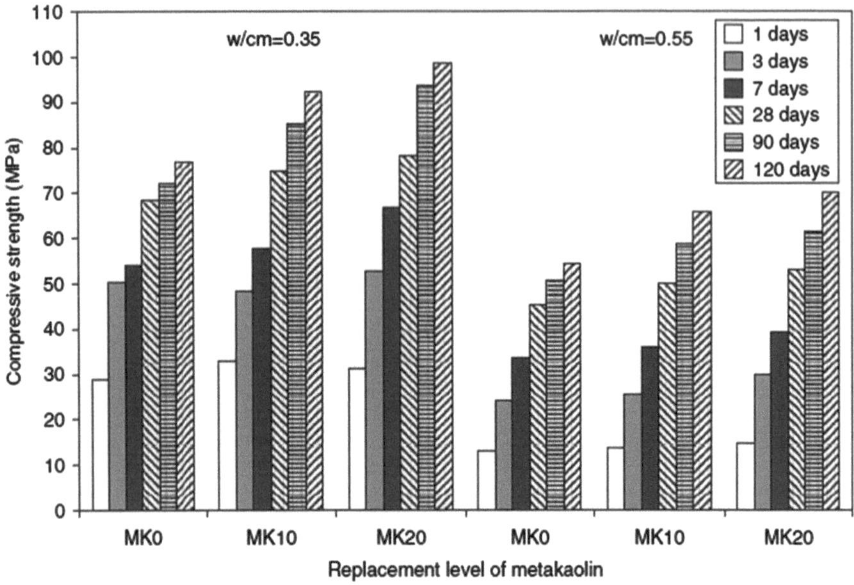

Fig. 5.8 Effect of metakaolin (MK) on the compressive strength development of concretes [33]

Wild et al. show that increasing the specific surface of MK from 12 to 15 m^2/g reduce the age at which maximum strength enhancement occurs in MK mortars, illustrating the effect of particle size on reaction rate. This increase in fineness also resulted in an increase in the optimum level of replacement of cement by MK, meaning that more of the cement could be replaced by this MK without the system suffering a lag due to dilution. Interestingly, this change in fineness did not influence the long-term (90 days) strength [25].

In quite a few studies, the optimization of the amount of metakaolin used as Portland cement replacement (mainly from the point of view of improvement of compressive strength) was a substantial part of research work. The results showed that while for normal concrete (typically with w/c 0.5 and fc 40–50 MPa) the optimal amount of metakaolin was 20 %, for high performance concrete (typically with w/c 0.3 and fc 80–100 MPa) the best results were achieved with 10 % of metakaolin [35].

The influence of curing temperature on the strength development in concretes has been investigated by some researchers [36]. It was shown that curing MK concrete at 50 °C results in increased early strength (7 days) compared to the strength of specimens cured at 20 °C. The acceleration in strength development due to the high curing temperature diminishes in the long-term (365 days).

The strength development of metakaolin has been compared with silica fume and shown in Fig. 5.10. It is clear that at the same replacement level, MK increased concrete strength at all ages to almost the same extent as SF did and in

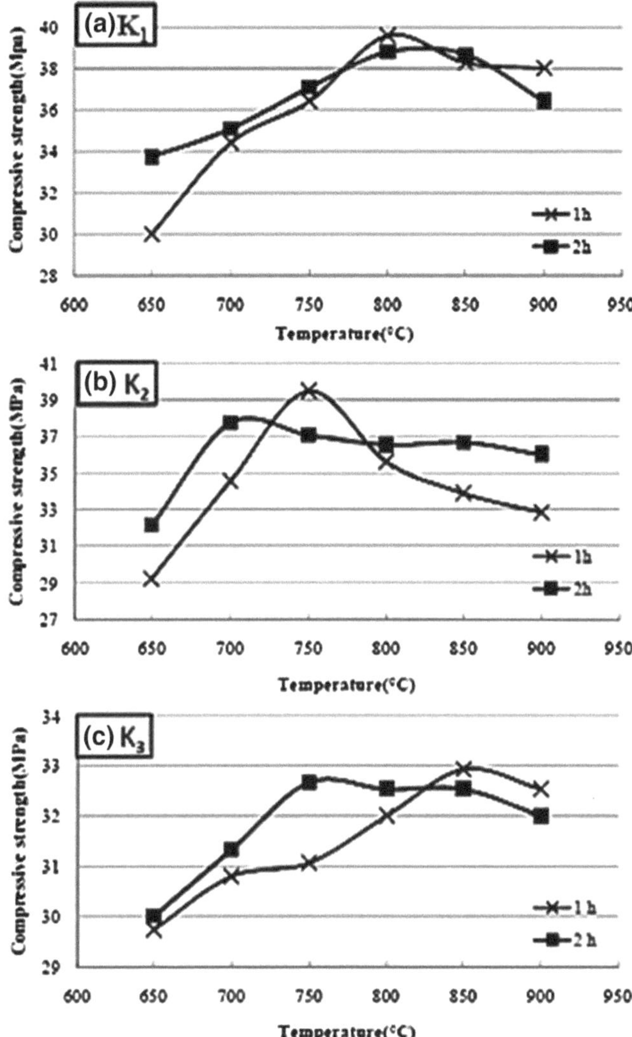

Fig. 5.9 Effect of temperature and time of thermal activation on 28 day compressive strength of mortars containing 15 % metakaolin, **a** K_1, **b** K_2, **c** K_3 [10]

early age MK provided slightly better results. Similar influences of MK on the strength of concrete have been reported by other researchers [22, 25].

The strength development pattern for Flexural and tensile strength is similar to that of compressive strength. In general, the splitting tensile strength increased with the increase in MK content at all ages. However, the increase in the splitting tensile strength was smaller compared to that obtained in the compressive strength. Some researchers have found similar trend for metakaolin concrete [3, 37]. As shown in

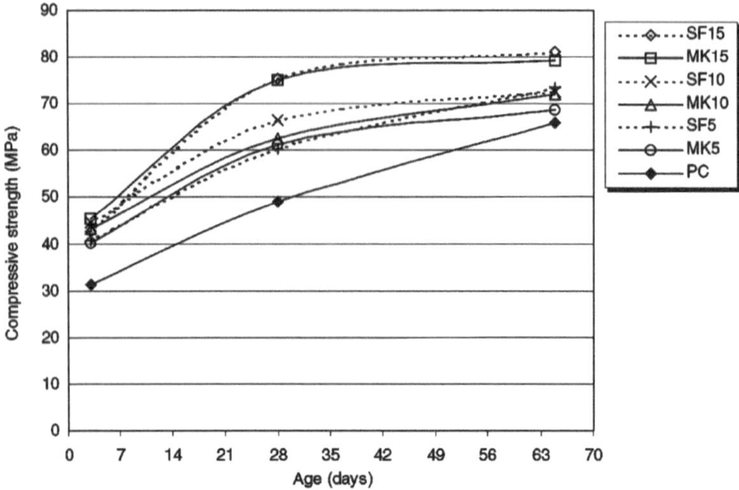

Fig. 5.10 Compressive strength developments of MK and SF concretes [2]

Fig. 5.11, a replacement ratio of 15 % is seen to improve the development of flexural strength, but such effect appears to reduce for 20 %. In other words, the most remarkable flexural strength were developed for replacement rates from 10 to 15 %, with poor improvement effect for a replacement rate of 15 % compared to 20 %.

Modulus of Elasticity

Modulus of elasticity (MOE) is greater in metakaolin concretes compared to ordinary concrete, most significantly at the lower w/cm. MOE generally increase with increasing MK content, although the rate of increase is lower than that for compressive strength. Table 5.5 summarized some published data on modulus of elasticity. This suggests that the metakaolin may have a greater effect on elastic modulus at lower w/cm. However, unlike the compressive strength data where the finer material produced greater strengths, no apparent effect of metakaolin fineness was observed in the elastic modulus data.

Effect on Volume Changes of Concrete

The replacement of cement with normal percentages of metakaolin does change the shrinkage. At early ages (24 h), autogenous shrinkage was reduced with the increase in MK content whereas at long-term, autogenous shrinkage increased with

Table 5.5 Increase in modulus of elasticity _MOE_ due to partial replacement with Metakaolin, as reported/measured at 28 Days of age [38]

Author/material	w/cm	Replacement (%)	MOE (GPa)	MOE increase over control (%)
Qian and Li [37]	0.38	10	33.2	11
Calderone et al. (1994)	0.40	10	38.9	18
MK235[a]	0.40	8	34.6	19
MK349[a]	0.40	8	34.4	18
Khatib and Hibbert (2004)	0.50	10	38.0	3
MK235[a]	0.50	8	30.7	5
MK349[a]	0.50	8	30.8	6
MK235[a]	0.60	8	27.7	8
MK349[a]	0.60	8	28.3	11

[a] Data from this investigation

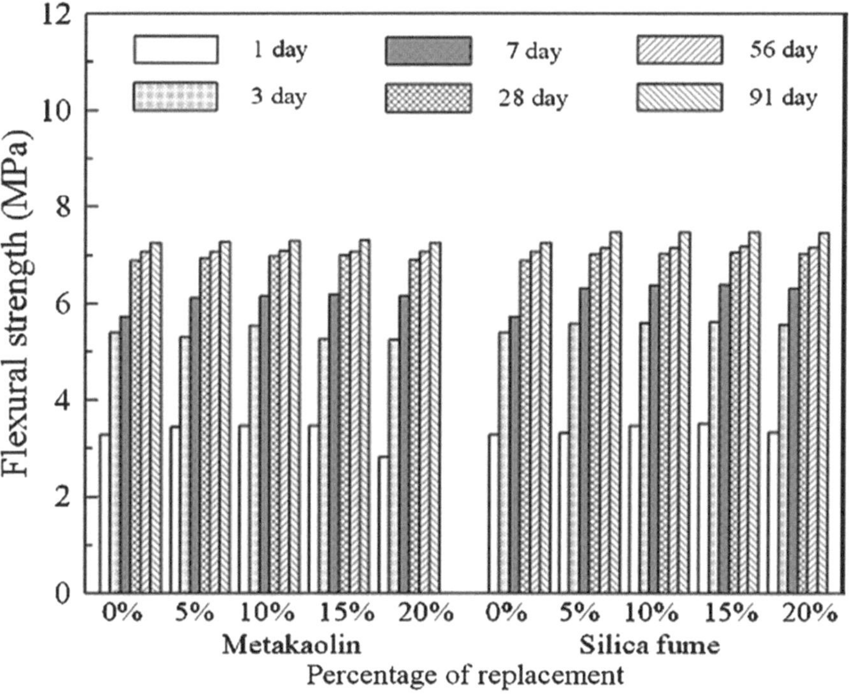

Fig. 5.11 Effect of metakaolin and silica fume on flexural strength development of concrete [24]

the increase in MK content (Fig. 5.12). This is clearly shown in Table 5.6 that there is difference between the shrinkage of normal Portland cement and cements containing metakaolin [39].

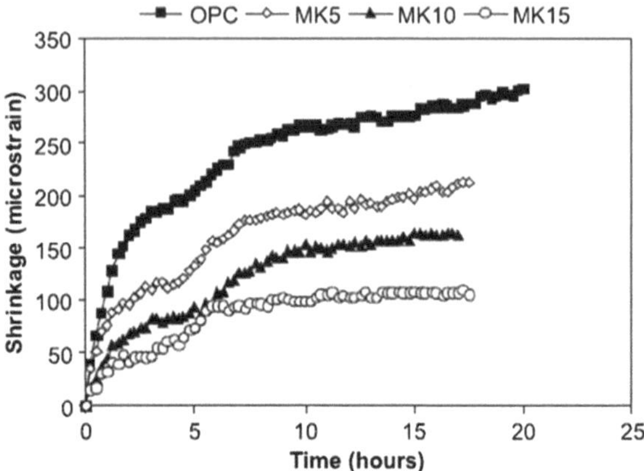

Fig. 5.12 Effect of metakaolin on the early ages autogenous shrinkage of concrete [39]

Table 5.6 200-day total and autogenous shrinkage of concrete [39]

	OPC	MK5	MK10	MK15
Total shrinkage from initial set (10^{-6})	861	713	618	516
Total shrinkage from 24 h (10^{-6})	558	499	455	410
Total autogeneous shrinkage from initial set (10^{-6})	445	485	419	327
Autogeneous shrinkage 24 h (10^{-6})	142	271	256	221
Percentage of A.S.–T.S. from initial set	52	68	68	63
Percentage of A.S.–T.S. from 24 h	25	54	56	54
Weight loss of the exposed specimens (%)	0.85	0.67	0.73	0.76

Chemical shrinkage increases with increase in MK content and reaches a maximum between 10 and 15 % MK and then decreases sharply for higher MK content [40]. The drying shrinkage of mortars and concretes vary with the variation of water and aggregate type and contents. The use of metakaolin reduce shrinkage and It is evident from the Fig. 5.13 that the higher the replacement of MK, the higher the reduction in the drying shrinkage, irrespective of w/cm ratio. The reason in the reduction of shrinkage owing to the use of MK is surely the decrease in shrinkage rate. Many researchers have found similar trend for metakaolin concrete investigated at different conditions [33].

Creep is related to many factors including the strength and the amount and time of loading. The strength of concretes containing metakaolin is usually higher than the control concrete and hence lower specific creep is expected for such concretes at all ages. This is clearly shown in Table 5.7 that Inclusion of MK reduced both total and basic creep of concrete as well as drying shrinkage, with a greater reduction in creep at higher replacement levels. The reduction in creep could be attributed to a denser pore structure, stronger paste matrix and improved paste aggregate interface of MK concrete mixtures as a result of the formation of additional hydrate phases from secondary pozzolanic reaction of MK and its filler effect [39].

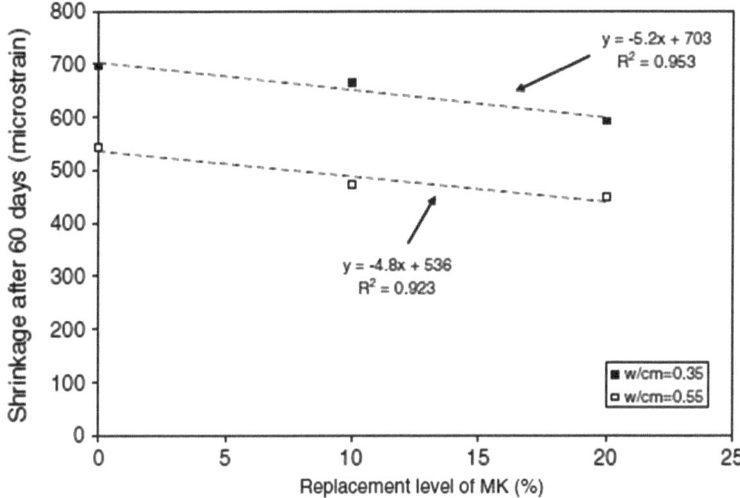

Fig. 5.13 Relationship between shrinkage after 60 days and replacement level of MK [33]

Table 5.7 Results of creep after 200 days [39]

Concrete mixture	Creep (10^{-6})		Stress (MPa)	Specific creep (10^{-6}/Mpa)		Drying creep (10^{-6}/MPa)
	Dry	Sealed		Dry	Sealed	
OPC	358	285	13.9	25.8	20.5	5.3
MK5	312	235	16.8	18.5	14.0	4.5
MK10	201	126	17.7	11.3	7.2	4.1
MK15	171	115	17.9	9.5	6.4	3.1

Microstructure, Porosity and Permeability

The evolution of porosity depends on certain characteristics of the SCM, such as particle size, chemical composition, mineralogy, and loss on ignition. In most cases, mortars and concrete containing pozzolanic SCMs have total porosity values equal to or less than that of PC concrete. However, MK because of its fineness and high pozzolanic reactivity has great potential to decrease concrete porosity. The porosity of cement pozzolan pastes have been determined by several methods such as mercury intrusion porosimetry, nitrogen adsorption, helium pycnometry and methanol displacement.

The results of mercury intrusion method (MIP) are given in Table 5.8. It can be deduced that the total porosity decreased substantially with increasing replacement level of MK. The magnitude of this reduction varies between 22 and 49 %, depending mainly on w/cm ratio and replacement level of MK [33].

Table 5.8 Results of mercury intrusion porosimetry test for the low and high w/cm ratio samples tested at 120 days [33]

W/cm ratio	MK (%)	Porosity (%)	Mean pore diameter (μm)	Median pore diameter (μm)
0.35	0	10.7	0.286	0.098
0.35	10	7.3	0.139	0.058
0.35	20	5.5	0.088	0.037
0.55	0	14.6	0.614	0.202
0.55	10	11.4	0.320	0.129
0.55	20	10.2	0.283	0.093

It was also noted that there was a considerable reduction in the mean (or median) pore diameter of the samples due to the inclusion of MK. The effect was particularly beneficial at 20 % MK content, where the lowest porosity and pore diameter were achieved.

Some researcher report that cement metakaolin paste had lower porosity and smaller average pore diameters than the control and the silica fume pastes at all ages tested [41, 42]. This indicates that MK is more effective than silica fume in the refinement of pore structure. These results are different than those reported by other researchers who found MK pastes to have greater porosity than controls at 28 days [42, 43].

The microstructure of the MK cement paste is more uniform and compact than that of the plain Portland cement paste. The reaction mechanism can be divided into physical and chemical aspects. The physical effect is that the ultra-fine particles fill the voids in cement, which makes the microstructure of cement paste denser. The chemical effect is the reaction of MK with the cement hydrates. The principal reaction is between the AS_2 and the CH, derived from cement hydration, in the water. This reaction forms additional cementitious aluminum containing CSH gel, together with crystalline products, which include calcium aluminate hydrates and alumino-silicate hydrates (i.e., C_2ASH_8, C_4AH_{13} and C_3AH_6) [24]. This is clearly shown in Fig. 5.14 which illustrates the results of SEM and EDS analysis obtained for plain Portland cement pastes without metakaolin and with 10 % metakaolin.

Effect of Metakaolin on Durability of Concrete

In addition to strength, MK incorporation is widely regarded as an effective means to increase concrete durability. This is achieved primarily in the ITZ, which is characterized by a higher porosity, a higher local w/cm, and differing mineralogical and chemical composition than the bulk paste. It has been suggested that these properties of the ITZ can be detrimental to some composite properties, including resistance to chloride and sulfate transport. MK, which has been shown to affect the chemistry and microstructure of the ITZ, may thus play a role in reducing ion transport and improving concrete durability.

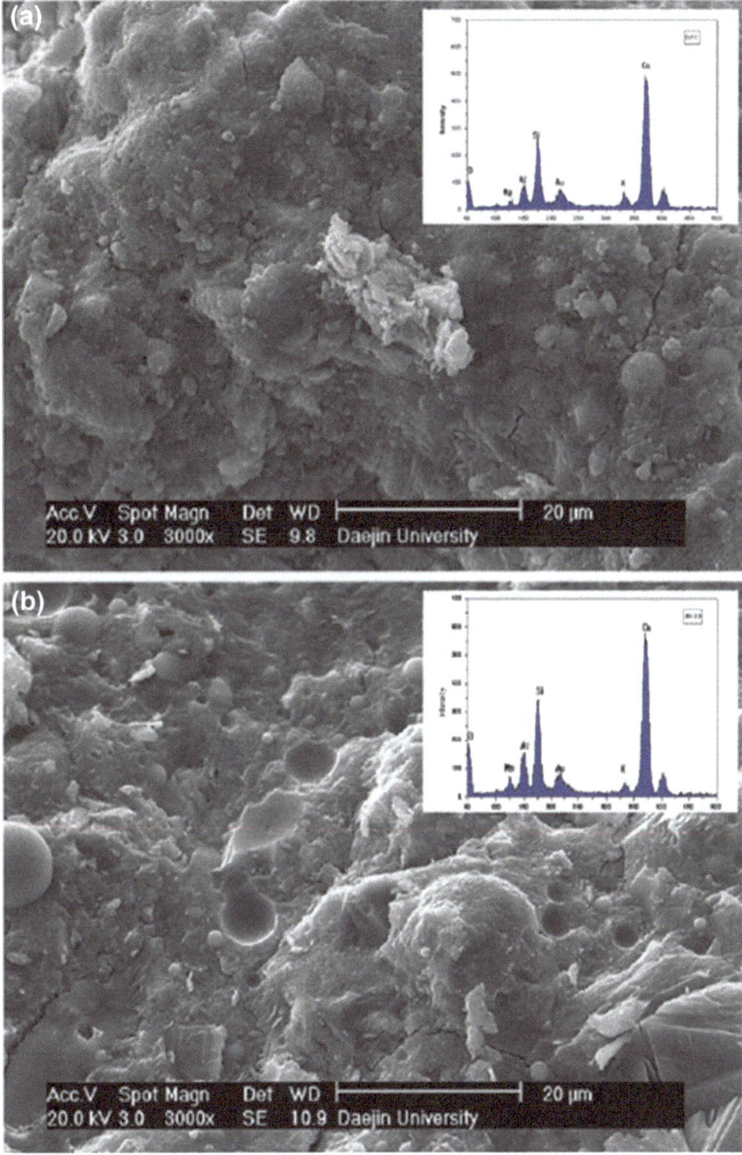

Fig. 5.14 SEM micrographs and EDS results of the hydrated pastes: **a** hydrated paste (ordinary Portland cement), **b** hydrated paste (10 % Metakaolin) [24]

Concrete durability depends mainly on the chemistry (cement hydration process), and the microstructure of the concrete. Metakaolin addition affects positively both factors; Metakaolin consumes rapidly and effectively the $Ca(OH)_2$ that is produced from the cement hydration process and additional to CSH, phases like

C_2ASH_8 (stratlingite), C_4AH_{13} and C_3ASH_6 (hydrogarnet) are produced. These pozzolanic products contribute to a total pore refinement [45–47]. The refined pore system results in a more compact concrete, through which transportation of the water and other aggressive chemicals is significantly impeded and therefore a decrease in the diffusion rate of harmful ions is reported [4, 44, 48, 49].

Effect of Metakaolin on Carbonation of Mortars and Concretes

The progress of carbonation in concrete depends on environmental conditions and concrete quality. Relative humidity, carbon dioxide concentration in the air, and temperature are the most important parameters affecting carbonation. Water cement ratio, permeability and moist curing duration are the main factors affecting the concrete quality and hence its carbonation depth.

It is generally believed that carbonation of concrete containing metakaolin is higher than that for the plain concrete. It is attributed to the lower portlandite content of the pozzolanic cement paste. However, some researchers have reported similar or lower carbonation for metakaolin concrete [35, 50].

As seen in Table 5.9, the carbonation extent manifested by the decrease of pH after 60 days in CO_2-rich environment was in metakaolin concrete (BM) substantially lower than in the plain concrete (BR).

In an accelerated carbonation measurement of concretes, which determined by the phenolphthalein method, the depth of carbonation of metakaolin concrete was found higher than control concrete. It is clearly seen that (see Fig. 5.15) the carbonation depth of metakaolin concrete is larger than the control mix, regardless of the type of admixture and age of concrete. It can be explained from the fact that the replacement of cement by MK decreases the content of portlandite in hydrate products due to pozzolanic reaction [24].

Table 5.9 Measured pH values of water leaches of concrete specimens Submerged for 60 days in various environments [35]

Environment	BR	BM
Air	12.33	12.10
Distilled water	12.40	12.08
$MgCl_2$	12.44	12.10
NH_4Cl	12.27	12.02
Na_2SO_4	12.40	12.03
HCl	12.40	12.01
CO_2	11.82	11.98

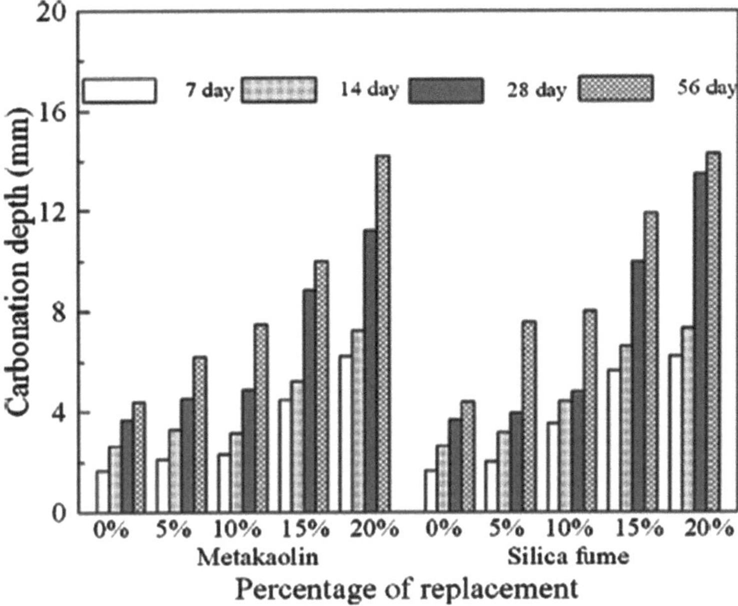

Fig. 5.15 Carbonation depth versus replacement level with metakaolin or silica fume [24]

Effect of Chloride Ions on Durability of Metakaolin Cement Mortars and Concretes

The resistance to chloride penetration is governed by two factors: firstly, the permeability of concrete mixtures, which is a function of the pore size distribution. Secondly, the chloride ion binding capacity of the cementitious paste, which increases with increasing CSH gel formation and increased alumina levels enabling more chloride to be fixed as Friedel's salt ($C_3A\_CaCl_2\_10H_2O$) [51]. The replacement of cement with metakaolin reduces the diffusion of chloride ions at all ages. Results of the rapid chloride penetration test (RCPT) of concretes containing metakaolin are shown in Table 5.10. It is clearly seen that both the MK concretes showed lower total ion penetration than the control. This may be related to the refined pore structure of these concretes and their reduced electrical conductivity.

Unfortunately, the results of the rapid chloride penetration test (RCPT) cannot be used to directly predict the rate of chloride penetration under field conditions, since they do not incorporate the mass transport coefficients that true service life models require. Bulk diffusion, or ponding, tests are necessary to determine apparent chloride diffusion coefficients. Many researchers report the results of bulk diffusion test to determine apparent chloride diffusion coefficients [53–55].

The results for chloride concentration profiles of metakaolin concrete (MK) and control(C) are presented in Fig. 5.16. The values of coefficient of diffusion (D) for C

Table 5.10 Chloride permeability of control and blended concretes [52]

Series	w/b ratio	Mix	Total charge passed (Coulombs)			
			3 days	7 days	28 days	90 days
1	0.3	Control	2461	2151	1035	931
		5 % MK	1327	1244	862	646
		10 % MK	417	347	199	135
		20 % MK	406	395	240	124
2	0.5	Control	5312	4054	2971	2789
		5 % MK	4215	3765	2079	1065
		10 % MK	1580	1247	918	752
		20 % MK	751	740	640	580

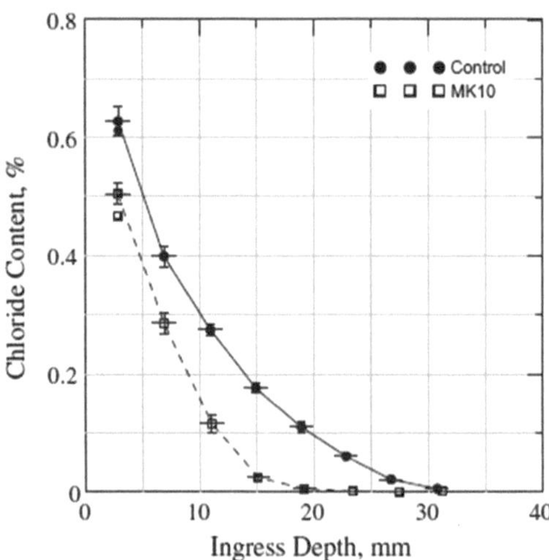

Fig. 5.16 Chloride ions ingress profile test result [30]

and MK10 were calculated using Fick's 2nd Law of diffusion as 3.2 ± 0.05 (10^{-12}) (m^2/s) and 1.7 ± 0.05 (10^{-12}) (m^2/s), respectively. This shows a 47 % reduction of ionic diffusivity of concrete obtained in the metakaolin concrete.

Sulfate Resistance of Mortars and Concretes Containing Metakaolin

Another aspect of durability that MK replacement affects is resistance to sulfate attack. The sulfate attack is generally attributed to the reaction of sulfate ions with calcium hydroxide and calcium aluminate hydrate to form gypsum and ettringite. The gypsum and ettringite formed as a result of sulfate attack is significantly more voluminous (1.2–2.2 times) than the initial reactants [56]. The formation of gypsum and ettringite leads to expansion, cracking, deterioration, and disruption of concrete structures. In addition to the formation of ettringite and gypsum and its subsequent expansion, the deterioration due to sulfate attack is partially caused by the degradation of calcium silicate hydrate (C–S–H) gel through leaching of the calcium compounds. This process leads to loss of C–S–H gel stiffness and overall deterioration of the cement paste matrix [57]. The sulfate attack chemical interaction is a complicated process and depends on many parameters including concentration of sulfate ions, ambient temperature, cement type and composition, water to cement ratio, porosity and permeability of concrete, and presence of supplementary cementitious materials [58]. MK is chemically different from many other SCMs in that it has very high alumina content. The reaction products that MK and CH form are C–S–H, and also include C_4AH_{13}, C_3AH_6, and C_2ASH_8. A clear correlation has been drawn between the tricalcium aluminate (C3A) content of a portland cement and its susceptibility to sulfate attack [59].

Figure 5.17 shows the effect of various percentages of metakaolin on the expansion of mortars for high and intermediate C_3A Portland cements. Expansion due to sulfate attack was found to decrease systematically with increasing MK content for both cement types. Metakaolin, by consuming $Ca(OH)_2$, has a large positive effect on mortar durability in the sulfate environment. In addition, the replacement of a portion of Portland cement with MK reduces the total amount of tricalcium aluminate hydrate in the cement paste matrix of concrete. Furthermore, the filler effect and the formation of secondary C–S–H by the pozzolanic reaction, although less dense than the primary C–S–H gel, is effective in filling and segmenting large capillary pores into small, discontinuous capillary pores through pore size refinement. Thus, the total permeability of concrete decreases and enhances the resistance of MK concrete to sulfate attack [60].

Compressive strength reduction of MK concrete decreased with increasing MK content. The decrease in strength indicates the expansion effect of the sulfate attack which results in the softening of the material and the formation of microcrack. The effects of curing on the sulfate resistance of MK concrete are shown in Fig. 5.18. Autoclaved concrete specimens showed superior sulfate resistance (as indicated by strength reduction) compared to the moist cured concrete specimens. The superior resistance of autoclaved MK concrete to sulfate attack compared to moist cured MK concrete is attributed to many factors. First, autoclaving curing using high pressure steam produce tricalcium aluminate hydrates more stable in the presence of sulfate ions than those formed in the moist cured concrete specimens [62]. The second factor is the reduction of pore volume of autoclaved concrete compared to the moist cured concrete. The reduction of pore volume causes a reduction in the permeability of autoclaved concrete and consequently a reduction in the sulfate ions intruding into autoclaved concrete. The third factor is the fast reduction of calcium hydroxide in the cement paste as a result of the pozzolanic reaction [60].

Entrained air increases and decreases the sulfate resistance of the MK and plain concrete, respectively. Incorporating entrained air bubbles may increase the uniformity and homogeneity of MK in the cement paste matrix of concrete mixtures. Therefore, the efficiency of MK effect (pozzolanic reaction and filler action) within

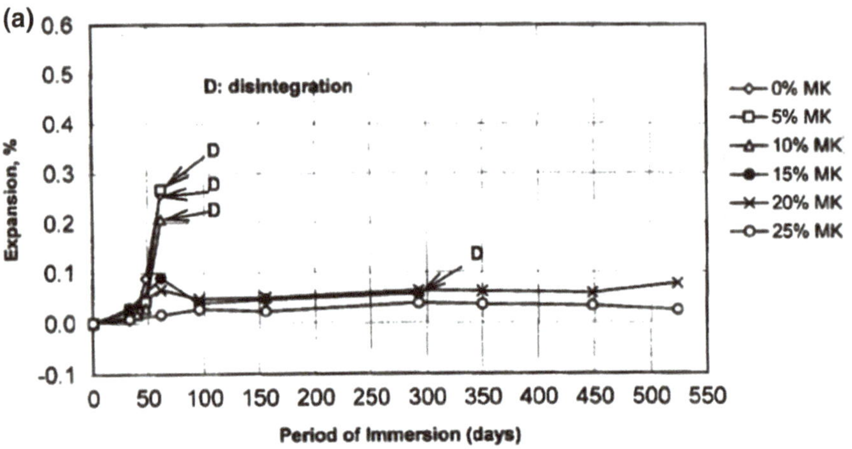

Fig. 5.17 Expansion of MK mortar bars versus sodium sulfate exposure time for **a** high and **b** intermediate C_3A Portland cements [61]

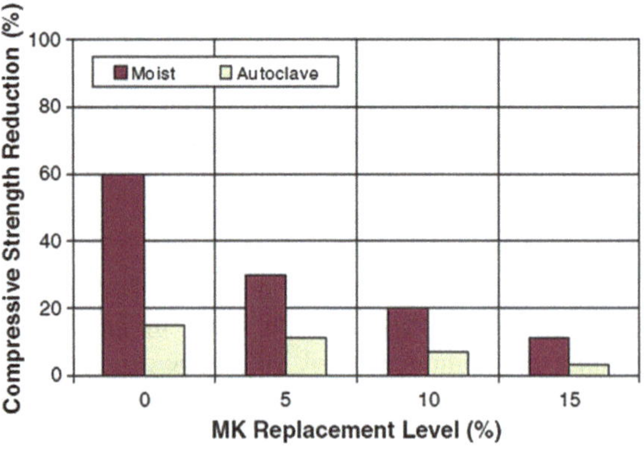

Fig. 5.18 Effect of moist and autoclaving curing on the compressive strength reduction of MK concrete [60]

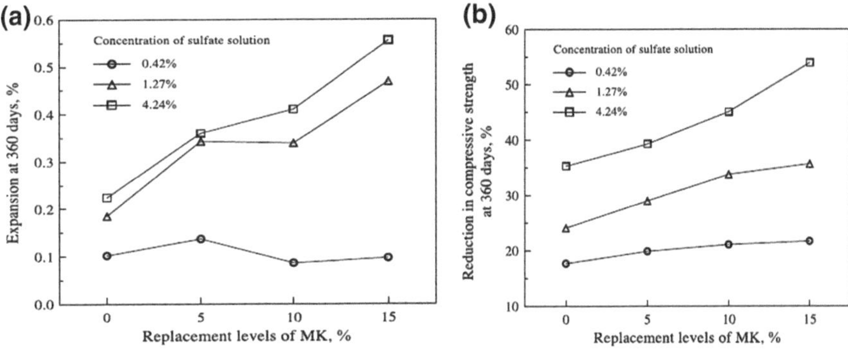

Fig. 5.19 Effect of solution concentrations and MK replacement levels on the **a** expansion and **b** Reduction in the compressive strength of mortar specimens exposed to magnesium sulfate solutions [64]

concrete specimens is increased. However, plain air entrained concrete showed decrease in the sulfate resistance due to increasing the porosity in the cement matrix, which increases the rate of sulfate ions intruding within concrete [60].

The sulfate attack of magnesium sulfate is considered to be more severe than that of sodium sulfate. Metakaolin usually increase the resistance of concretes and mortars in sodium sulfate exposure, but increased metakaolin replacement levels play a negative role on the performance of mortar or paste specimens when exposed to magnesium sulfate solution. This is clearly seen in Fig. 5.19. This likely significantly contributed to gypsum formation as magnesium sulfate attack. Tensile stresses developed during gypsum formation may lead to significant expansion [63]. Another possible reason may be the decalcification of primary and secondary C–S–H gel following the formation of M–S–H gel [64].

Effect of Metakaolin on Suppressing the Alkali Aggregate Reaction

Alkali-silica reactivity (ASR) is another aspect of durability that MK replacement affects in controlling expansion due to alkali-silica reaction (ASR).

The best protection against ASR is achieved by:

1. Avoiding susceptible aggregate
2. Reducing water content (to reduce concrete permeability)
3. Specifying a low-alkali cement
4. Reducing the overall alkali content of the concrete by limiting the cement content.

Secondary reactions involving metakaolin promote refinement and densification of the concrete microstructure, reducing permeability and limiting the availability of water. In addition, when metakaolin is used as partial replacement for cement, its effect is to dilute the cement, which reduces the alkali content of the concrete system and the pH of the pore solution, thereby increasing the solubility of calcium and promoting the formation of non-expanding gel in place of swelling N(K)–S–H. The potential for using metakaolin to control alkali-silica expansion in concrete has been reported by many researchers [65, 66]. It has been reported that between 1962 and 1979, about 260,000 tones of metakaolin was used in construction of four dams in Brazil where local aggregates were highly alkali-active; results showed that ASR was successfully mitigated where metakaolin was used in concrete.

Potential for alkali reactivity is measured according to the accelerated mortar bar method (ASTM C 1260). This method has been shown reliable for evaluating the effectiveness of SCMs in suppressing ASR. The effect of metakaolin on reducing the alkali aggregate reaction is shown in Fig. 5.20. It is clearly seen that expansion reductions as much as 50 % lower in metakaolin specimens.

The suitable method for assessing the efficacy of pozzolans and slag in reducing expansion due to ASR is the concrete prism method (ASTM C 1293). Current experience suggests that a 2-year testing period is sufficient for the evaluation of mineral admixtures, with an expansion limit criterion of 0.04 %.

In an investigation shown in Fig. 5.21 for the control of alkali aggregate reaction of concrete prisms containing of two reactive aggregates, Portland cement Type I blending with 10–15 % metakaolin is the appropriate level of replacement, depending on the nature of the aggregate. The mechanism by which MK may suppress ASR expansion is entrapment of alkalis by the supplementary hydrates and a consequent decrease in the pH of the pore solution. This was supported by sampling pore solutions of paste samples over a two-year period and titrating to determine OH- concentration [66].

The effect of metakaolin on alkali carbonate reaction has not been fully investigated and needs more research.

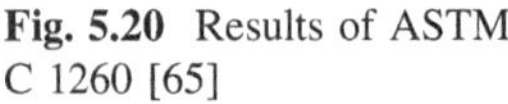

Fig. 5.20 Results of ASTM C 1260 [65]

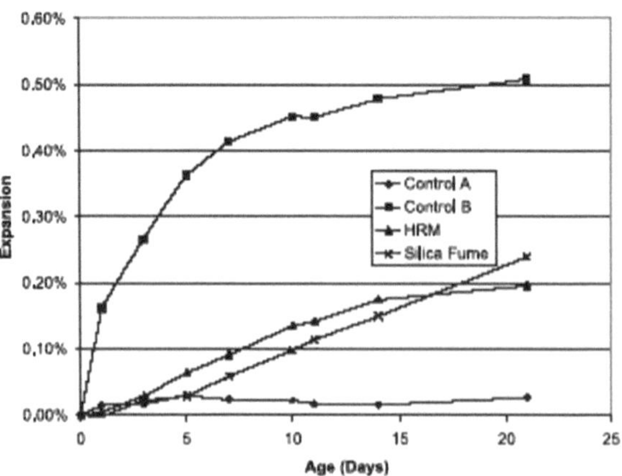

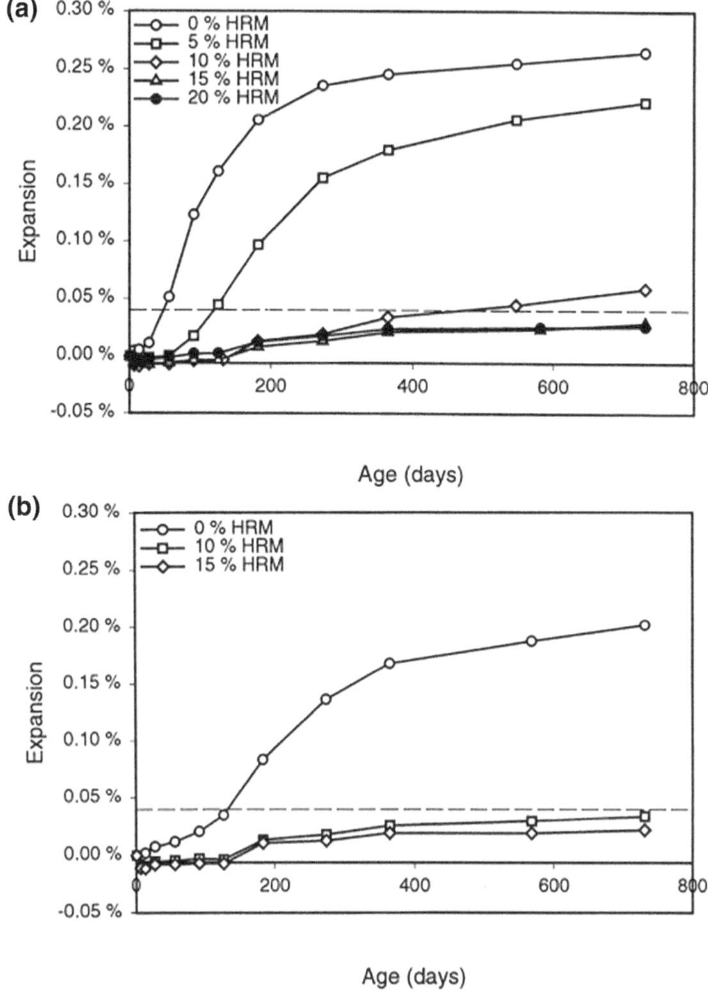

Fig. 5.21 Expansion of concrete prisms containing metakaolin, cement A, and **a** Sudbury or **b** Spratt aggregate [66]

Freezing and Thawing of Concretes Containing Metakaolin

It is imperative for development of resistance to deteriora-tion from cycles of freezing and thawing that a concrete have adequate strength and entrained air. For concrete containing supplementary cementing materials to provide the same resistance to freezing and thawing cycles as a concrete made using only Portland cement as a binder, four conditions for both concretes must be met:

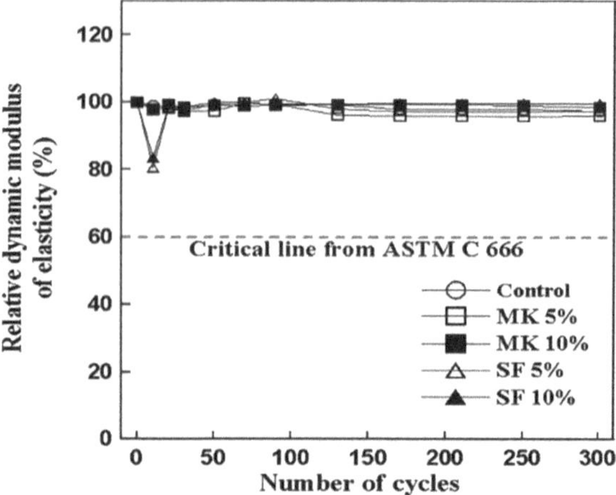

Fig. 5.22 Accelerated freezing and thawing test results [24]

1. They must have the same compressive strength.
2. They must have adequate entrained air content with proper air-void characteristics.
3. They must be properly cured.
4. They must be air-dried for one month prior to expo-sure to saturated freezing conditions.

Researches investigating the effect of metakaolin on freezing—thawing action show similar or better performance as for plain cements with similar strengths and air bubble contents. Figure 5.22 shows the resistance of concrete to freezing and thawing which was assessed by accelerated freezing and thawing cycle test until 300 cycles according to ASTM C 666. Then the relative dynamic modulus of elasticity was calculated. It is seen that the relative dynamic modulus of elasticity remains quasi-constant for the five types of concrete until 300 cycles. This can be explained by the low water to binder ratio of the high performance concretes and the entrained air content.

The effect of metakaolin on the frost resistance of concrete subjected to 100 freezing and thawing cycles is depicted in Table 5.11. The frost resistance of the concrete containing metakaolin (BM) was excellent; 100 freeze–thaw cycles had no effect on both compressive and bending strengths [35].

Freeze–thaw resistance of mortar specimens containing MK as a partial PC replacement under three point loading was examined and MK was found to

Table 5.11 Frost resistance [35]

Material	Frost resistance coefficient K	
	As the ratio of compressive strengths(-)	As the ratio of bending strengths(-)
BR	1	0.8
BM	1	1

improve the durability of the mortars tested. This was attributed to the pore refinement produced by the pozzolanic activity [67].

Application of Metakaolin in Mortars and Concretes

The first documented use of MK was in 1962, when it was incorporated in the concrete used in the Jupia Dam in Brazil. It has been commercially available since the mid-1990s. Where conventional silica fume leaves a dark coloration to concrete, pure white HRM does not. HRM can be used in combination with white cement to get a high performance white concrete. The HRM permits placement of high-performance concrete next to conventional concrete, because the admixture does not darken the concrete. This is desirable for architectural treatments. For slab placements, or areas where good finish ability (perhaps by hand) is needed, HRM can offer additional benefits. In addition, early-age strength developments, control efflorescence, lower permeability, higher resistance to sulfate attack and alkali aggregate reaction and reduce drying shrinkage are the technical improvements in the application of cement containing natural pozzolans.

At this time, HRM's widest commercial use is in precast elements of high-rise skyscrapers along the booming Pacific Rim. But in April 1996, based on its lab tests, the New York State Department of Transportation approved the use of Engelhard MetaMax HRM as a substitute for silica fume in bridge deck applications.

In the USA metakaolin have been used in the construction Con Edison dock and Reconstruction of Erie Canal (for the refacing of lock walls and abutments and extension of a dam apron) and I-87 overpass (for the decks of new four-lane, twin bridges that carry I-87 over the Hudson River near south Glen Falls). Also, the Pennsylvania Turnpike Commission is using metakaolin concrete to construct bridge and viaduct parapets on their Highways of Hope. The white concrete increases the visibility of the parapets and improves safety, especially at night and in wet conditions (Fig. 5.23).

Fig. 5.23 Metakaolin was used on **a** I-87 overpass and **b** Reconstruction of erie canal

In the major rebuild of the Highway 39 Bridge over the Canadian Pacific Railway tracks east of Estevan in Saskatchewan, a metakaolin was used.

References

1. T. Ramlochan, M. Thomas, K.A. Gruber, The Effect of metakaolin on ALKALI Silica reaction in concrete. Cem. Concr. Res. **30**, 339–344 (2000)
2. J.T. Ding, Z.J. Li, Effects of metakaolin and silica fume on properties of concrete. ACI Mater. J. **99**(4), 393–398 (2002)
3. J. Bensted, P. Barnes, *Structure and performance of cements* (Spon, New York, 2002), p. 565
4. J.A. Kostuch, V. Walters, T.R. Jones, High performance concretes incorporating metakaolin: a review. ed. by R.K. Dhir and M.R. Jones. *Concrete 2000: Economic and Durable Concrete through Excellence* (E&FN Spon, London, 1993), pp. 1799–1811
5. T.R. Jones, Metakaolin as a pozzolanic addition to concrete. ed. by J. Bensted, P. Barnes. *Structure and performance of cements* (Spon Press, London, 2001)
6. W.D. Kingery, D.R. Uhlmann, H.K. Bowen, *Introduction to ceramics*, 2nd edn. (Wiley, New York, 1976)
7. G.W. Brinkley, *Ceramic fabrication processes*. (Technology Press & Wiley, Cambridge & New York, 1958)
8. J. Ambroise, M. Murat, J. Pera, Hydration reaction and hardening of calcined clays and related minerals, V Extension of the research and general conclusions. Cem. Concr. Res. **15**, 261–268 (1985)
9. J. Bensted, P. Barnes, *Structure and performance of cements*, 2nd edn. (Spon Press, New York, 2002)
10. F. Moodi, A.A. Ramezanianpour, A.Sh. Safavizadeh, *Evaluation of the Optimal Process of Thermal Activation of Kaolins* (Sientia Iranica, 2011)
11. S. Salvador, Pozzolanic properties of flash-calcined kaolinite: a comparative study with soak-calcined products. Cem. Concr. Res. **25**(1), 102–112 (1995)
12. J.M. Justice, *Evaluation of Metakaolins for Use as Supplementary Cementitious Materials*. MSc Thesis, Georgia institute of technology, Georgia, USA (2005)
13. S. Salvador, T.W. Davies, Modeling of combined heating and dehydroxylation of kaolinite particles during flash calcination; production of metakolin. Processing Adv. Mater **9**, 128–135 (1994)
14. C. Nita, V.M. John, D.M.R. Cleber, H. Savastano Jr, M.S. Takeashi, Effect of metakaolin on the performance of PVA and cellulose fibers reinforced cement, in *Proceedings of the 9th international inorganic-bonded composite materials conference; 2004 October 11–13th*, ed. by A.A. Moslemi (University of Idaho, Moscow, 2004), pp. 1–48
15. B.B. Sabir, S. Wild, j. Bai, Metakaolin and cacined clays as Pozzolans for concrete: a review. Cem Conc Comp **23**(6), 441–454 (2001)
16. R. Siddique, J. Klaus, Influence of metakaolin on the properties of mortar and concrete: a review. Appl. Clay Sci. **43**, 392–400 (2009)
17. A.H. Asbridge, G.V. Walters, T.R. Jones, *Ternary blended concretes OPC/GGBFS/ Metakaolin* (Concrete Across Borders, Denmark, 1994), pp. 547–557
18. M. Oriol, J. Pera, Pozzolanic activity of metakaolin under microwave treatment. Cem. Concr. Res. **25**(2), 265–270 (1995)
19. M.F. Rojas, J. Cabrera, The effect of temperature on the hydration rate and stability of the hydration phases of metakaolin–lime–water systems. Cem. Concr. Res. **32**, 133–138 (2002)
20. C.S. Poon, L. Lam, S.C. Kou, Y.L. Wong, R. Wong, Rate of pozzolanic reaction of metakaolin in high-performance cement pastes. Cem. Concr. Res. **31**, 1301–1306 (2001)

21. S. Mindess, F.J. Young, D. Darwin, *Concrete*, 2nd edn. (Prentice Hall, Upper Saddle River, 2003)
22. M.A. Caldarone, K.A. Gruber, R.G. Burg, High-reactivity Metakaolin: a new generation mineral admixture. Concr. Int. **16**(11), 37–40 (1994)
23. M.A. Caldarone, K.A. Gruber, *High reactivity Metakaolin, a mineral admixture for high-performance concrete, concrete under severe conditions: environment and loading*, ed. by K. Sakai, N. Banthia, O.E. Gjorv, Proceedings of the International Conference on Concrete under Severe Conditions, CONSEC 1995, Sapporo, Japan, Aug. 1995, V. 2 (E&FN Spon: Chapman & Hall, New York, 1995), pp. 1015–1024
24. H.S. Kim, S.H. Lee, H.Y. Moon, Strength properties and durability aspects of high strength concrete using Korean metakaolin. Constr. Build. Mater. **21**(6), 1229–1237 (2007)
25. S. Wild, J.M. Khatib, A. Jones, Relative strength, pozzolanic activity and cement hydration in superplasticised metakaolin concrete. Cem. Concr. Res. **26**(10), 1537–1544 (1996)
26. L. Courard, A. Darimont, M. Schouterden, F. Ferauche, X. Willem, R. Degeimbre, Durability of mortars modified with metakaolin. Cem. Concr. Res. **33**(9), 1473–1479 (2003)
27. J.J. Brooks, M.A.M. Johari, M. Mazloom, Effect of admixtures on the setting times of high-strength concrete. Cem. Concr. Compos. **22**(1), 293–301 (2000)
28. G. Batis, P. Pantazopoulou, S. Tsivilis, E. Badogiannis, The effect of metakaolin on the corrosion behavior of cement mortars. Cem Conc Comp in press (2004)
29. E. Moulin, P. Blanc, D. Sorrentino, Influence of key cement chemical parameters on the properties of metakaolin blended cements. Cem. Concr. Compos. **23**(6), 463–469 (2001)
30. M. Shekarchi, A. Bonakdar, M. Bakhshi, A. Mirdamadi, B. Mobasher, Transport properties in metakaolin blended concrete. Constr. Build. Mater. **24**(11), 2217–2223 (2010)
31. S. Wild, J.M. Khatib, Portlandite consumption in metakaolin cement pastes and mortars. Cem. Concr. Res. **27**, 137–146 (1997)
32. D.P. Bentz, E.J. Garboczi, Simulation studies of the effects of mineral admixtures on the cement paste-aggregate interfacial zone. ACI Mater J **88**(5), 518–529 (1991)
33. E. Güneyisi, M. Gesoğlu, K. Mermerdas, Improving strength, drying shrinkage, and pore structure of concrete using metakaolin. Mater Struct **41**(5), 937–949 (2008)
34. J.M. Khatib, Metakaolin concrete at a low water to binder ratio. Constr. Build. Mat. **22**, 1691–1700 (2008)
35. E. Vejmelkova, M. Pavlikova, M. Keppert, Z. Keršner, P. Rovnanikova, M. Ondracek, M. Sedlmajer, R. Cerny, High performance concrete with Czech metakaolin: Experimental analysis of strength, toughness and durability characteristics. Constr. Build. Mat. **24**(8), 1404–1411 (2010)
36. B.B. Sabir, The effects of curing temperature and water/binder ratio on the strength of metakaolin concrete, in Sixth CANMET/ACI International Conference on Fly ash, Silica Fume, Slag and Natural Pozzolans in Concrete, Supplementary volume. Bangkok, Thailand, pp. 493–506 (1998)
37. X. Qian, Z. Li, The relationships between stress and strain for high-performance concrete with metakaolin. Cem. Concr. Res. **31**, 1607–1611 (2001)
38. J.M. Justice, K.E. Kurtis, Influence of metakaolin surface area on properties of cement-based materials. J. Mater. civil ASCE/September2007
39. J.J. Brooks, M.A.M. Johari, Effect of metakaolin on creep and shrinkage of concrete. Cement Concr. Compos. **23**(6), 495–502 (2001)
40. S. Wild, J. Khatib, L.J. Roose, Chemical and autogenous shrinkage of Portland cement-metakaolin pastes. Adv. Cem. Res. **10**(3), 109–119 (1998)
41. C.S. Poon, S.C. Kou, L. Lam, Pore size distribution of high performance metakaolin concrete. J. Wuhan Univ Technol. Mat. Sci. Edn. **17**(1), 42–46 (2002)
42. J. Ambroise, S. Maximilien, J. Pera, Properties of metakaolin blended cements. Adv. Cem. Based Mater. **1**(4), 161–168 (1994)
43. M. Frías, J. Cabrera, Pore size distribution and degree of hydration of metakaolin-cement pastes. Cem. Concr. Res. **30**(4), 561–569 (2000)

44. J.M. Khatib, S. Wild, Pore size distribution of metakaolin paste. Cem. Concr. Res. **26**(10), 1545–1553 (1996)
45. J.P. Skanly, *Materials science of concrete, I* (American Ceramic Society Inc., Westerville, 1989)
46. N.J. Coleman, C.L. Page, Aspects of the pore solution chemistry of hydrated cement pastes containing metakaolin. Cem. Concr. Res. **27**, 147–154 (1997)
47. E. Badogiannis, S. Tsivilis, Exploitation of poor Greek kaolins: durability of metakaolin concrete. Cement Concr. Compos. **31**(2), 128–133 (2009)
48. P.K. Mehta, J.M. Monteiro, Effect of aggregate, cement and mineral admixtures on the microstructure of the transition zone, ed. by S. Mindess, P.S. Shah, Bonding in cementitious composites. Materials Research Society, Pittsburgh pp. 65–75 (1987)
49. P. Bredy, M. Chabannet, J. Pera, Microstructure and porosity of metakaolin blended cements. Proc. Mater. Res. Soc. Symp. **137**, 431–436 (1989)
50. J. Bai, S. Wild, B.B. Sabir, Sorptivity and strength of air-cured and water-cured PC–PFA–MK concrete and the influence of binder composition on carbonation depth. Cem. Concr. Res. **32**, 1813–1821 (2002)
51. H.E.H. Seleem, A.M. Rashad, B.A. El-Sabbagh, Durability and strength evaluation of high-performance concrete in marine structures. Constr. Build. Mater. **24**(6), 878–884 (2010)
52. C.S. Poon, S.C. Kou, L. Lam, Compressive strength, chloride diffusivity and pore structure of high performance metakaolin and silica fume concrete. Constr. Build. Mat. **20**(10), 858–865 (2006)
53. K.A. Gruber, T. Ramlochan, A. Boddy, R.D. Hooton, M.D.A. Thomas, Increasing concrete durability with high-reactivity metakaolin. Cem. Conc. Comp. **23**(6), 479–484 (2001)
54. A.H. Asbridge, C.L. Page, M.M. Page, Effects of metakaolin, water/binder ratio and interfacial transition zones on the microhardness of cement mortars. Cem. Concr. Res. **32**(9), 1365–1369 (2002)
55. A. Boddy, R.D. Hooton, K.A. Gruber, Long-term testing of the chloride penetration resistance of concrete containing high-reactivity metakaolin. Cem. Concr. Res. **31**(5), 759–765 (2001)
56. R.D. Hooton, Influence of silica fume replacement of cement on physical properties and resistance to sulfate attack, freezing and thawing, and alkali-silica reactivity. ACI Mater. J. **90**(2), 143–151 (1993)
57. P.K. Mehta, Mechanics of sulfate attack on Portland cement concrete another look. Cem. Concr. Res. **13**(3), 401–406 (1983)
58. P.J. Tumidajski, G.W. Chan, K.E. Philipose, An effective diffusivity for sulfate transport into concrete. Cem. Concr. Res. **25**(6), 1159–1163 (1995)
59. K.E. Kurtis, P.J.M. Monteiro, S.M. Madanat, Empirical models to predict concrete expansion caused by sulfate attack. ACI Mater. J. **97**(2), 156–162 (2000)
60. Nabil M. Al-Akhras, Durability of metakaolin concrete to sulfate attack. Cem. Concr. Res. **36**, 1727–1734 (2006)
61. J.M. Khatib, S. Wild, Sulphate resistance of metakaolin mortar. Cem. Concr. Res. **28**(1), 83–92 (1998)
62. A.M. Neville, *Properties of concrete*, 4th edn. (Addison Wesley Longman Limited, London, 1995)
63. B. Tian, M.D. Cohen, Does gypsum formation during sulfate attack on concrete lead to expansion Cem. Concr. Res. **30**(1), 117–123 (2000)
64. S.T. Lee, H.Y. Moon, R.D. Hooton, J.P. Kim, Effect of solution concentrations and replacement levels of metakaolin on the resistance of mortars exposed to magnesium sulfate solutions. Cem. Concr. Res. **35**, 1314–1323 (2005)

65. W. Aquino, D.A. Lange, J. Olek, The influence of metakaolin and silica fume on the chemistry of alkali-silica reaction products. Cem. Concr. Compos. **23**(6), 485–493 (2001)
66. T. Ramlochan, M. Thomas, K.A. Gruber, The effect of metakaolin on alkali-silica reaction in concrete. Cem. Concr. Res. **30**(3), 339–344 (2000)
67. C. Girodet, M. Chabannet, J.L. Bosc, J. Pera, *Influence of the type of cement on the freeze-thaw resistance of the nortar phase of concrete*, ed. by M.J. Setzer, R. Auberg, Proceedings of the International RILEM Workshop on the Frost Resistance of Concrete. (E & FN Spon, London, 1997) pp 31–40

Chapter 6
Rice Husk Ash

Introduction

Rice-husk (RH) is an agricultural by-product material. It constitutes about 20 % of the weight of rice. It contains about 50 % cellulose, 25–30 % lignin, and 15–20 % of silica. When rice-husk is burnt rice-husk ash (RHA) is generated. On burning, cellulose and lignin are removed leaving behind silica ash. The controlled temperature and environment of burning yields better quality of rice-husk ash as its particle size and specific surface area are dependent on burning condition. For every 1000 kg of paddy milled, about 200 kg (20 %) of husk is produced, and when this husk is burnt in the boilers, about 50 kg (25 %) of RHA is generated. Completely burnt rice-husk is grey to white in color, while partially burnt rice-husk ash is blackish [1–3].

Rice-husk ash (RHA) is a very fine pozzolanic material. The utilization of rice husk ash as a pozzolanic material in cement and concrete provides several advantages, such as improved strength and durability properties, reduced materials costs due to cement savings, and environmental benefits related to the disposal of waste materials and to reduced carbon dioxide emissions. Reactivity of RHA is attributed to its high content of amorphous silica, and to its very large surface area governed by the porous structure of the particles [1, 2]. Generally, reactivity is favored also by increasing fineness of the pozzolanic material. However, Mehta [1] has argued that grinding of RHA to a high degree of fineness should be avoided, since it derives its pozzolanic activity mainly from the internal surface area of the particles.

Production

Rice is the seed of the monocot plants Oryza sativa or Oryzaglaberrima. As a cereal grain, it is the most important staple food for a large part of the world's human population, especially in East and South Asia, the Middle East, Latin America, and the West Indies. It is the grain with the second-highest worldwide production, after corn.

A. A. Ramezanianpour, *Cement Replacement Materials*,
Springer Geochemistry/Mineralogy, DOI: 10.1007/978-3-642-36721-2_6,

Rice cultivation is well-suited to countries and regions with low labor costs and high rainfall, as it is labor-intensive to cultivate and requires ample water. Rice can be grown practically anywhere, even on a steep hill or mountain. Although its parent species are native to South Asia and certain parts of Africa, centuries of trade and exportation have made it commonplace in many cultures worldwide.

Rice is normally grown as an annual plant, although in tropical areas it can survive as a perennial and can produce a ratoon crop for up to 30 years. World production of rice has risen steadily [4].

Rice Husk

Rice hulls (or rice husks) are the hard protecting coverings of grains of rice. In addition to protecting rice during the growing season, rice husks can be put to use as building material, fertilizer, insulation material, or fuel.

Rice husks are the coating for the seeds, or grains, of the rice plant. To protect the seed during the growing season, the husk is made of hard materials, including opaline silica and lignin. The husk is mostly indigestible to humans.

One practice, started in the seventeenth century, to separate the rice from husks, it to put the whole rice into a pan and throw it into the air while the wind blows. The husks are blown away while the rice fell back into the pan. This happens because the husk isn't nearly as dense as the rice. These steps are known as winnowing. During the milling process, the husks are removed from the grain to create brown rice, the brown rice is then milled further to remove the bran layer to become white rice [1, 5].

Usage of Rice Husk

Some of the current and potential applications are listed below.

Chemistry

Rice hulk can be used to produce mesoporos molecular sieves (e.g., MCM), [6, 7] which are applied as catalysts for various chemical reactions, as a support for drug delivery system and as adsorbent in waste water treatment.

Pet Food Fiber

Rice husks are the outermost covering of the rice and come as organic rice husks and natural rice husks. Rice husks are an inexpensive byproduct of human food processing, serving as a source of fiber that is considered a filler ingredient in cheap pet foods [8].

Building Materials

Rice husks are a Class A insulating material because they are difficult to burn and less likely to allow moisture to propagate mold or fungi. It has been found out that when burned, rice husk produces significant amounts of silica. For these reasons it provides excellent thermal insulation.

Fertilizer

Rice husks are organic material and can be composted. However, their high lignin content can make this a slow process. Sometimes earthworms are used to accelerate the process. Using vermicomposting techniques, the husks can be converted to fertilizer in about four months.

Sic Production

Rice husks are a low-cost material from which silicon carbide "whiskers" can be manufactured. The SiC whiskers are then used to reinforce ceramic cutting tools, increasing their strength tenfold [9].

Fuel

With proper techniques, rice husks can be burned and used to power steam engines [10].

Factors Influencing the Use of Rice Husk

Low cost building blocks

Ordinary Portland cement (OPC) is expensive and unaffordable to a large portion of the world's population. Since OPC is typically the most expensive constituent of concrete, the replacement of a proportion of it with RHA offers improved concrete affordability, particularly for low-cost housing in developing countries.

The potential for good but inexpensive housing in developing countries is especially great. Studies have been carried out all over the world, such as in Guyana, Kenya and Indonesia on the use of low cost building blocks. Portland cement is not affordable in Kenya and a study showed that replacing 50 % of Portland cement with RHA was effective, and the resultant concrete cost 25 % less [5, 11].

Using a concrete mix containing 10 % cement, 50 % aggregate and 40 % RHA plus water, an Indonesian company reported that it produced test blocks with an average compressive strength of 12 N/mm^2. This compares to normal concrete blocks, without RHA, which have an average compressive strength of 4.5–7 N/mm^2 or high strength concrete blocks which have a compressive strength of 10 N/mm^2. Higher strength concrete with RHA allows lighter weight products to be produced, such as hollow blocks with enhanced thermal insulation properties, which provide lighter walls for steel framed buildings. It also leads to reduced quantities of cement and aggregate [5, 11].

Rice Husk Ash Production

On average 20 % of the rice paddy is husk, giving an annual total production of 120 million tonnes. In the majority of rice producing countries much of the husk produced from the processing of rice is either burnt or dumped as waste.

Production of rice is dominated by Asia, where rice is the only food crop that can be grown during the rainy season in the waterlogged tropical areas.

Larger rice mills such as the Patum rice mill in Thailand produce 320,000 t/year, and already utilize husk for cogeneration. In developed countries, where the mills are typically larger, disposal of the husks is a big problem. Burning in open piles is not acceptable on environmental grounds, and so the majority of husk is currently going into landfill. The cost of this erodes the profit of the milling company. This has led to many research programs into potential end uses of both husk and ash, primarily in the USA.

As an example, in Iran, where rice production has had a dramatic increase over the past ten years (becoming the most important crop since 2001), the main use of rice husk is as fuel in different process. But the use of this fuel generates a huge volume of ash which has no immediate useful application and is usually dumped into water streams causing pollution and contamination of springs. As a result, the use of RHA has aroused great interest.

Rice husks are one of the largest readily available but most under-utilised biomass resources, being an ideal fuel for electricity generation. The calorific value varies with rice variety, moisture and bran content but a typical value for husks with 8–10 % moisture content and essentially zero bran is 15 MJ/kg.

The treatment of rice husk as a 'resource' for energy production is a departure from the perception that husks present disposal problems. The concept of generating energy from rice husk has great potential, particularly in those countries that are primarily dependent on imported oil for their energy needs. For these countries, the use of locally available biomass, including rice husks is of crucial importance.

Rice husk is unusually high in ash compared to other biomass fuels—close to 20 %. The ash is 92–95 % silica (SiO_2), highly porous and lightweight, with a very high external surface area. Its absorbent and insulating properties are useful to many industrial applications, and the ash has been the subject of many research studies.

If a long term sustainable market and price for rice husk ash (RHA) can be established, then the viability of rice husk power or co-generation plants are substantially improved. A 3 MW power plant would require 31,000 tonnes of rice husk per year, if operating at a 90 % capacity factor. This would result in 5580 tonnes of ash per year. Revenue from selling the ash for beneficial use would decrease the pay-back period for the capital needed to build the project. Many more plants in the 2–5 MW range can become commercially viable around the world and this biomass resource can be utilized to a much greater extent than at present [5, 11].

Rice husk ash has many applications due to its various properties. It is an excellent insulator, so has applications in industrial processes such as steel foundries, and in the manufacture of insulation for houses and refractory bricks. It is an active pozzolan and has several applications in the cement and concrete industry. It is also highly absorbent, and is used to absorb oil on hard surfaces and potentially to filter arsenic from water.

The chemical composition of RHA depends on temperature and burning time, but the variations in the components are not significant. The ash from open-field burning (or from non-controlled combustion in industrial furnaces) usually contains a higher proportion of non-reactive silica minerals such as cristobalite and tridymite, and it should be ground into very fine particles to develop pozzolanic activity. In addition, highly pozzolanic ash can be produced by means of controlled combustion when silica is kept in non-crystalline form. Such silica can react when added to cement in the presence of water (with calcium hydroxide) resulting in cementitious compounds. Most research confirms the fact that burning temperature is a critical point in the production of amorphous reactive ash.

RHA is a general term describing all types of ash produced from burning rice husks. In practice, the type of ash varies considerably according to the burning technique. Two forms predominate in combustion and gasification. The silica in the ash undergoes structural transformations depending on the temperature regime it undergoes during combustion. At 550–800 °C amorphous silica is formed and at greater temperatures, crystalline silica is formed.

These types of silica have different properties and it is important to produce ash of the correct specification for the particular end use. There are health issues associated with the use of crystalline ash, inhalation of which can lead to a number of diseases, the most common being silicosis. This affects the potential markets for this type of ash [10, 12].

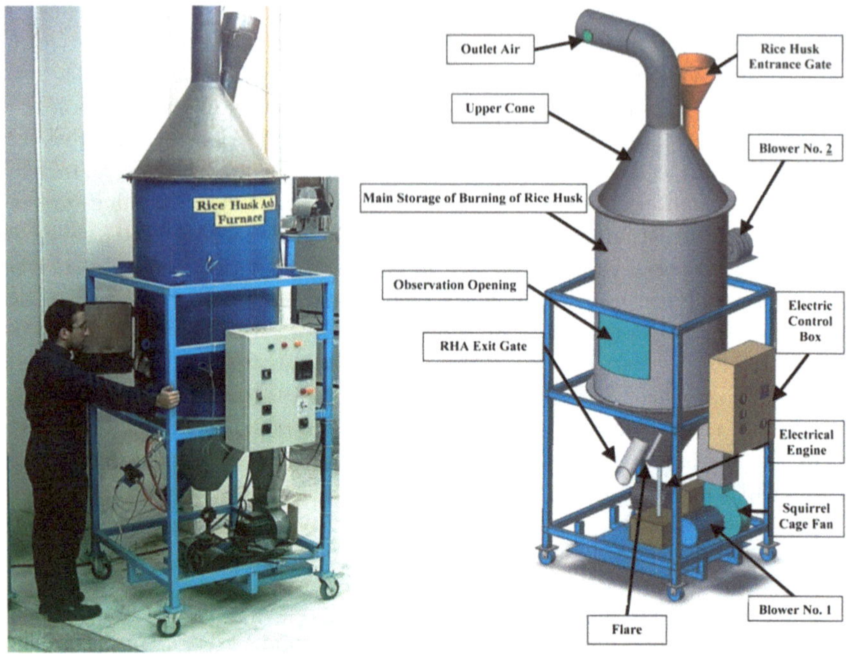

Fig. 6.1 Photograph and schematic shape of rice husk ash furnace (Ramezanianpour et al. [12])

Rice Husk Ash Optimization

From literature review it can be concluded that burning rice husks at temperature below 700 °C produces rice husk ashes with high pozzolanic activity.

In an investigation by author, ashes used for evaluation of properties of RHA concrete were produced by burning rice husks from North of Iran in the furnace. In order to supply typical RHA, a special furnace was designed and constructed in the Amirkabir University of Technology. A view of the furnace and produced RHA are shown in Fig. 6.1 [10].

This furnace was built in pilot size having the ability to control the conditions of combustion. The furnace was designed with a possibility for controlling the temperature and the rate of burning. Therefore, the furnace can be used to produce rice husk ashes with various un-burnt carbon contents.

The highest temperature in the furnace was maintained below 750 °C. This temperature was recorded in the fire zone where the rice hush was burnt. The measured temperatures were 550, 600, 650, 700 and 750 °C. Furthermore, time of burning was another variable parameter that was investigated at 30, 60 and 90 min. In addition, an ash sample was processed at temperature of 1100 °C for a few minutes. The temperatures can be varied by regulating the blowing air. Therefore ashes with various contents of un-burnt carbon were obtained.

Table 6.1 XRF analysis of RHA (Ramezanianpour et al. [12])

Oxide elements	Percentage
SiO_2	88
Al_2O_3	0.04
Fe_2O_3	0.1
CaO	0.6
MgO	0.5
Na_2O	0.03
K_2O	1.7

To identify and quantify the major and minor elements present in the samples of obtained rice husk ash, X-Ray Fluorescence (XRF) analysis was carried out. The results are given in Table 6.1. The chemical compositions of the RHA indicate that the material is mainly composed of SiO_2. From the results of this investigation Fig. 6.2 has been drawn out. Figure 6.2a can be used to obtain silicon dioxide content from temperature and time duration of burning ashes and Fig. 6.2b shows variation of LOI value versus temperature and time. Figure 6.3 shows a photo of rice husk, high carbon RHA, optimum RHA and RHA with crystalline silica.

Then in order to determine the crystalline compounds present in the various rice husk ash specimens, X-ray Diffraction (XRD) test was carried out. Figure 6.4 shows XRD patterns of the ashes. The ash patterns were denoted as RHA-Temp.-Min. respectively. The figure shows that silica in the rice husk initially exists in the amorphous form, but will not remain porous and amorphous, when combusted for a prolonged period at a temperature above 650 °C, or during less than a few minutes at 1100 °C, under oxidizing conditions. It means that the reactivity of rice husk ash is generally decreased by the increase of burning temperature and the heating duration. Burning rice husks at temperature below 650 °C produces amorphous crystals of rice husk ashes. Combination of 650 °C temperature and 60 min burning time seems to present the optimized solution resulting in non-crystallize RHA and lowest burning time.

The results of pozzolanic activity test are shown in Table 6.2. Results demonstrate high pozzolanic activity index of RHA over that of the control in accordance with ASTM C-311/ASTM C-618 test method. On the other hand, produced rice husk ash is a high reactive pozzolanic material, and entirely satisfies other requirements.

The form of silica obtained after combustion of rice husk depends on the temperature and duration of combustion of rice husk. Mehta [1] suggested that essentially amorphous silica can be produced by maintaining the combustion temperature below 500 °C under oxidizing conditions for prolonged periods or up to 680 °C with a hold time less than 1 min. However, Yeoh et al. [13] reported that RHA can remain in the amorphous form at combustion temperatures of up to 900 °C if the combustion time is less than 1 h, while crystalline silica is produced at 1000 °C with combustion time greater than 5 min. Using X-ray diffraction, Chopra et al. [14] observed that at burning temperatures up to 700 °C, the silica was in an amorphous form. The effect of different burning temperatures and the

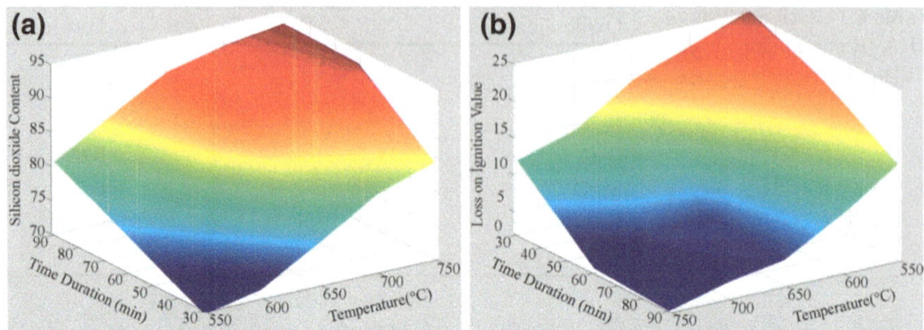

Fig. 6.2 Estimated variation of **a** SiO_2 and **b** LOI versus temperature and time duration (Ramezanianpour et al. [12])

chemical composition of rice husk (Taiwan RHA) were studied by Hwang and Wu [15]. It was observed that at 400 °C, polysaccharides begin to depolymerize. Above 400 °C, dehydration of sugar units occurs. At 700 °C, the sugar units decompose. At temperatures above 700 °C, unsaturated products react together and form a highly reactive carbonic residue. The X-ray data and chemical analyses of RHA produced under different burning conditions given by Hwang and Wu [15] showed that the higher the burning temperature, the greater the percentage of silica in the ash. K, S, Ca, Mg as well as several other components were found to be volatile.

Della et al. [16] reported that a 95 % silica powder could be produced after heat-treatment at 700 °C for 6 h. And specific surface area of particles was increased after wet milling from 54 to 81 m^2/g.

Physical and Chemical Properties of RHA

Physical Properties

RHA is a very fine material. The average particle size of rice-husk ash ranges from 5 to 10 μm. Some physical properties of rice husk ashes reported by a number of authors are given in Table 6.3.

Chemical Composition

Rice husk ash is very rich in silica content. Silica content in RHA is generally more than 80–85 %. Chemical composition of RHA reported by few authors is given in Table 6.4. For RHA to be used as pozzolan in cement and concrete, it

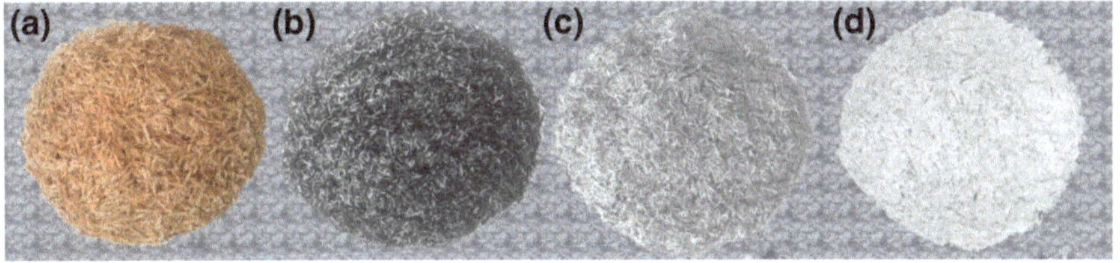

Fig. 6.3 **a** Rice husk. **b** High carbon RHA. **c** Optimum RHA. **d** RHA with crystalline silica. (Ramezanianpour et al. [12])

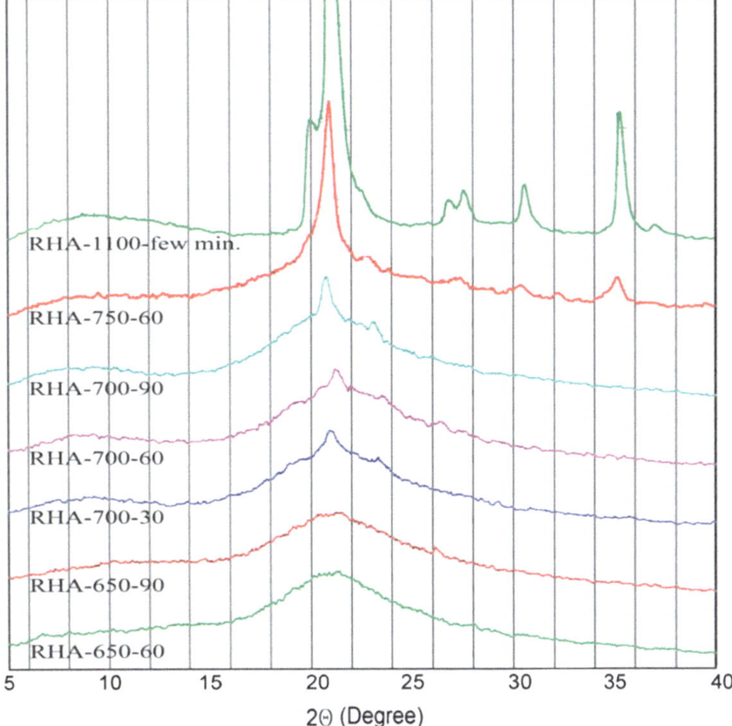

Fig. 6.4 Results of XRD on rice husk ash samples for different burning time and temperature (Ramezanianpour et al. [12])

should satisfy requirements for chemical composition of pozzolans as per ASTM C618.

The combined proportion of silicon dioxide (SiO_2), aluminium oxide ($A1_2O_3$) and iron oxide (Fe_2O_3) in the ash should be not be less than 70 %, and LOI should not exceed 12 % as stipulated in ASTM requirement.

Table 6.2 Comparison in chemical and physical specifications of produced RHA with ASTM standard C618-03 (Ramezanianpour et al. [12])

	ASTM	RHA results
Chemical requirements		
$SiO_2 + Al_2O_3 + Fe_2O_3$, min., %	70	89.9
SO_3, max., %	4	0.15
Moisture content, max., %	3	0.23
Loss on ignition (LOI), max., %	6	5.9
Physical requirements		
Fineness: amount retained when wet-sieved on 45 μm sieve, max., %	34	8
Strength activity index (20 % RHA) at 3-day, min. % control	–	102
Strength activity index (20 % RHA) at 7-day, min. % control	75	106
Strength activity index (20 % RHA) at 28-day, min. % control	75	110

Table 6.3 Physical properties of RHA

Property	Value			
	Mehta (1992)	Zhang et al. (1996)	Feng et al. (2004)	Bui et al. (2005)
Mean particle size (μm)	–	–	7.4	5
Specific gravity	2.06	2.06	2.10	2.10
Fineness: passing 45 μm (%)	99	99	–	–

Pozzolanic Activity

Rice husk contains considerable amount of SiO_2. Well burnt and well-ground rice husk ash is very active and considerably improves the strength and durability of cement and concrete. This pozzolanic material with good and consistent properties can be obtained only by burning rice husk under well-defined conditions. The sensitivity of burning conditions is the primary reason that prevents the widespread use of this material as pozzolan (Hewlett [17], Real et al. [18]). It has been reported that acid leaching of the husk helps to obtain relative pure silica with high specific surface area (Inoue and Hara [19]). It is possible to enhance the pozzolanic properties and decrease the sensitivity of burning conditions of this material.

Zhang et al. [20] studied the effect of incorporation of RHA on the hydration, microstructure and interfacial zone between the aggregate and paste. Based on the investigation, they concluded that: (i) calcium hydroxide [$Ca(OH)_2$] and calcium silicate hydrates [C–S–H] were the major hydration and reaction products in the RHA paste. Because of the pozzolanic reaction, the paste incorporating RHA had lower $Ca(OH)_2$ content than the control Portland cement paste; and (ii) incorporation of the RHA in concrete reduced the porosity and the $Ca(OH)_2$, amount in the interfacial zone; the width of the interfacial zone between the aggregate and the cement paste compared with the control Portland cement composite was also

Table 6.4 Chemical composition of RHA

Constituents	Precentage		
	Mehta (1992)	Zhang et al. (1996)	Bui et al. (2005)
Silica (SiO_2)	87.2	87.3	86.98
Alumina (Al_2O_3)	0.15	0.15	0.84
Iron Oxide (Fe_2O_3)	0.16	0.16	0.73
Calcium Oxide (CaO)	0.55	0.55	1.4
Magnesium Oxide (MgO)	0.35	0.35	0.57
Sodium Oxide (Na_2O)	1.12	1.12	0.11
Potassium Oxide (K_2O)	3.68	3.68	2.46
Sulfur Oxide (SO_3)	0.24	0.24	–
LOI	8.55	8.55	5.14

reduced. However, the porosity of the rice-husk ash composite in the interfacial zone was higher than that of the silica fume composite.

Yu et al. [21] reported that improvement of concrete properties up on addition of RHA may be attributed to the formation of more C–S–H gel and less portlandite in concrete due to the reaction between RHA and the Ca+ , OH− ions or $Ca(OH)_2$.

Cisse and Laquerbe [22] concluded that pozzolanic activity of rice-husk ash was responsible for the better strength and performance of sandcrete blocks in comparison with classic mortar blocks. In addition, the use of rice-husk ash enabled production of lightweight sandcrete with insulating properties at reduced cost.

Jauberthie et al. [23] reported that concentration of silica is high on the external face of the husk, much weaker on the internal face and practically non-existent within the husk. An analysis of these zones is given in Table 6.5.

These results confirm that the presence of amorphous silica is concentrated at the surfaces of the rice husk and not within the husk itself. Amorphous silica concentrated on the interior and exterior surfaces of the uncalcinated husk promote a pozzolanic action on the surface of the husk and therefore enable its use in cement/concrete.

Feng et al. [24] investigated the effect of hydrochloric acid pretreatment and heating course on the pozzolanic activity of rice husk ash. Three methods were used to estimate the pozzolanic activity: (i) Rapid evaluation method—conductivity measurement, the conductivity method proposed by Luxan et al. [25]. The greater the change in the conductivity of the saturated $Ca(OH)_2$ solution added with rice husk ash, the more active rice husk ash is. This method is valid to evaluate the pozzolanic activity of rice husk ash [4]; (ii) measuring the rate of consumption of $Ca(OH)_2$ that reacted with rice husk ash. Rice husk ash/ $Ca(OH)_2$ = 1:1, W/C = 1.0, 20 °C, the $Ca(OH)_2$ used was analytical grade $Ca(OH)_2$; and (iii) comparing the strength of mortar made with and without rice husk ash. Hydrochloric acid-treated rice husk was prepared by immersing rice husk in 1 N HCl aqueous solution. The husks were washed repeatedly with water

Table 6.5 Analysis of rice husk (Jauberthie et al. [23])

Element	Micro analysis of external surface of rice husk		Micro analysis of interior surface of rice husk		Micro analysis of internal surface of rice husk	
	% (by weight)	% (by atomic)	% (by weight)	% (by atomic)	% (by weight)	% (by atomic)
C	6.91	11.11	62.54	69.54	30.2	40.93
O	47.93	57.84	35.19	29.38	42.53	43.27
Si	45.16	31.05	2.27	1.08	27.27	15.8
Total	100	100	100	100	100	100

until hydrochloric acid was undetected in the filtrate and then air-dried at room temperature. Two kinds of rice husk ash were obtained by heating rice husk in a batch furnace under oxidizing atmosphere by which rice husk ash can be produced about 300 g at a time. One kind of rice husk ash (ADR) was obtained by heating hydrochloric acid treated rice husk, and another kind of rice husk ash (RHA) was obtained by heating untreated rice husk.

The heating temperature ranged from 350 to 1100 °C and the maintaining time was 4 h, and then the rice husk ash was ground in a ball mill for 60 min by adding grinding agent. Based on the study, they concluded that: (i) with hydrochloric acid pretreatment of rice husks, the pozzolanic activity of rice husk ash was stabilized and enhanced as well; the sensitivity of the pozzolanic activity of the rice husk ash to burning conditions was reduced. The pozzolanic activity of ADR (pretreated) was slightly affected by the change of maintaining time, but the maintaining time had a great affect on the pozzolanic activity of RHA (no pretreatment); (ii) the two kinds of rice husk ashes had the same rate of lime consumption at a very rapid rate in the initial period of reaction. The mechanism of reaction was consistent with diffusion control and ADR had a faster reaction rate with lime than RHA. The main reaction product was s C–S–H gel; (iii) during the first 12 h, RHA and ADR showed similar behavior in the increase of hydration heat of the cement. The pozzolanic activity of ADR is higher than that of RHA; and (iv) because of the high of amorphous SiO_2 content in ADR with high activity, a significant increase in the strength of ADR specimen was observed compared with the strength of control mortar and that made with RHA. The cement mortar added with ADR had lower $Ca(OH)_2$ content after 7 days. The pore size distribution of the mortar with ADR showed a tendency to shift towards the smaller pore size.

Agarwal [26] studied the accelerated pozzolanic activity of various siliceous materials, like rice husk ash, silica fume, fly ash, quartz, precipitated silica, and metakaolin. The compressive strength of accelerated tests was compared with cubes cured in water at 7 and 28 days. Maximum activity was observed in case of RHA (<45 μm), followed by quartz and silica fume.

Effects of Rice Husk Ash on the Properties of Fresh Concrete

Workability

Addition of fine particles of rice husk ash to the concrete mixtures increases the cohesiveness of the mixture and makes it slightly stiffer. Result is the lower slump in concrete mixtures containing rice husk ash. In order to maintain the workability, it is recommended to use water reducing admixtures in RHA concrete mixtures.

Ikpong and Okpala [27] studied the variation in workability of concrete of designed strengths with the incorporation of RHA. Cement was partially replaced with 0, 20, 25, and 30 % of RHA. 28-day designed strength of concretes was 20, 25, 30, and 40 N/mm^2. Slump and compacting factor results are given in Table 6.6. Based on the results, they concluded that: (i) to attain the same level of workability, the mixes containing rice husk ash required higher water content than those containing only ordinary Portland cement as the binder. This was reflected in the water-cement ratios of the three mixes (0, 30 and 40 % RHA contents) for each design strength, which have been calculated on the assumption of a constant cement content by weight (using the cement content of the mix containing 0 % RHA); and (ii) with minor variations, this higher water content of mixes containing RHA when compared with the conventional mixes was found to increase with the design strength of the mixes.

For the mixes having 30 % RHA content the percentage increases were 3.2, 4.4, 8.5 and 5.6 for the design strengths of 20, 25, 30, and 40 N/mm^2 respectively. And for the mixes having 40 % RHA content the percentage increases, were 6.3, 8.7, 10.2 and 13.0 for the corresponding design strengths.

Ismail and Waliuddin [28] reported the effect of rice husk ash (RHA) passing #200 and #325 sieves as 10–30 % partial replacement of cement on the workability of high-strength concrete (HSC). Control mixture had proportion of

Table 6.6 W/C ratio, slump and compaction factor values of mixes (Ikpong and Okpala [27])

Design strength (N/mm^2)	RHA (%)	W/C ratio	Slump (mm)	Compacting factor
20	0	0.80	40	0.926
	30	0.83	33	0.93
	40	0.85	35	0.92
25	0	0.69	35	0.89
	30	0.72	35	0.87
	40	0.75	37	0.92
30	0	0.59	50	0.92
	30	0.64	55	0.90
	40	0.65	35	0.89
40	0	0.54	50	0.90
	30	0.57	45	0.87
	40	0.61	45	0.91

Table 6.7 Fresh concrete properties (Ismail and Waliuddin [28])

Mix type	RHA (%)		W/(C + RHA)	Slump (mm)	Density (kg/m^2)
	#200	#325			
A	-	-	0.24	70	2425
Aa-10	10	-	0.31	30	2405
Aa-20	20	-	0.33	60	2400
Aa-30	30	-	0.36	30	2398
Ab-10	-	10	0.30	30	2403
Ab-20	-	20	0.32	45	2396
Ab-30	-	30	0.34	32	2390

1:1.07:1.90 with cement content of 571 kg/m^3. Slump and density results are given in Table 6.7.

It can be seen that slump and density values were reduced with the increase in RHA content. However, fineness of RHA did not have significant effect on both of these properties.

Bui et al. [29] investigated the influence of rice husk ash on the slump of concrete mixtures. Two types (PC30 and PC40) of ordinary Portland cement were used. Cement PC 30 and PC 40 had Blain specific surface area of 2700 and 3759 cm^2/g, respectively. Three water-to-binder ratios (0.30, 0.32, and 0.34) were used. Rice husk ash was used to replace 10, 15 and 20 % by mass of PC. The superplasticizer was added to all mixtures for obtaining high workability. In the mixtures with water to binder ratio of 0.34, the amount of superplasticizer was kept constant to investigate the influence of RHA on workability. Figure 6.5 shows the influence of RHA content on the slump of gap-graded mixtures made with water to binder ratio of 0.34. It is clear that slump decreased with the increase in RHA content for same level of superplasticizer.

Air-Entrainment

Zhang and Malhotra [20] studied the effect of RHA on the AEA requirement of concrete mixtures. They used RHA as partial replacement of cement in varying percentages (0, 5, 8, 10 and 15 %). Results are shown in Fig. 6.6. They concluded that AEA requirement increased with the increase in RHA content possibly because of high specific surface area of RHA in comparison to cement.

Consistency and Setting Times

Singh et al. [30] examined the effect of lignosulfonate (LS) and $CaCl_2$ on the setting times, consistence and pozzolanic reaction of blended cement made with

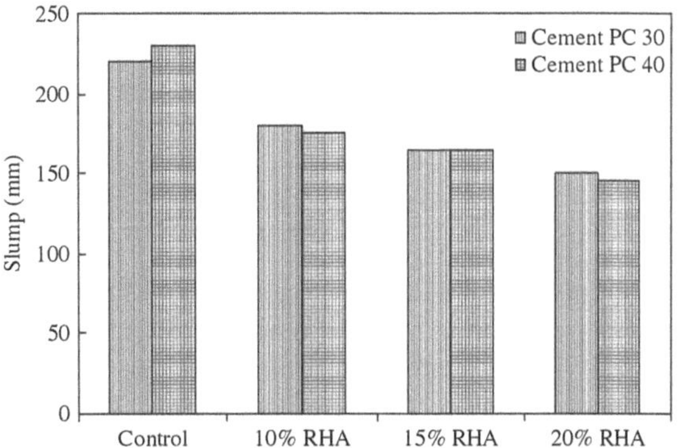

Fig. 6.5 Slump variation of gap-graded concretes made with different fineness at constant superplasticizer content (Bui et al. [29])

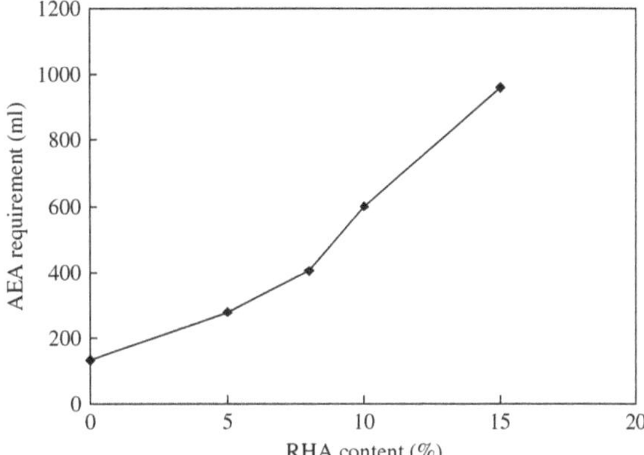

Fig. 6.6 Relationship between requirement of air-entraining admixture and RHA content (Zhang and Malhotra [20])

RHA. Pastes of 10 % RHA-blended Portland cement were made with the addition of 2 % $CaCl_2$, 1 %LS, and (1 %LS + 2 %$CaCl_2$). Both the initial and final setting times were enhanced considerably in the presence of 1 % LS indicating that it prolonged the setting behavior. It showed that 1 % LS acted as a retarder. The 2 % $CaCl_2$, on the other hand, reduced both initial and final setting times showing that it acted as an accelerator in the hydration of RHA-blended OPC. However, in the presence of a mixture of 1 % LS and 2 % $CaCl_2$, the setting time values were higher than that in the presence of 2 % $CaCl_2$ and lower than that in the presence of 1 % LS. The effect of admixtures on the consistency of 10 % RHA-blended OPC is given in Table 6.8.

From the table, it is clear that in the presence of 1 % LS, the water/binder ratio was reduced considerably as is generally expected in the presence of a superplasticizer. In the presence of 2 % $CaCl_2$ (an accelerator), the water/binder ratio was lower than that of control but higher than that in the presence of 1 % LS. However, in the presence of a mixture of 1 % LS and 2 % $CaCl_2$, the water/binder ratio was the lowest. This indicated that water reduction was enhanced in the presence of a mixture of a superplasticizer (1 % LS) and an accelerator (2 % $CaCl_2$).

Table 6.8 Variation of water consistency with admixtures (Singh et al. [30])

Composition	Water-binder
90 % OPC + 10 % RHA	0.42
90 % OPC + 10 % RHA + 2 % $CaCl_2$	0.40
90 % OPC + 10 % RHA + 1 % LS	0.38
90 % OPC + 10 % RHA + 2 % $CaCl_2$ + 1 % LS	0.37

Jaturapitakkul and Roongreung [31] studied the normal consistency and setting times of cementing material made from rice husk ash (RHA) and calcium carbide residue (CCR). Calcium Carbide Residue (CCR) is a by-product of acetylene production process. CCR consists mainly of calcium hydroxide, $Ca(OH)_2$, and is obtained in a slurry form. Chemical composition of CCR was 51.94 % CaO, 3.36 % SiO_2, 2.56 % Al_2O_3, 0.33 % Fe_2O_3, 0.46 % MgO, 0.03 % K_2O, 0.22 % SO_3 and 41.72 % LOI. The major compound in RHA was SiO_2 (78.22 %). The normal consistency of cement paste was 23.9 % while those of CCR-RHA pastes were between 43.7 % and 62.0 %, depending on RHA content. Paste 80C20R (80 % of CCR and 20 % of RHA) had normal consistency of 43.7 % and increased to 62.0 % in paste 20C80R (20 % of CCR and 80 % of RHA). The normal consistency of CCR-RHA pastes needed more water than that of cement paste because of the high porosity of the two materials as well as the high LOI of CCR. The higher the RHA content in the paste, the higher was the water requirement to maintain the same normal consistency. Initial and final setting times of the cement paste were 107 and 195 min, respectively. For CCR-RHA pastes, the shortest setting time occurred in paste 50C50R and the longest setting time was paste 80C20R. The initial and final setting times of paste 50C50R were 345 and 635 min, respectively, while those of paste 80C20R, the longest setting times, were 502 and 680 min, respectively. The initial setting time of the new cementing material was about 3.2–4.6 times longer than that of the cement paste. The results suggested that the ratio of CCR to RHA of 50–50 % by weight was the suitable ratio to obtain the highest pozzolanic reaction. In CCR-RHA paste, the $Ca(OH)_2$ from CCR reacted with SiO_2 from RHA to form CSH while Portland cement reacted with water and formed CSH directly. The long setting times of calcium carbide residue-rice husk ash paste was because the pozzolanic reaction between pozzolan and lime was usually much slower than the hydration of cement.

Effects of Rice Husk Ash on the Mechanical Properties of Hardened Concrete

Compressive and Tensile Strengths

In many investigations on the compressive and tensile strengths of mortars and concretes containing rice husk ash, there has been a significant increase when compared with plain mortars and concretes.

Ikpong and Okpala [27] studied the variation in strength of medium workability concrete with the incorporation of RHA. Cement was partially replaced with 0, 20, 25 and 30 % of RHA. 28-day designed strength of concretes was 20, 25, 30 and 40 N/mm^2. Compressive strength results up to the age of 90 days are given in Table 6.9.

The compressive strength continued to increase with age for each of the mixes. The control mix (0 % RHA) attained a higher strength than the OPC/RHA mixes at all age. At the age of 28-days all the OPC/RHA concretes, except the designed mix of 40 N/mm^2, had attained their 28-day design strengths. For the designed mix of 40 N/mm^2, the mix having 30 % RHA content reached 98.5 % of its design strength, while the one having 40 % RHA content reached 86.5 %. Furthermore even at the early age of 7 days most of the OPC/RHA concretes had attained over 60 % of their 28-day design strengths. The worst case which was the designed mix of 40 N/mm^2 containing 40 % RHA, attained 46 % of its 28-day design strength.

Zhang and Malhotra [20] investigated the influence of 10 % RHA inclusion as partial replacement of cement on the compressive strength of concrete and compared it with the compressive strength of concrete containing 10 % silica fume (SF). Water to-cementitious material ratio was maintained at 0.40. Compressive strength results up to the age of 730 days are shown in Fig. 6.7. It can be seen that RHA concrete, in general, achieved higher strength than control concrete mixture but lower than that of silica fume concrete. At 28-days, the RHA concrete had a compressive strength of 38.6 MPa compared with 36.4 MPa for the control concrete and 44.4 MPa for SF concrete. At 180 days, RHA concrete exhibited compressive strength of 48.3 MPa compared with 44.2 MPa for the control concrete and 50.2 MPa for SF concrete.

Ismail and Waliuddin [28] examined the effect of rice husk ash (RHA) passing #200 and #325 sieves as 10–30 % partial replacement of cement on the compressive strength of high-strength concrete (HSC). Control mixture had mixture

Table 6.9 Compressive strength development with age (Ikpong and Okpala [27])

Design strength (N/mm^2)	RHA (%)	Compressive strength (N/mm^2)			
		7-day	14-day	28-day	90-day
20	0	17.5	23.9	33.1	37.8
	30	15.8	18.2	29.1	36.1
	40	3.27	15.2	27.2	33.9
25	0	22.9	27.2	35.4	39.2
	30	18.7	23.8	31.1	37.5
	40	14.8	21.0	28.6	34.0
30	0	28.4	37.6	43.5	46.3
	30	22.6	32.0	38.3	43.2
	40	16.7	27.5	31.8	37.1
40	0	32.4	38.8	44.2	47.6
	30	24.0	33.5	39.4	44.9
	40	1.5	29.3	34.6	38.9

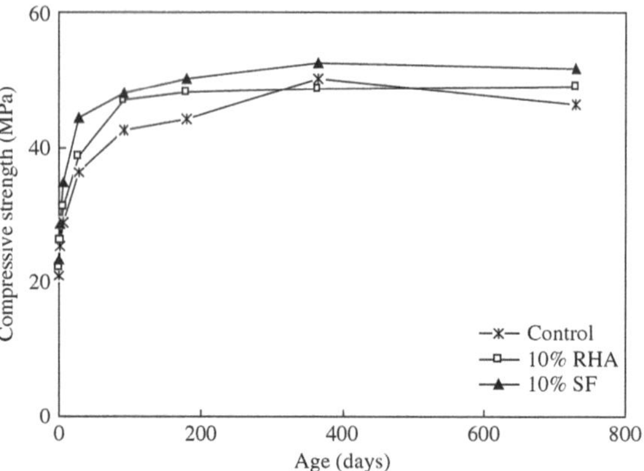

Fig. 6.7 Development of compressive strength of concrete with RHA and silica fume (Zhang and Malhotra [20])

proportion of 1:1.07:1.90 with cement content of 571 kg/m^3. Strength tests were conducted up to the age of 150 days, and results are given in Table 6.10. Based on the test results, they concluded that: (i) it is possible to get high strength concrete economically using RHA; (ii) optimum replacement of cement by RHA was around 10–20 % with finely ground RHA; (iii) rate of hydration in concrete made with part replacement of cement by RHA was slow as compared to concrete with OPC. This fact was very dominant during the initial 3 days of age of concrete. This rate of slow hydration also affects the 150 day strength of concrete made by part replacement of cement by RHA in the mix; and (iv) the lower strength of concrete made with part replacement of cement by RHA was because of higher w/c ratios. Though the w/c ratio for samples Ab-10 and Ab-20 was 23 and 28 % higher than w/c ratio in specimen-A, the 28 day strength was 98 % and 97 % of the concrete of Mix-A where no replacement was made.

Zhang and Malhotra [20] studied the mechanical properties of concrete made with 10 % RHA and 10 % silica fume (SF). Tests were conducted at the age of 28 days, and results are given in Table 6.11. These results indicated that splitting tensile strength, flexural strength and modulus of elasticity of control and concrete incorporating RHA and SF were comparable.

Table 6.10 Compressive strength of RHA concrete (Ismail and Waliuddin [28])

Mix type	RHA (%)		w/(C + RHA)	Compressive strength (MPa)			
	#200	#325		3-day	7-day	28-day	150-day
A	–	–	0.24	54.3	62.3	72.4	85.0
Aa-10	10	–	0.31	46.2	56.0	68.1	71.1
Aa-20	20	–	0.33	35.3	46.8	57.3	57.4
Aa-30	30	–	0.36	31.5	39.3	47.7	48.8
Ab-10	–	10	0.30	47.0	61.0	71.0	72.4
Ab-20	–	20	0.32	46.7	56.0	70.2	70.3
Ab-30	–	30	0.34	43.1	51.7	63.0	63.2

Table 6.11 Mechanical properties of concrete mixtures (Zhang and Malhotra [20])

Mix	RHA (%)	SF (%)	W/Cm ratio	Spliting tensile strength (MPa)	Flexural strength (MPa)	Modulus of elasticity (GPa)
1	0	0	0.40	2.7	6.3	29.6
2	10	–	0.40	3.5	6.8	29.6
3	–	10	0.40	2.8	7.0	31.1

Wada et al. [32] demonstrated that RHA mortar and concrete exhibited higher compressive strength than the control mortar and concrete. They have further reported excellent strength development at the early stages even without steam curing for RHA mortar and concrete.

Singh et al. [30] investigated the effect of lignosulfonate (LS) and $CaCl_2$ on the compressive strength of blended cement made with RHA. Pastes of 10 % RHA blended Portland cement were made with the addition of 2 % $CaCl_2$, 1 % LS, and (1 % LS + 2 % $CaCl_2$). Compressive strength was determined at the age of 7 and 28 days, and results are shown in Fig. 6.8. The figure shows that in the presence of 2 % $CaCl_2$, compressive strength was higher at 7 days and lower at 28 days of hydration than that of the control. In the presence of 1 % LS, the values were lower than that of control on all days of hydration. However, in the presence of a mixture of 1 % LS and 2 % $CaCl_2$, the compressive strength was maximum at 28 days of hydration. The results exhibited very clearly that when a mixture of a superplasticizer (1 % LS) and an accelerator (2 % $CaCl_2$) was added to the RHA blended OPC, the superplasticizer reduced the water demand and the accelerator enhanced the compressive strength. Thus the role of the two admixtures becomes supplementary to each other.

Nehdi et al. [33] made a comparative study of the compressive strength of concrete mixtures made with: (i) 0 % rice husk ash; (ii) Egyptian rice husks tested

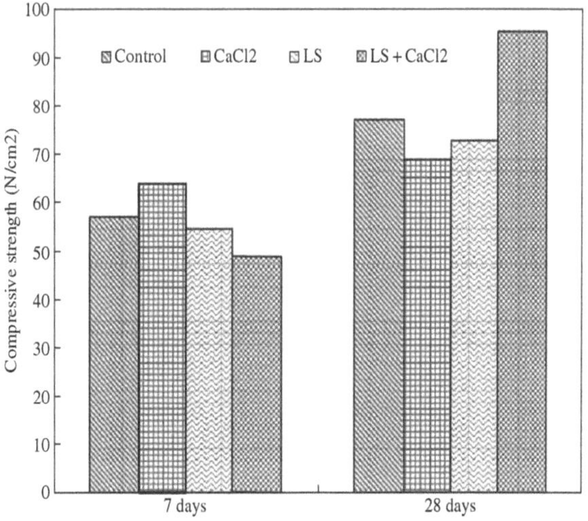

Fig. 6.8 Effect of admixtures on the compressive strength of blended cement and 28 days (Singh et al. [30])

in a Torbed reactor at combustion temperatures between 750 and 830 C plus external jet of air. These were categorized as EG-RHA (A), EG-RHA (B) and EG-RHA (C); (iii) a raw rice husk ash (RAW-US-RHA) and a high quality RHA (US-RHA) from USA, produced using fluidized bed technology; and (iv) silica fume (SF). Physical properties of cementitious materials are given in Table 6.12. Three different percentages (7.5, 10, and 12.5 %) of Egyptian and US rice husk ashes, SF, and two percentages (7.5 and 10 %) of raw US rice husk ashes were used. Compressive strength test results at the age of 28 days are shown in Fig. 6.9. It can be seen from this figure that the strength of all mixtures incorporating RHA and SF additions outperformed that of the reference mixture.

Bui et al. [29] investigated the compressive strength of concrete mixtures made with two types (PC30 and PC40) of ordinary Portland cement having Blain specific surface area of 2,700 and 3,759 cm^2/g, respectively. Rice husk ash was used to replace 10, 15 and 20 % by mass of PC. Based on the results, they concluded that: (i) rice husk ash (RHA) can be used as a highly reactive pozzolanic material to improve the microstructure of the interfacial transition zone (ITZ) between the cement paste and the aggregate in high performance concrete; (ii) in addition to the pozzolanic reactivity of RHA (chemical aspect), the particle grading (physical aspect) of cement and RHA mixtures exerted significant influences on the blending efficiency; (iii) relative strength increase (relative to the concrete made with plain cement, expressed in %) was higher for coarser cement; and (iv) simulation results demonstrated that the favorable results for coarser cement (i.e., the gap-graded binder) reflected improved particle packing structure companied by a decrease in porosity and particularly in particle spacing.

Rodriguez de Sensale [34] studied the compressive and splitting tensile strengths of concretes with rice-husk ashes (RHA).

Test results of strengths are given in Table 6.13. Based on the results obtained, he concluded that: (i) at lower age (7 days), concretes with UY-RHA exhibited higher compressive strength that concretes with USA-RHA; (ii) at higher age (91 days), the RHA concrete had higher compressive strength in comparison with that of concrete without RHA, and the highest values of compressive strengths were achieved in concretes with 20 % USA-RHA; (iii) long term compressive

Table 6.12 Physical properties of cementitious materials (Nehdi et al. [33])

Property	Type 1 cement	RHA-A (750 °C)	RHA-B (830 °C)	RHA-A (750 °C) + air	US-RHA	SF
Specific gravity	3.17	2.05	2.05	2.05	2.06	2.16
Passing 45 μm (%)	83.9	98.2	97.7	97.9	98.4	97.2
Mean partial size (μm)	13.4	7.15	7.63	7.41	6.8	0.16
Water requirement (%)	–	99	95	100	104	114
Pozzolanic activity index (%) at 7 days		88	105	117	114	94
Pozzolanic activity index (%) at 28 days		122	117	144	109	102

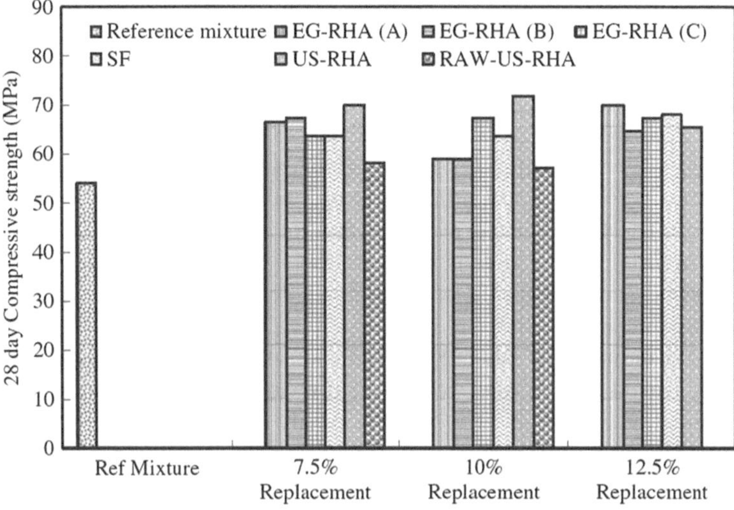

Fig. 6.9 Compressive strength for various concrete mixtures made with rice husk ashes from Egypt and US (Nehdi et al. [33])

Table 6.13 Compressive and splitting tensile strength results of RHA concretes (Rodriguez de Sensale [34])

w/(c + RHA)	RHA		Compressive strength (MPa)			Splitting tensile strength (MPa)
	Type	%	7-day	28-day	91-day	
0.32	–	0	48.4	55.5	60.6	3.63
	UY	10	51.1	60.4	67.3	3.57
		20	44.3	54.8	62.7	3.54
	USA	10	39.5	51.4	64.5	3.62
		20	30.5	47.4	68.5	3.54
0.40		0	35.8	42.3	45.6	–
	UY	10	41.1	50.4	54.9	–
		20	27.9	40.7	51.4	–
	USA	10	29.7	40.8	51.5	–
		20	23.6	39.4	57.3	–
0.50		0	24.6	32.9	35.9	2.85
	UY	10	24.1	31.5	35.5	2.32
		20	24.9	34.9	37.9	2.63
	USA	10	22.7	34.5	44.4	2.92
		20	20.8	35.9	52.9	3.00

strength of the concretes with UY-RHA was not as high as the one obtained with USA-RHA, which also increased with the increase in RHA content; and (iv) the results of splitting tensile revealed the significance of the filler and pozzolanic effect for the concretes with RHA. On the one hand, the results were consistent with the compressive strength development at 28 days for the USA-RHA. On the other hand, in the concretes with UY-RHA, lower splitting tensile strengths were observed, which can be due to the fact that with residual RHA, the filler effect of the smaller particles in the mixture was higher than the pozzolanic effect. The results are compared with those of the concrete without RHA, with splitting tensile strength and air permeability. It is concluded that residual RHA provides a positive effect on the compressive strength at early age, but the long term behavior of the concretes with RHA produced by controlled incineration was more significant. Results of splitting tensile and air permeability reveal the significance of the filler and pozzolanic effect for the concretes with residual RHA and RHA produced by controlled incineration.

Sakr [35] examined the role of RHA inclusion as partial replacement of cement, on the mechanical properties of heavyweight concrete. The aggregates used were special natural heavy weight mineral ores, mainly ilmenite and baryte. Tests were conducted for compressive strength, indirect tensile strength, flexural strength, bond strength, and modulus of elasticity at the age of 28 days. They concluded that: (i) with the increase in RHA content up to 15 %, generally, all the mechanical properties increased. Beyond 15 %, decrease in strength properties values was observed.

Saraswathy and Song [36] reported the compressive, splitting tensile and bond strength of concrete in which cement was partially replaced with RHA. 28-day results of compressive strength, splitting tensile strength and bond strength are given in Table 6.14. Based on the results obtained, they concluded that: (i) compressive strength increased with increase in RHA contents, blending percentage and with age. At 28 days, all the rice husk ash replaced concretes exhibited higher compressive strength than the control concrete; (ii) inclusion of RHA (up to 25 %) as partial replacement of cement did not affect the splitting tensile strength of concrete. At 25 % replacement level, a slight decrease in splitting tensile strength was observed; and (iii) all the rice husk replaced concretes showed higher bond strength values than the conventional concrete, indicating that replacement of rice husk ash does not affect the bond strength properties.

Chindaprasirt et al. [37] determined the compressive strength development of blended cements containing rice husk ash and fly ash. Class F lignite fly ash and

Table 6.14 Twenty-eight-day strengths of RHA concretes (Saraswathy and Song [36])

RHA (%)	Compressive strength (Mpa)	Splitting tensile strength (Mpa)	Bond strength (Mpa) at 0.25 mm slip
0	36.45	4.49	3.32
5	37.49	4.57	4.11
10	37.43	4.65	4.31
15	37.38	4.92	3.79
20	37.71	4.60	3.43
25	39.55	4.58	4.07
30	37.80	3.67	3.87

Table 6.15 Compressive strength of blended cements (Chindaprasirt et al. [37])

Mix	w/b ration	Compressive strength (Mpa)			
		7 day	28 day	90 day	180 day
OPC	0.55	44	51	57	60
FA20	0.53	32	45	57	57
FA40	0.51	29	46	62	77
RHA20	0.68	31	54	61	62
RHA40	0.80	17	32	43	53

RHA were used at replacement dosages of 20 and 40 % by weight of cement. RHA contained high silica content of 90 % and low loss on ignition (LOI) of 3.2 %. The Blaine fineness of RHA was 140,00 cm^2/g. Table 6.15 shows water-to-binder (W/B) ratios and compressive strengths of the mortar mixes. Use of fly ash and RHA resulted in a reduction of 7 and 28 day compressive strength. The reduction of early strength is typical of the fly ash mixes. For RHA mixes, the low initial strength was due to the high water-to-binder of the mixes. For 20 % RHA replacement level, although the W/B ratio was increased, the strength at 28 days was higher than that of PC mix, suggesting that RHA was quite reactive.

Although possessing high fineness, RHA contribution to the strength development was limited by the high water demand associated with its high surface areas.

Ramezanianpour et al. [12] studied the variation in strength of concrete containing RHA. A total of four concrete mixtures were made; one corresponding to a control concrete (CTL) and three others with 7, 10 and 15 % RHA replaced with cement by weight. Table 6.16 shows the physical and chemical characteristics of RHA and cement. Concrete cubes of 100 × 100 × 100 mm dimension were cast for compressive strength, while two 150 × 300 mm cylinder concrete specimens were prepared for the tensile strength test.

The results of pozzolanic activity test are shown in Table 6.17. Results demonstrate high pozzolanic activity index of RHA over that of the control in accordance with ASTM C-311/ASTM C-618 test method. On the other hand, produced rice husk ash is a high reactive pozzolanic material, and entirely satisfies other requirements.

Results of the compressive strengths of concretes are given in Fig. 6.10. In general, the RHA concrete had higher compressive strengths at various ages and up to 90 days when compared with the control concrete. The results show that it was possible to obtain a compressive strength of as high as 46.9 MPa after 28 days. In addition, strengths up to 63.2 MPa were obtained at 90 days.

Figure 6.11 shows that concrete containing RHA has a greater splitting tensile strength than that the control concrete at all ages. It is clear that, as the amount of RHA increases, the tensile strength increases up to 20 %. For instance, at 90 days the 15 % RHA concrete had a compressive strength of 5.62 MPa compared with 4.58 MPa for the control concrete.

Table 6.16 Physical and chemical characteristics of cement and RHA (Ramezanianpour et al. [12])

	Physical tests		Chemical analyses,(%)								Bogue composition, (%)			
	Specific gravity	Blaine (cm^2/gram)	SiO_2	Al_2O_3	Fe_2O_3	CaO	MgO	Na_2O	K_2O	LOI	C_3S	C_2S	C_3A	C_4AF
RHA	2.15	3600	89.61	0.4	0.22	0.91	0.42	0.07	1.58	5.91	–	–	–	–
Cement	3.21	3200	21.50	3.68	2.76	61.50	4.80	0.12	0.95	1.35	51.1	23.1	5.1	8.4

Table 6.17 Comparison in chemical and physical specifications of produced RHA with ASTM standard C618-03 (Ramezanianpour et al. [12])

	RHA results
Chemical requirements	
$SiO_2 + Al_2O_3 + Fe_2O_3$, min., %	89.9
SO_3, max., %	0.15
Moisture content, max., %	0.23
Loss on ignition (LOI), max., %	5.9
Physical Requirements	
Fineness: Amount retained when wet-sieved on 45 μm sieve, max., %	8
Strength activity index (20 % RHA) at 3-day, min. % control	102
Strength activity index (20 % RHA) at 7-day, min. % control	106
Strength activity index (20 % RHA) at 28-day, min. % control	110

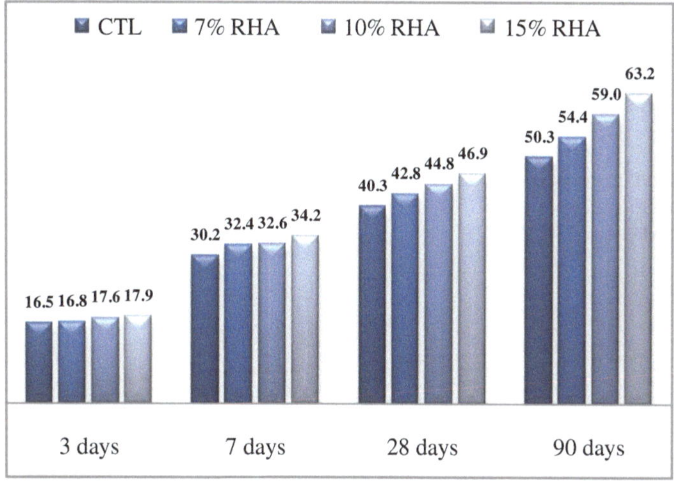

Fig. 6.10 Compressive strength (MPa) at various ages for control (CTL) and RHA mixtures (Ramezanianpour et al. [12])

Effect on Volume Changes of Concrete

Drying Shrinkage

Shrinkage of concrete mixtures incorporating rice husk ashes have been investigated by only a few researchers.

Zhang and Malhotra [20] studied the drying shrinkage strain of concretes made with 10 % RHA and 10 % silica fume (SF). Tests were conducted up the age of 448 days. Figure 6.12 shows the drying shrinkage strain of concretes after 7 days of initial curing in lime-saturated water. These results indicated that RHA concrete

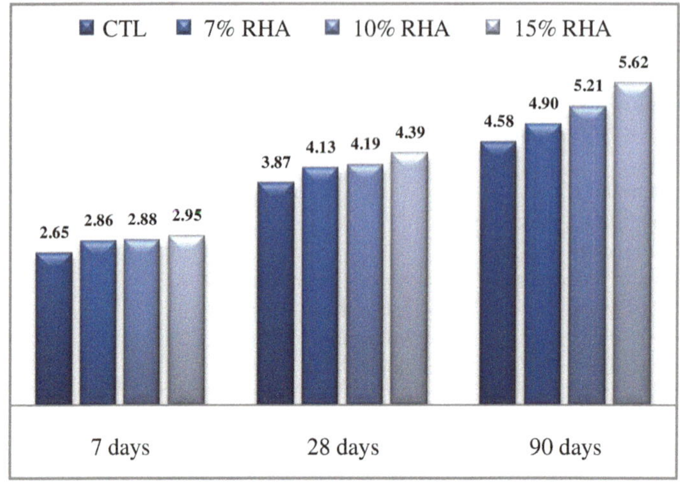

Fig. 6.11 Tensile strength (MPa) at various ages for control (CTL) and RHA mixtures (Ramezanianpour et al. [12])

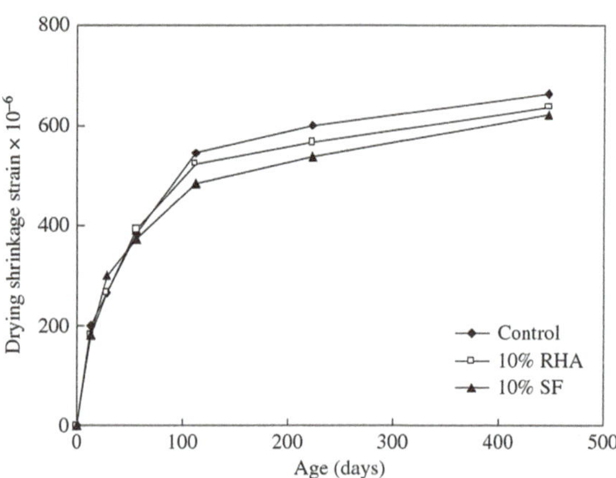

Fig. 6.12 Drying shrinkage of RHA and SF concretes (Zhang and Malhotra [20])

had a drying shrinkage of 638×10^{-6} after 448 days, which was similar to the strains for the control and silica fume concretes.

Microstructure, Porosity and Permeability

Microstructure of concrete will be affected by the incorporation of rice husk ash in the mixture. The major influence is the refinement of the pore structure of cement paste.

Rice husk ash addition to the concrete mixture also improves the interfacial transition zone between cement paste and aggregates. The content of calcium hydroxide is reduced in the ITZ due to the pozzolanic reaction of silica fume. This will reduce the porosity and permeability of concretes at the interface region and enhances the bonding of paste and aggregates.

Permeability of concrete is also affected by the incorporation of rice husk ash in concrete mixtures. It is even more important than the effect on compressive strength. Saraswathy and Song [36] investigated the effect of partial replacement of cement with rice husk ash (RHA) on the porosity and water absorption of concrete. Cement was replaced with 0, 5, 10, 15, 20, 25, and 30 % RHA. Proportion of control (without RHA) mix was 1:1.5:3 with w/c ratio of 0.53. Porosity and water absorption test was carried out as per ASTM C642-97. For determination of effective porosity and coefficient of water absorption, specimens of size 83 mm diameter and 50 mm thick were cast with and without rice husk ash and cured for 28 days in distilled water. After the curing period was over the specimens were dried in an oven at 105 ± 5 C for 48 h in order to evaporate the moisture content present in the concrete.

Table 6.18 shows the results of porosity and coefficient of water absorption. Based on the results, they concluded that: (i) porosity values decreased with the increase in RHA content because small RHA particles improved the particle packing density of the blended cement, leading to a reduced volume of larger pores; and (ii) coefficient of water absorption for rice husk ash replaced concrete at all replacement levels was found to be less when compared to control concrete.

Rodriguez de Sensale [34] studied the air-permeability of concretes with two sources of rice-husk ash; a residual RHA from rice paddy milling industry in Uruguay and another produced by controlled incineration from the USA. Two (10 and 20 %) replacement percentages of cement by RHA, and two water/cementitious material ratios (0.32 and 0.50) were used. The percentage of reactive silica contained in the USA RHA was 98.5 % and in the UY-RHA was 39.55 %. Air-permeability for concrete was determined with the "Torrent permeability tester" method (Torrent [38]; Torrent and Frenzer [39]). 150 × 300 mm cylinders were used for measurement of air-permeability at the age of 28 days. The particular features of the Torrent method are a two-chamber vacuum cell and a pressure

Table 6.18 Porosity and water absorption of rice husk replaced Concrete (Saraswathy and Song [36])

RHA (%)	Porosity (%)	Water absorption coefficient (m^2/s)
0	18.06	3.5571×10^{-10}
5	18.18	6.7587×10^{-11}
10	18.82	1.0302×10^{-11}
15	13.82	1.0644×10^{-11}
20	13.54	1.2122×10^{-10}
25	13.04	1.4548×10^{-10}
30	11.89	1.3030×10^{-10}

Table 6.19 Permeability results of RHA concretes (Rodriguez de Sensale [34])

w/(C + RHA)	RHA		Permeability coefficient (m^2)
	Type	%	
0.32	–	0	1.08×10^{-16}
	UY	10	0.23×10^{-16}
		20	0.05×10^{-16}
	USA	10	0.08×10^{-16}
		20	0.0×10^{-16}
0.50		0	28.20×10^{-16}
	UY	10	71.82×10^{-16}
		20	9.10×10^{-16}
	USA	10	26.36×10^{-16}
		20	14.20×10^{-16}

regulator, which ensures that air flows at right angles to the surface and is directed towards the inner chamber; this allows the calculation of the permeability coefficient Kt on the basis of a simple theoretical model. By comparing the results (Torrent and Frenzer [39]) of gas permeability measured by the Torrent permeability tester (Kt) and oxygen permeability obtained for the Cembureau method (Ko), the following relation is presented: Ko = 2.5 $K^{0.7}$t where Ko and Ktare expressed in 10–16 m^2. Permeability results of RHA concretes are given in Table 6.19. It can be seen from the table that: (i) for a particular water-cementitious ratio, permeability of UY-RHA concrete was more than that of USA-RHA concrete; and (ii) with the increase in water-cementitious ratio, permeability increased for both types of RHA. The results of air permeability revealed the significance of the filler and pozzolanic effect for the concretes with RHA. On one hand, the results are consistent with the compressive strength development at 28 days for the USA RHA. On the other hand, in the concretes with UY RHA, lower air permeability was observed, which can be due to the fact that with residual RHA, the filler effect of the smaller particles in the mixture is higher than the pozzolanic effect.

Effect of Rice Husk Ash on Durability of Concrete

Durability of concrete is usually enhanced by the addition of rice husk ash. Rice husk ash can improve the chemical and physical properties of concrete. It also reduces the permeability of concretes leading to improvement of durability.

Carbonation of Concretes Containing Rice Husk Ash

Similar to other pozzolanic materials, rice husk ash consumes the calcium hydroxide of the cement paste. This may increase the risk of carbonation in the

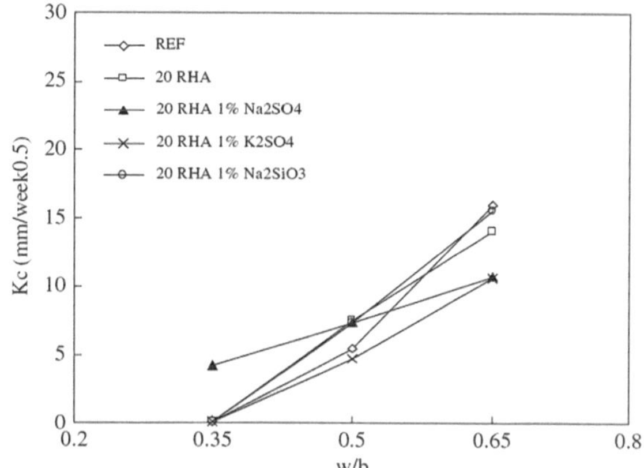

Fig. 6.13 Carbonation coefficients with reference to the w/b ratios (Gastaldini et al. [40])

rice husk ash mortar and concrete. On the other hand rice husk ash addition will reduce the permeability of concrete which may result in lower carbonation. These two contradictory effects of silica fume on concrete carbonation have been reported by a few researchers.

Gastaldini et al. [40] examined the role of chemical activators on the carbonation of concrete containing 20 % of rice husk ash as partial replacement of cement.

Water/binder ratios used were 0.35, 0.50 and 0.65 and binder/aggregate ratios were 1:3.75, 1:5.25 and 1:6.9. Potassium sulfate (K_2SO_4), sodium sulfate (Na_2SO_4) and sodium silicate (Na_2SiO_3) were used as chemical activators in concentrations of 1 % by weight of cement. The top surface of the specimens was sealed and they underwent a preconditioning cycle as required by RILEM TC116-PCD [41]. They were then placed in a controlled atmosphere chamber with 5 % CO_2, at 23 $\pm$ 1 C and RH 65 $\pm$ 1 %. The carbon dioxide penetration depth was measured at different exposure times, 4, 8 and 12 weeks by means of the phenolphthalein test carried out on the transversely split section of the cylinders specimens using the RILEM CPC18 [42] method. Figure 6.13 shows the changes in these coefficients for the w/b ratios used. For all mixtures investigated, Kc increased with the increase in w/b ratio because of the increase in concrete porosity and the lower concentration of cement. For the same w/b ratio (0.35, 0.50 and 0.65), the lowest carbonation coefficients were seen in the mixture with RHA and 1 % K_2SO_4 showed. The values obtained were lower than those in the reference concrete.

Effect of Chloride Ions on Mortars and Concretes Containing Rice Husk Ash

The use of rice husk ash in concrete slightly reduces the alkalinity of pore water and will cause a reduction in the threshold value of chloride necessary for the initiation of corrosion of steel bars. On the contrary, the use of silica fume reduces the permeability and chloride diffusion of concrete.

Zhang and Malhotra [20] investigated the chloride-ion penetration resistance of concretes made with 10 % RHA and 10 % silica fume (SF). Details of the concrete mixture along with the chloride-ion penetration test conducted as per ASTM C1202 are given in Table 6.20. It can be seen that the use of RHA and SF has drastically reduced the chloride-ion penetration at both ages. These values were less than 1000 coulombs. As per ASTM C1202, when charge passed through concrete is less than 1000 coulombs, the concrete has very high resistance to chloride-ion penetration.

Nehdi et al. [33] made a comparative study of the rapid chloride permeability of concrete mixtures made with: (i) 0 % rice husk ash; (ii) Egyptian rice husks; EG-RHA (A), EG-RHA (B) and EG-RHA (C); (iii) a raw rice husk ash (RAW-US-RHA) and a high quality RHA (US-RHA) from USA, produced using fluidized bed technology; and (iv) silica fume (SF). Physical properties of cementitious materials have already been given in Table 6.12. Three different percentages (7.5, 10, and 12.5 %) of Egyptian and US rice husk ashes, SF, and two percentages (7.5 and 10 %) of raw US rice husk ashes were used. Chloride permeability tests were carried out according to ASTM C1202 test method. According to this standard, permeability categorization is as: Very Low (0–1000 Coulombs); Low (1000–2000 Coulombs); Moderate (2000–3500 Coulombs). The 28 day test results are shown in Fig. 6.14. Based on the results, they concluded that: (i) non-ground RHA did not significantly change the rapid chloride penetrability classification of concrete; and (ii) finely ground RHA reduced the rapid chloride penetrability of concrete from a moderate rating to low or very low ratings depending on the type and addition level of RHA. Such reductions are comparable to those achieved by SF.

Coutinho [43] determined the rapid chloride permeability of concrete with partial replacement of RHA in various percentages (10, 15, and 20 %) and when using controlled permeability formwork (CPF). Controlled permeability formwork (or CPF) is one of the few techniques developed recently for directly improving the concrete surface zone. This technique reduces the near-surface water/binder ratio and reduces the sensitivity of concrete to poor site curing. CPF consists of using a textile liner on the usual formwork, allowing air bubbles and surplus water to drain out but retaining the cement particles and so enabling the water–cement ratio of the outer layer to become very low and the concrete to hydrate to a very dense surface skin as the filter makes enough water available at the right time to activate optimum hydration. So CPF enhances durability by providing an outer concrete layer which is richer in cement particles, with a lower water/binder ratio,

Table 6.20 Test results of resistance of concrete to chloride-ion penetration (Zhang and Malhotra [20])

Mix no.	Type of concrete	W/Cm	Unit weight (kg/m^3)	Comperssive strength (MPa)	Chloride-ion resistance (Coulombs)	
					28 days	90 days
CO-D	Control	0.40	2320	36.5	3175	1875
R10-D	10 % RHA	0.40	2320	45.5	875	525
SF10-D	10 % SF	0.40	2320	42.5	410	360

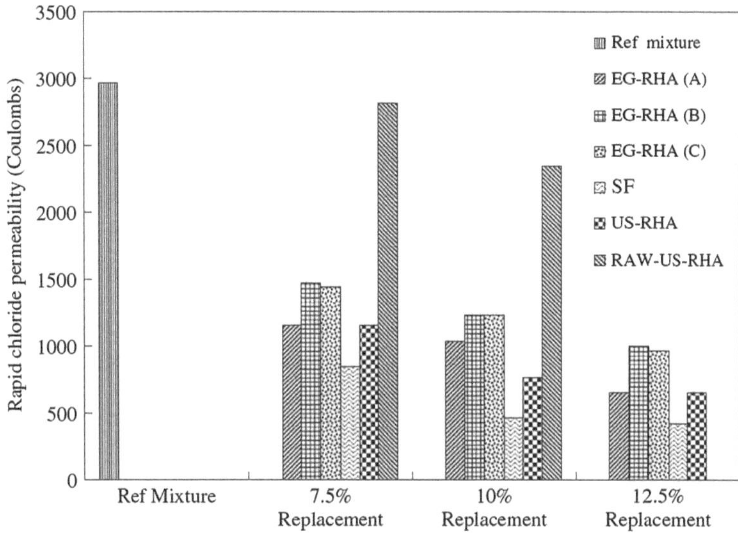

Fig. 6.14 Rapid chloride permeability at 28 days for various concrete mixtures made with rice husk ashes from Egypt and US (Nehdi et al. [33])

less porous and so much less permeable than when ordinary formwork is used. Resistance to chloride penetration was assessed with the AASHTO T277-83 test method up to the age of 100 days. Table 6.21 shows the average of 3 test specimens for each series. It is evident from these results that inclusion of RHA significantly reduced the charge passed. Furthermore, when CPF was used, it greatly reduced the permeability of concrete mixtures.

Saraswathy and Song [36] investigated the effect of partial replacement of cement with rice husk ash (RHA) on the chloride permeability of concrete. Proportion of control (without RHA) mix was 1:1.5:3 with w/c ratio of 0.53. Cement was replaced with 0, 5, 10, 15, 20, 25, and 30 % RHA. Test was conducted according to ASTM C1202-94 test method. Concrete specimens with and without rice husk ash were cast and allowed to cure for 28 days. Rapid chloride

Table 6.21 Rapid chloride permeability results (Coutinho [43])

Type of mixture	Average charge passed (Coulombs)
Control	2349.3
Control + CPF	1916.3
10 % RHA	435.0
10 % RHA + CPF	384.7
15 % RHA	322.0
15 % RHA + CPF	245.0
20 % RHA	260.0
20 % RHA + CPF	202.0

permeability (charge passed) through concretes made with 0, 5, 10, 15, 20, 25, and 30 % RHA were 1161, 1108, 653, 309, 265, 213, and 273 coulombs, respectively.

It can be seen that replacement of rice husk ash drastically reduced the Coulomb values. As the replacement level increased, the chloride penetration decreased. According to ASTM C1202 criteria, RHA reduced the rapid chloride penetrability of concrete from a low to very low ratings from higher to lower replacement levels.

Gastaldini et al. [40] studied the influence of chemical activators (K_2SO4, Na_2SO_4, Na_2SiO_3) on the chloride-ion permeability of concrete made with 20 % of rice husk ash as partial replacement of cement. Water/binder ratios used were 0.35, 0.50 and 0.65 and binder/aggregate ratios were 1:3.75, 1:5.25 and 1:6.9. Potassium sulfate (K_2SO_4), sodium sulfate (Na_2SO_4) and sodium silicate (Na_2SiO_3) were used as chemical activators in concentrations of 1 % by weight of cement. Chloride-ion penetration was measured according to ASTM C1202 at the age of 28 and 91 days. Results of chloride-ion penetration are given in Table 6.22. Based on the results they concluded that: (i) at 28 days, RHA concrete exhibited significant reduction in the total charge passed. This reduction was about 42, 51 and 27 % for w/b = 0.35, 0.50 and 0.65, respectively. At 91 days, the same w/b ratios showed reductions of 65, 68 and 59 %; (ii) concrete mixtures with activators showed lower total charge passed values when compared with the mixture without activator.

The mixtures activated with K_2SO_4 showed the best results. At 28 days, the mixture activated with Na_2SiO_3 showed the lowest charge passed values. Overall, the best results at 91 days were seen in the sample activated with K_2SO4; and (iii) at 91 days, all mixtures with chemical activators showed very low charge passed values (100–1000 C), even for w/b ratios as high as 0.65 which can be rated as very low in accordance with ASTM C1202.

Table.6.22 Chloride-ion permeability of RHA concretes (Gastaldini et al., [38])

Mixture	w/b ratio	Total charge passed(C)	
		28 days	91 days
REF (Control)	0.35	1227	1288
	0.50	3166	2136
	0.65	3681	2866
20 RHA	0.35	999	452
	0.50	1557	692
	0.65	2677	1176
20 RHA 1% Na_2So_4	0.35	933	515
	0.50	1393	630
	0.65	2001	760
20 RHA 1% K_2SO_4	0.35	820	326
	0.50	1312	552
	0.65	2242	818
20 RHA 1% Na_2So_3	0.35	704	342
	0.50	914	578
	0.65	1470	732

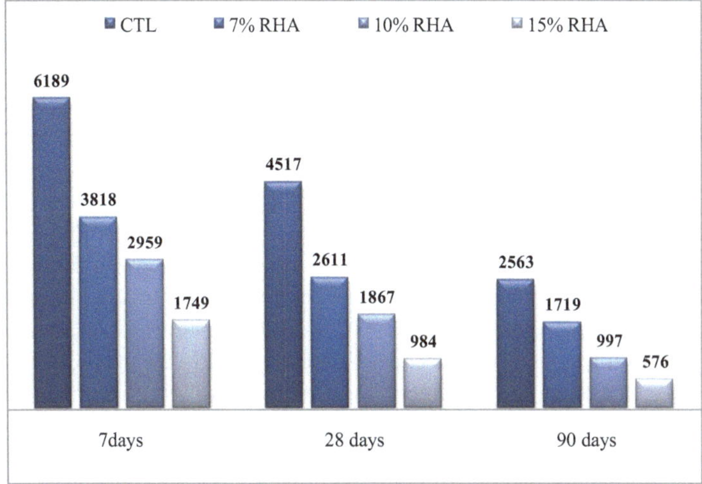

Fig. 6.15 Resistance to chloride ion penetration (coulomb) at various ages for control (CTL) and RHA mixtures (Ramezaniampour et al. [44])

Ramezanianpour et al. [44], investigated the chloride-ion penetration resistance of concretes made with 7, 10 and 15 % RHA replaced with cement by weight. Details of the concrete mixture along with the chloride-ion penetration test conducted as per ASTM C1202 are given in Fig. 6.15.

The results show that using RHA drastically enhances resistance to chloride penetration compared to control concrete on average, around 4 $\sim$ 5 times higher for the 15 % RHA. At 7 days, the control concrete showed the highest value of 6189 coulombs while the charge passed through the 15 % RHA concrete was 1749 coulombs. With a continuous moist-curing of up to 91 days, the charge passed through all concretes; was reduced. The charge for the 15 % RHA concrete was reduced to 576 coulombs, which was well below that of the control concrete (2563 coulombs). According to ASTM C 1202, when the charge passed through concrete is below 1000 coulombs, it is categorized as a very high resistance concrete to chloride ion penetration.The chloride permeability of the concrete specimens incorporating 15 % RHA was "very low", while that of the concrete specimens with 0, 7, 10 % RHA were "moderate", "low" and "low", respectively, as per ASTM C 1202 criteria.

Corrosion Resistance

Saraswathy and Song [36] investigated the corrosion performance of concrete made with 0, 5, 10, 15, 20, 25, and 30 % RHA as partial replacement of cement. Proportion of control (without RHA) mix was 1:1.5:3 with w/c ratio of 0.53. Corrosion performance was evaluated by impressed voltage test and open circuit potential measurements. For impressed voltage test, cylindrical concrete specimens of 50 mm diameter and 100 mm height were cast with centrally embedded rebar of 12 mm diameter and 100 mm height. After 28 days of curing, the specimens were subjected to impressed voltage test. In this technique, the concrete specimens were immersed in 5 % NaCl solution and embedded steel in concrete was made anode with respect to an external stainless steel electrode serving as cathode by applying a constant positive potential of 12 V to the system from a DC source. The variation of current was recorded with time. For each specimen, the time taken for initial crack and the corresponding maximum anodic current flow was recorded. For open circuit potential test, specimens of size 100 × 100 × 100 mm were cast. 12 mm diameter rebar of 120 mm length were embedded at a cover of 25 mm from one side of the specimen. After casting the specimens were subjected to water curing for 28 days. After 28 days of curing the cubes were taken out and dried for 24 h and subjected to alternate wetting and drying in 3 % NaCl solution. One cycle consisted of seven days immersion in 3 % NaCl solution and seven days drying in open atmosphere. The tests were continued over a period of 200 days. Open circuit potential measurements were monitored with reference to saturated calomel electrode (SCE) periodically with time according to ASTM C876 test method. Impressed voltage results were expressed in terms of time to cracking in hours. Time of cracking was 42, 72, and 74 h for concretes made with 0, 5, and 10 % RHA. However, no cracking was observed for concretes with 15, 20, 25, and 30 % RHA even after 144 h of exposure. It can be observed from the results that there was no cracking in concretes made with 15, 20, 25 and 30 % rise husk even after 144 h of exposure. Whereas in ordinary Portland cement concrete, the specimen cracked even after 42 h of exposure in 5 % NaCl solution. The concrete specimens containing 5 and 10 % rise husk ash also failed within 72 and 74 h of exposure. This indicated that the replacement of rice husk ash refined the pores and thereby the permeability and corrosion were reduced.

Open circuit potential measurements values (OCP) according to ASTM C876-97 were lesser than –275 mV versus saturated calomel electrode (SCE), was considered to be passive in condition. They concluded that all the rice husk ash replaced concretes had shown less negative potential than −275 mV even up to 100 days of exposure indicating the passive condition of the rebars. Beyond 100 days of exposure all the systems showed a more negative potential than −275 mV versus SCE irrespective of the replacement ratio showing the active condition of rebars.

Sulfate Resistance of Mortars and Concretes Containing Rice Husk Ash

The performance of rice husk ash mortars and concretes in sulfate solutions was better than the normal ones. This resistance depends on the replacement level of rice husk ash by Portland cement and the type of sulfate.

Sakr [35] investigated the role of RHA on the sulfate resistance of heavyweight concrete. The aggregates used were special natural heavy weight mineral ores, mainly ilmenite and baryte. They were used as the fine and coarse aggregates for the heavy weight concrete. Ilmenite, barite and gravel concrete mixtures were made with 0 and 15 % RHA. For the sulfate resistance test, 100 mm cubes were immersed in a 5 % $MgSO_4$ solution for different periods (1, 3, and 6 months) and the loss in compressive strength due to sulfate attack was determined. They concluded that: (i) reductions in compressive strength of the gravel, baryte, and ilmenite control mixtures were 8.5, 7.0, and 8.0 %, respectively, after immersion in 5 % $MgSO_4$ for 28 days, while the reductions after 90 days of immersion were 16, 14, and 14 %, respectively; (ii) reduction percentages of the compressive strength of gravel, baryte and ilmenite concrete incorporating 15 % of RHA when immersed in a sulfate solution for 28 days were 5, 6, and 6 %, respectively; and (iii) decrease rate of the effect of sulfate ions on compressive strength was generally increased as the immersion time in the sulfate solution increased.

Chindaprasirt et al. [37] determined the sulfate resistance of mortars made from ordinary Portland cement containing fly ash and ground rice husk ash (RHA). Class F fly ash and RHA were used at replacement dosages of 20 and 40 % by weight of cement. The Blaine fineness of PC, FA, and RHA was 2900, 2600, and 14,000 cm^2/g, respectively. Mortars were made with sand-to-binder ratio of 2.75 and adjusted water contents to achieve similar flow of 110 $\pm$ 5 %. The test for sulfate-induced expansion was done following the procedures described in ASTM C1012 with 5 % sodium sulfate solution. The expansion of the PC prism was much larger than those made with the blended cements. The accelerating expansion pattern of the PC mortar was observed from 120 days onward. There was no obvious accelerating expansion pattern shown by blended cement mortar prisms. FA20, FA40 and RHA20 mixes show a "linear" pattern of expansion after about 120 days in sulfate solution. RHA40 mix, however, shows a very small expansion even after immersion for 360 days. Figure 6.16 shows the pH levels of the sodium sulfate solution. The highest level of the pH of sulfate solution was approximately 12.5 for all the mortars and was observed within the first 7 days of immersion, indicating that a substantial amount of the calcium hydroxide was leached out and thus increased the pH of the solution. The fresh solution pH was 7.0–7.5. After one day of immersion, the pH level was found to be more than 12. Even Khatri et al. [45] have reported that the pH of 12–12.5 was obtained within a few hours of immersion. The pH of the solution increased slightly as the immersion period continued until the solution was replaced by a fresh solution. At 90 and 180 days, the pH levels of the sulfate solutions were significantly lower and different. The pH value of the solution with PC mortar was the highest of 11.0 followed by those of FA20, FA40, RHA20 and RHA40 with 10.7, 10.5, 10.1 and 9.5, respectively.

The expansion of the mortar bar is sensitive to the pH level of the solution (Cao et al. [46]). At pH 12–12.5 only ettringite formation can take place and at pH of 8.0–11.5 gypsum formation and decalcification occur (Khatri et al. [45]). Shi and Stegemann [47] have reported that dissolution of $Ca(OH)_2$ and calcium sulfoaluminates, and the decalcification of CSH with a high C/S ratio in hardened PC

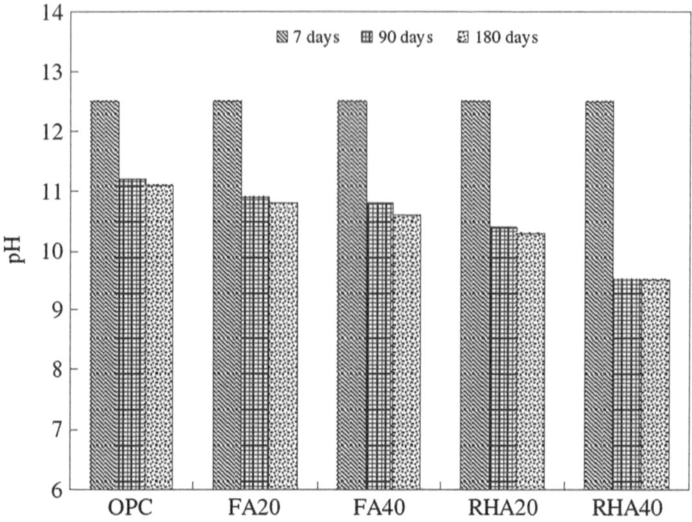

Fig. 6.16 pH level of the sodium sulfate solutions immersed with mortar bars (Chindaprasirt et al. [37])

paste results in a very porous layer whereas the decalcification of the low C/S ratio of CSH results in a protective layer of silica gel. For FA and RHA blended cement, C/S ratio of CSH would have been lower as a result of the pozzolanic reaction. FA and RHA mortars thus showed better resistance to the sulfate attack in comparison to PC mortar with RHA, being more reactive and showing better resistance to sulfate attack.

Ramezanianpour et al. [48] investigated the role of RHA on the sulfate resistance of concrete. The compressive strengths computed as an average of three cubic specimens subjected to a 5 % Na_2SO_4 solution are presented in Table 6.23. As seen in that table, when subjected to continuous sulfate exposure all specimens showed an increase in compressive strength up to 2 months. In cubic specimens subjected to a 5 % Mg_2SO_4 solution all specimens showed an increase in compressive strength up to 2 months except control specimens which was not significant. This was attributed to the hydration of calcium silicates and to the pozzolanic reactions of blended cements. However, in following weeks we expect,

Table 6.23 Compressive strength of specimens after 2 months sulfate exposure (Mpa) (Ramezanianpour et al. [48])

	28 days curing	2 months in 5 % Na_2So_4	2 months in 5 % $MgSo_4$
CTRL	41.92	41.58	45.5
7 %	44.61	49.78	50.28
10 %	46.19	51.40	55.49
15 %	47.46	47.74	52.16

the compressive strength of the OPC mortars began to decrease drastically for all specimens. This was an expected result, as ettringite formation leading to an expansion, cracking and drastic reduction in the strength and gypsum formation leading to a reduction of stiffness and strength were the major form of deterioration of mortars containing cements with a high C3A content.

Effect of Rice Husk Ash on Suppressing the Alkali Aggregate Reaction

Very fine particles of rice husk ash react with the alkalis in cement paste to form alkali silicates. This reduces the available alkalis in the pore solution and would prevent attack on reactive siliceous aggregates. Mixtures containing rice husk ash are also less permeable and prevent the penetration of water necessary for the alkali aggregate reaction. Therefore it is expected to have less expansion in concrete mixtures due to alkali aggregate reaction when rice husk ash is used. The amount of rice husk necessary for prevention of alkali aggregate reaction varies for different aggregates.

Hasparyk et al. [49] studied the expansion of mortar bars made with different levels of cement replacement with rice husk ash (RHA). Two types of reactive aggregates (quartzite and basalt) were used. Percentages of RHA were 0, 4, 8, 12, and 15 %. Tests were conducted in accordance with ASTM C 1260 test method. Expansion at 16 and 30 days are given in Table 6.24. It can be seen from these results that inclusion of RHA was very effective in controlling the expansion of mortar bars at the age of 16 and 30 days.

Ramezanianpour et al. [50] studied the expansion of mortar bars made with different levels of cement replacement with rice husk ash (RHA). The expansion of the concrete prism in accordance with ASTM C 1293 [51] is shown in Fig. 6.17. Up to 175 days, mixtures containing 7 and 10 % RHA reduced the expansion at about 52 and 33 %, respectively, when compared with the reference concrete prisms. However, specimens containing 15 % RHA show highest expansion at about 0.045 %. Figure 6.18 presents the expansion rate of concrete prisms up to

Table 6.24 Expansion of mortar bars containing RHA (Hasparyk et al. [49])

Replcement (%)	Expansion (%)			
	Quartzite		Basalt	
	16 days	30 days	16 days	30 days
0	0.28	0.53	0.84	1.10
4	0.16	0.36	0.76	0.99
8	0.21	0.45	0.32	0.81
12	0.09	0.25	0.08	0.33
15	0.06	0.15	0.04	0.25

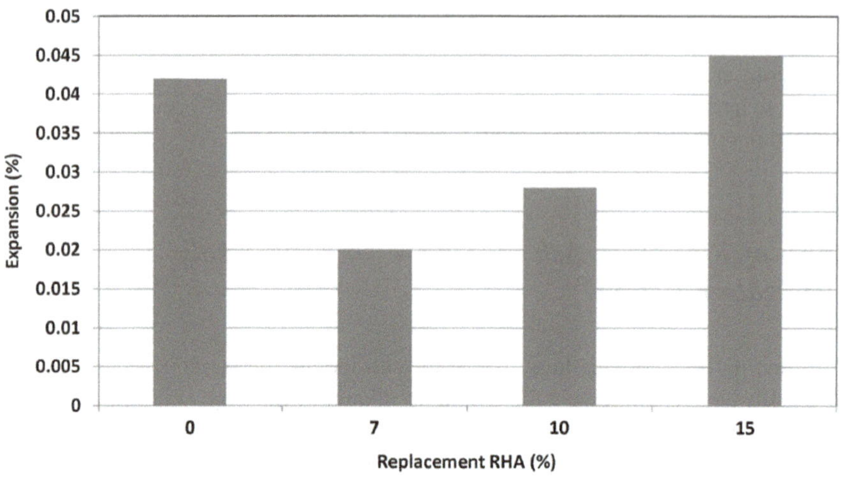

Fig. 6.17 Expansion of concrete prisms (ASTM C 1293 [51]) at 175 days (Ramezanianpour et al. [50])

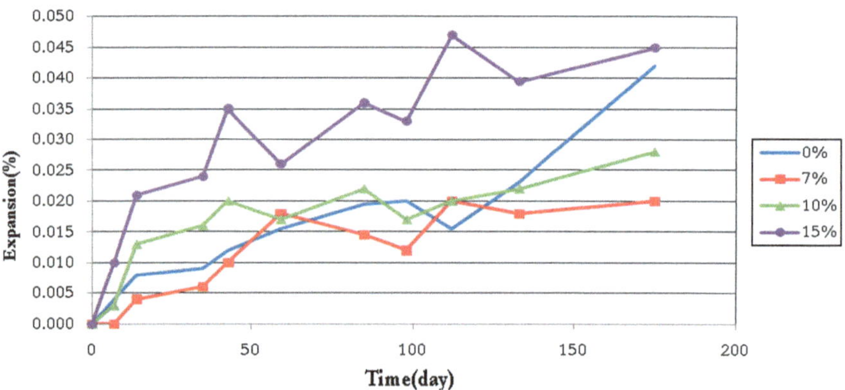

Fig. 6.18 Effect of RHA content on expansion of concrete prism (ASTM C 1293 [51]), (Ramezanianpour et al. [50])

175 days. The results show that the replacement of RHA to control alkali aggregate reaction has an optimum amount which seems to be between 7 and 10 %. This indicates that increasing the amount of RHA, causes an increase in the expansion.

Since total RHA alkali equivalent content (1.11 %) and also RHA soluble alkali (0.78 %) are both higher than cement alkali equivalent content (0.75 %), the increase of RHA would lead to the alkali increase of the entire system which was about 0.05 % in specimens containing 15 % RHA.

Table 6.25 Test results of concrete to freezing and thawing cycles (Zhang and Malhotra [20])

Mix no.	Type of concrete	Compressive strength (Mpa)	Change at the end of 300 freezing/thawing cycles				Durability factor (%)
			Weight	Length	Pulse velocity	Resonant frequency	
CO-D	Control	36.5	0.08	0.006	−0.55	−0.84	98.0
R10_D	10 % RHA	45.5	0.02	0.001	0.01	−0.86	98.3
SF10-D	10 % SF	42.8	0.12	0.001	0.19	0.47	101.0

Freezing and Thawing Resistance of Concretes Containing Rice Husk Ash

The published data on the freezing and thawing of mortars and concretes containing rice husk ash indicate that similar or slightly better performance has been observed.

Zhang and Malhotra [20] studied the freezing and thawing resistance of concretes made with 10 % RHA and 10 % silica fume (SF) in accordance with ASTM C666 Procedure A. Details of the concrete mixture along with the results of freezing/thawing are given in Table 6.25. The control, RHA, and silica fume concretes showed excellent performance in the freezing and thawing test. The RHA concrete had a durability factor of 98.3 and very small changes in length, mass, pulse velocity, and resonant frequency after 300 cycles of freezing and thawing.

Deicing Salt Scaling Resistance of Mortars and Concretes Containing Rice Husk Ash

Zhang and Malhotra [20] reported the scaling resistance of concretes made with 10 % RHA and 10 % silica fume (SF) in accordance with ASTM C 672 test method. Scaling resistance results are given in Table 6.26. The visual evaluation of test slabs showed that the performance of the RHA concrete was similar to that of the control concrete but marginally better than SF concrete. For both control and

Table 6.26 Test results of deicing salt scaling (Zhang and Malhotra [20])

Mix no.	Type of concrete	W/C ratio	Compressive strength (Mpa)	Visual rating	Total scaling residue (kg/m^3)
CO-D	Control	0.40	36.5	2	0.3
R10-D	10 % RHA	0.40	45.5	2	0.6
SF10-D	10 % SF	0.40	42.8	3	0.8

RHA concrete, no coarse aggregate was visible after 50 cycles whereas for SF concrete some coarse aggregates were visible. All the three concretes had a total mass of scaling residue of equivalent to or less than 0.8 kg/m^3 after 50 cycles in the presence of deicing salts.

The control, RHA, and silica fume concretes showed excellent performance in the freezing and thawing test. The RHA concrete had a durability factor of 98.3 and very small changes in length, mass, pulse velocity, and resonant frequency after 300 cycles of freezing and thawing.

Nehdi et al. [33] investigated the effect of RHA and SF on the deicing scaling resistance of concrete. Concrete mixtures were made with 10 % RHA and 10 % SF as partial replacement of cement. Tests were conducted up to 50 cycles according to ASTM C672 test method. Results showed that RHA concrete specimens (visual rating 2) performed similar to the control concrete mixture (visual rating 2) but little better than SF concrete mixtures (visual rating 3). These results were in agreement with those reported by Zhang and Malhotra ([20]).

References

1. P.K. Mehta, The chemistry and technology of cement made from rice husk ash, in *Proceedings UNIDO/ESCAP/RCTT Workshop on Rice Husk Ash Cements*, Peshawar, Pakistan, Regional Centre for Technology Transfer, Bangalore (India), pp. 113–22, Jan 1979
2. D.J. Cook, Development of microstructure and other properties in rice husk ash—OPC systems, in *Proceedings of the 9th Australasian Conference on the Mechanics of Structures and Materials*, University of Sydney, Sydney, pp. 355–360 (1984)
3. P.K. Mehta, Rice husk ash—a unique supplementary cementing material, in *Proceedings of the International Symposium on Advances in Concrete Technology*, Athens, Greece, pp. 407–430 (1992)
4. M.F.M. Zain, M.N. Islam, F. Mahmud, M. Jamil, Production of rice husk ash for use in concrete as a supplementary cementitious material. Constr. Build. Mater. **25**(2), 798–805 (2011)
5. A.A. Ramezanianpour, F. Gafarpour, M.H. Magedi, *The use of rice husk ash in the building industry* (Building and Housing Research Center Press, Tehran, 1995). Winter
6. S. Chiarakorn, T. Areerob, N. Grisdanurak, Influence of functional silanes on hydrophobicity of MCM-41 synthesized from rice husk. Sci. Technol. Adv. Mater. **8**(1–2), 110–115 (2007)
7. J. Chumee, N. Grisdanurak, A. Neramittagapong, J. Wittayakun, Characterization of platinum–iron catalysts supported on MCM-41 synthesized with rice husk silica and their performance for phenol hydroxylation. Sci. Technol. Adv. Mater. **10**(1), 015006 (2008)
8. Erron Peuse, Project engineer of concrete canoe project, Michigan 49916, 2005
9. R.V. Krishnarao, Y.R. Mahajan, Formation of SiC whiskers from raw rice husks in argon atmosphere. Ceram. Int. **22**(5), 353–358 (1996)
10. D.O. Albina, Emissions from multiple-spouted and spout-fluid fluidized beds using rice husks as fuel. Renew. Energy **31**(13), 2152–2163 (2006)
11. A.A. Ramezanianpour, F. Gafarpour, M.H. Magedi, *Use of rice husk in the production of masonary cements* (Building and Housing Research Center Press, Tehran, 1997)
12. A.A. Ramezanianpour, M. Mahdikhani, Gh. Ahmadi, The effect of rice husk ash (RHA) on mechanical properties and durability of sustainable of concretes. Int. J. Civil Eng. **7**(2), 83–91 (2009)

13. A.K. Yeoh, R. Bidin, C.N. Chong, C.Y. Tay, The relationship between temperature and duration of burning of rice-husk in the development of amorphous rice-husk ash silica, in *Proceedings of UNIDO/ESCAP/RCTT, Follow-up Meeting on Rice-Husk Ash Cement*, Alor Setar, Malaysia (1979)
14. S.K. Chopra, S.C. Ahluwalia, S. Laxmi, Technology and manufacture of rice-husk ash masonry (RHAM) cement, in *Proceedings of ESCAP/RCTT Workshop on Rice-Husk Ash Cement*, New Delhi (1981)
15. C.L. Hwang, D.S. Wu, Properties of cement paste containing rice husk ash. Am. Concr. Inst. SP **114**, 733–765 (1989)
16. V.P. Della, I. Kuhn, D. Hotza, Rice husk ash as an alternate source for active silica production. Mater. Lett. **57**, 818–821 (2002)
17. P.C. Hewlett, *Chemistry of Cement and Concrete* (Wiley, New York, 1998), pp. 471–601
18. C. Real, M.D. Alcala, J.M. Criado, Preparation of silica from rice husks. J. Am. Ceram. Soc. **79**(8), 2012–2016 (1996)
19. K. Inoue, N. Hara, Thermal treatment and characteristics of rice husk ash. Inorg. Mater. **3**, 312–318 (1996)
20. M.H. Zhang, V.M. Malhotra, High-performance concrete incorporating rice husk ash as a supplementary cementing material. ACI Mater. J. **93**(6), 629–636 (1996)
21. Q. Yu, K. Sawayama, S. Sugita, M. Shoya, Y. Isojima, The reaction between rice-husk ash and $Ca(OH)_2$ solution and the nature of its product. Cem. Concr. Res. **29**(1), 37–43 (1999)
22. I.K. Cisse, M. Laquerbe, Mechanical characterization of filler sandcretes with rice husk ash additions: study applied to Senegal. Cem. Concr. Res. **30**, 13–18 (2000)
23. R. Jauberthie, F. Rendell, S. Tamba, I. Cisse, Origin of the pozzolanic effect of rice husks. Constr. Build. Mater. **14**, 419–423 (2000)
24. Q. Feng, H. Yamamichi, M. Shoya, S. Sugita, Study on the pozzolanic properties of rice husk ash by hydrochlortic acid pretreatment. Cem. Concr. Res. **34**, 521–526 (2004)
25. M.P. Luxan, M. Mndruga, J. Seavedra, Rapid evaluation of pozzolanic activity of natural products by conductivity measurement. Cem. Concr. Res. **19**, 63–68 (1989)
26. S.K. Agarwal, Pozzolanic activity of various siliceous materials. Cem. Concr. Res. **36**(9), 1735–1739 (2006)
27. A.A. Ikpong, D.C. Okpala, Strength characteristics of medium workability ordinary Portland cement-rice husk ash concrete. Build. Environ. **27**(1), 105–111 (1992)
28. M.S. Ismail, A.M. Waliuddin, Effect of rice husk ash on high strength concrete. Constr. Build. Mater. **10**(7), 521–526 (1996)
29. D.D. Bui, J. Hu, P. Stroeven, Particle size effect on the strength of rice husk ash blended gap-graded Portland cement concrete. Cement Concr. Compos. **27**, 357–366 (2005)
30. N.B. Singh, V.D. Singh, S. Rai, S. Chaturvedi, Effect of lignosulfonate, calcium chloride and their mixture on the hydration of RHA-blended Portland cement. Cem. Concr. Res. **32**, 387–392 (2002)
31. C. Jaturapitakkul, B. Roongreung, Cementing material from calcium carbide residue-rice husk ash. J. Mater. Civ. Eng. **15**(5), 470–475 (2003)
32. I. Wada, T. Kawano, N. Mokotomaeda, Strength properties of concrete incorporating highly reactive rice-husk ash. Trans. Jpn. Concr. Inst. **21**(1), 57–62 (1999)
33. M. Nehdi, J. Duquette, A. Damatty El, Performance of rice husk ash produced using a new technology as a mineral admixture in concrete. Cem. Concr. Res. **33**, 1203–1210 (2003)
34. G. Rodriguez de Sensale, Strength development of concrete with rice-husk ash. Cem. Concr. Compos. **28**, 158–160 (2006)
35. K. Sakr, Effects of silica fume and rice husk ash on the properties of heavy weight concrete. J. Mater. Civ. Eng. **18**(3), 367–376 (2006)
36. V. Saraswathy, Song Ha-Won, Corrosion performance of rice husk ash blended concrete. Constr. Build. Mater. **21**(8), 1779–1784 (2007)
37. P. Chindaprasirt, P. Kanchanda, A. Sathonsaowaphak, H.T. Cao, Sulfate resistance of blended cements containing fly ash and rice husk ash. Constr. Build. Mater. **21**(6), 1356–1361 (2007)

38. R. Torrent, Gas permeability of high-performance concretes-site and laboratory test. High performance concrete and performance and quality of concrete structures, in *Proceedings Second CANMET/ACI International*, Gramado, pp. 291–308 (1999)
39. R. Torrent, G.A. Frenzer, A method for rapid determination of the coefficient of permeability of the covercrete, in *Proceedings of the International Symposium on Non-Destructive Testing in Civil Engineering, (NDT-CE)*, pp. 985–92 (1995)
40. A.L.G. Gastaldini, G.C. Isaia, N.S. Gomes, J.E.K. Sperb, Chloride penetration and carbonation in concrete with rice husk ash and chemical activators. Cem. Concr. Compos. **29**, 176–180 (2007)
41. RILEM Recommendations of TC116-PCD, Tests for gas permeability of concrete. Mater. Struct **32**(217), 163–179 (1999)
42. RILEM CPC18, Measurement of hardened concrete carbonation depth. Mater. Struct. **21**(126), 453–455 (1988)
43. J.S. Coutinho, The combined benefits of CPF and RHA in improving the durability of concrete structures. Cem. Concr. Compos. **25**, 51–59 (2003)
44. A.A. Ramezanianpour, M. Mahdikhani, K. Zarrabi, Gh. AhmadiBeni, The optimization of quality of rice husk ash (RHA) and durability of RHA concretes in chloride-ion environment, in *Proceeding of International Conference on Sustainability in the Cement and Concrete Industry*, Lillehammer, Norway, (2007)
45. R.P. Khatri, V. Sirivivatnanon, J.L. Yang, Role of permeability in sulfate attack. Cem. Concr. Res. **27**, 1179–1189 (1997)
46. H.T. Cao, L. Bucea, A. Ray, S. Yozghatlian, The effect of cement composition and pH of environment on sulfate resistance of Portland cements and blended cements. Cem. Concr. Compos. **19**, 161–171 (1997)
47. C. Shi, J.A. Stegemann, Acid corrosion resistance of different cementing materials. Cem. Concr. Res. **30**, 803–808 (2000)
48. A.A. Ramezanianpour, P. Pourbeik, F. Moodi, Gh. Ahmadibeni, Sulfate resistance of concretes containing rice husk ash, in *Proceedings of the 1st International Conference on Concrete Technology*, Tabriz, Iran, 6–7 Nov 2009
49. N.P. Hasparyk, P.J.M. Monteiro, H. Carasek, Effect of silica fume and rice husk ash on alkalisilica reaction. ACI Mater. J. **97**(4), 486–491 (2000)
50. A.A. Ramezanianpour, K. Zarrabi, M. Mahdikhani, Mitigation of alkali aggregate reaction (AAR) of concretes containing rice husk ash (RHA), in *13th International Conference on Alkali-Aggregate Reactions in Concrete*, Trondheim, Norway (2008)
51. ASTM C 1293-95, Standard test method for concrete aggregate by deterioration of length change of concrete due to alkali-silica reaction, 1995 annual book of ASTM standard, 04.02 concrete and aggregates, American society for testing and materials, Philadelphia, PA, USA (1995)

Chapter 7
Limestone

Introduction

One of the materials being introduced to blended cements as a constituent is limestone or calcium carbonate ($CaCO_3$). This has led to the production of Portland-limestone cement, i.e. cements that have been interground with limestone. Most Portland cement specifications allow the use of up to 5 % limestone. Beyond that, Portland-limestone cements are categorized based on the percentage of limestone added to the cement. Portland-limestone cements consisting of limestone from 5 % up to about 40 % are being produced and used in various countries around the world, with the most commonly used cement type used in Europe being CEM II/A composite cement with 5–20 % limestone. Also, different standards have stated specifications in regards to the amount of limestone used in Portland-limestone cements.

Perhaps the main advantage of producing Portland-limestone cements is that by introducing limestone into cement, the total volume of cement would increase, or in other words, the amount of clinker required to produce a certain amount of cement would decrease. This would result in a substantial amount of energy saving in the production of cement as the consumption of natural raw materials and the fuel needed for production of clinker would be reduced. Moreover, it would contribute to sustainable development due to the reduction in greenhouse gas emissions, mostly CO_2, resulted from release of CO_2 from limestone in the pyroprocessing of clinker. On this basis, the future world production of Portland-limestone cement is expected to increase. Nevertheless, it should be noted that all the aforementioned benefits could be achieved provided that Portland-limestone cement has similar performance to Portland cement, and has no adverse effects on the properties of concrete.

The properties of Portland-limestone cements have been the subject of numerous studies. Researchers have studied the effect of using Portland-limestone cement with various limestone contents on fresh properties, mechanical properties, and durability of concrete. Overall, the data found in the literature seems to be inconsistent in some specific areas, especially in cases where the limestone content of Portland-limestone

A. A. Ramezanianpour, *Cement Replacement Materials*,
Springer Geochemistry/Mineralogy, DOI: 10.1007/978-3-642-36721-2_7,

cement is greater than 5 %. Data reported in the literature is apparently affected by the quality and particle size distribution of the limestone used. Also whether the limestone was interground, blended, or added at the mixer seems to have an influence on the results. Hence, it is important that these factors be taken into consideration when interpreting the data.

Production and Application

The use of Portland-limestone cements has been in practice for a considerable period of time in several countries. It seems that European countries have been the leaders in producing and using Portland-limestone cement. According to Schmidt [1], Heidelberg Cement, the main cement producer in Germany, used to produce an energy saving cement which contained 20 % interground limestone and was used for special applications as early as 1965. In 1979, a new standard introduced in France permitted the incorporation of up to 35 % of slag, fly ash, natural or artificial pozzolans, and limestone in a new type of cement called CPJ [2]. Later on, a specific cement designated as PKZ which consisted of 85 ± 5 % clinker and 15 ± 5 % limestone was specified in the 1987 draft of the European standard EN 197 [1]. This type of Portland-limestone cement, along with other types of blended cements, was reported to be commonly used throughout Germany by 1990 [1]. In addition to that, the British Standard BS 7583 allowed use of up to 20 % limestone in cement in the United Kingdom in 1992.

In the current version of the European standard EN 197-1 (2000), all the 27 common types of cement recognized by the standard are allowed to contain up to 5 % minor additional constituents (MAC), the most typical of which are limestone and cement raw meal. Moreover, four of the designated 27 types correspond to Portland-limestone cements which allow higher amounts of limestone in two replacement level ranges, namely CEM II/A-L and CEM II/A-LL with 6–20 % limestone, as well as CEM II/B-L and CEM II/B-LL with 21–35 % limestone in addition to the 5 % MAC. Portland-limestone cements have been in use in North American countries as well. In Canada, the use of up to 5 % ground limestone in Type 10 Portland cement (Type GU in the new standard) has been permitted in the Canadian cement standard CSA A5 since 1983. In the US, ASTM C150 allowed up to 5 % limestone to be used in all types Portland cements in 2004. Since then, CSA has also allowed addition of limestone to all types of cement.

Overall, based on the Cement Standards of the World [3], more than 25 countries allow the use of between 1 and 5 % limestone in their P ("Portland") cements. Also, many countries allow up to 35 % replacement in PB ("Portland composite") cements. According to Hawkins et al. [4], several countries have modified their standards to permit limestone in some amount, including Australia, Italy, New Zealand, and the United Kingdom since 1991. Taken from the latest edition of Cement Standards of the World, Fig. 7.1 represents how the production

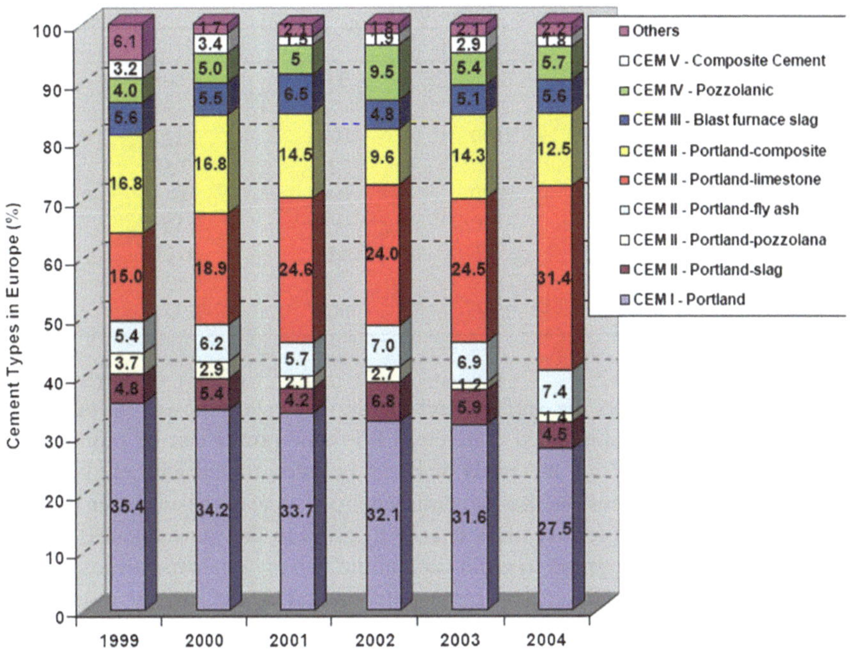

Fig. 7.1 Cement types commercialized in Europe (according to CEMBUREAU [3])

of Portland-limestone cement has been increasing over time in recent years in Europe.

Limestone is a rock that contains at least 50 % calcium carbonate in the form of calcite by weight. All limestones contain at least a few percent other minerals. These can be small particles of quartz, feldspar, clay minerals, pyrite, siderite and other minerals. It can also contain large nodules of chert, pyrite of siderite. A part of calcium molecules if being replaced by magnesium, it is known as magnesium limestone or dolomite limestone.

Chemically, limestones are calcareous rocks principally of calcic minerals with minor amounts of alumina, ferric and alkaline oxides.

Table 7.1 shows typical limestone chemical compositions.

Physical Properties

Physically limestones are quite impervious, hard, compact, fine to very fine grained calcareous rocks of sedimentary nature. The density of Limestones varies between 2.5 and 2.7 kg/cm^3. Its water absorption is usually less than 1 %. They have a very low porosity and hence a high compressive strength of about 60–170 MPa.

Table7.1 Typical limestone chemical compositions

Limestone	A	B	C	D	E
SiO_2	4.00	13.60	2.00	12.05	2.96
Al_2O_3	0.77	2.50	0.80	3.19	0.79
Fe_2O_3	0.30	0.90	0.20	1.22	0.30
CaO	51.4	43.4	52.9	43.5	52.3
MgO	1.30	3.20	0.90	1.68	1.30
SO_3	0.10	0.10	0.20	0.56	0.03
LOI	42.0	35.6	42.5	36.21	42.18
Na_2O	0.01			0.12	0.04
K_2O	0.02	0.60	0.20	0.72	0.23

The fineness and specific surface area of the Limestone power influence the properties of fresh and hardened concrete. Limestone power can be produced by grinding separately the raw materials or it can be intergrinded with clinker.

Intergrinding of Limestone has several benefits. Limestone is a softer material than clinker and therefore takes less energy to grind to the same fineness. Figure 7.2 shows the energy required to grind each of the two materials to various specific surface areas. As the content of Limestone increases, the energy required to produce the same fineness decreases as shown in Fig. 7.3 [5].

The particle size distribution of any one constituent is affected by grindability of others (Schiller and Ellerbrock 1992). Voglis et al. [6] found that Portland Limestone cement (PLC) gave a wider particle size distribution than that of cement interground with fly ash (PFC) or natural Pozzolan (PPC) as shown in Fig. 7.4 when designed to give equivalent compressive strengths. In addition, Limestone increased the grinding time required to obtain the target compressive strength (40 MPa), but also increased the fineness considerably. Vuk et al. [7] have found that Blaine surface area increased with 5 % of Limestone addition, for the same residue on the 90 μm sieve.

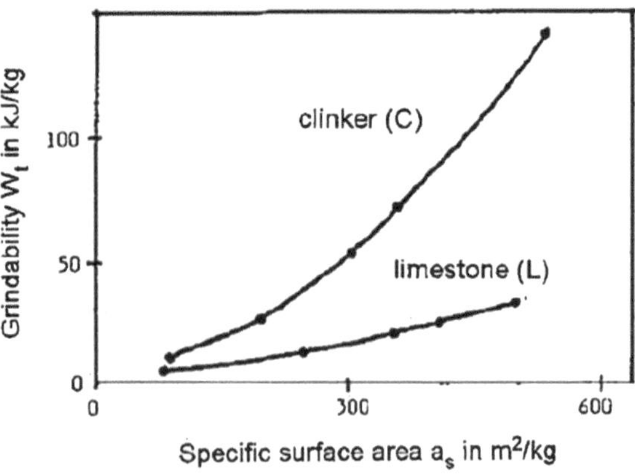

Fig. 7.2 Grindability of clinker and Limestone (Opoczky 1992)

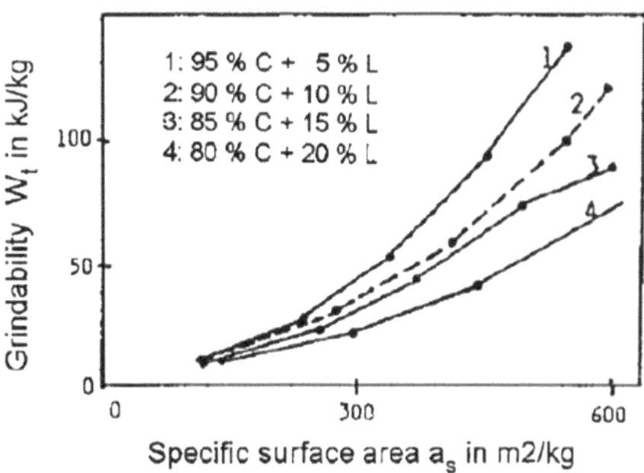

Fig. 7.3 Grindability of Limestone cement mixtures (Opoczky 1992)

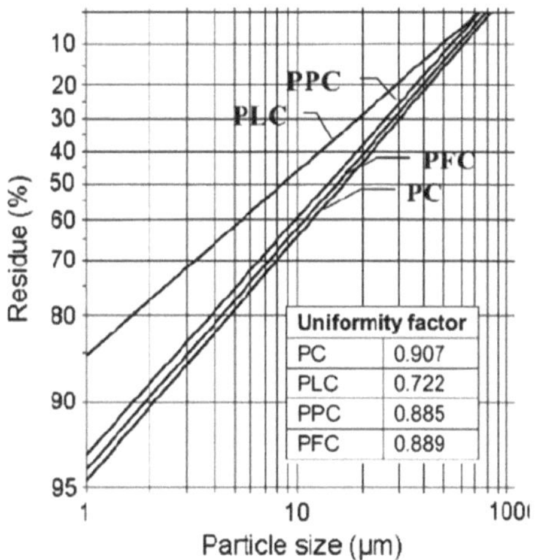

Fig. 7.4 Particle size distributions of interground cement mixtures (Voglis et al. 2005)

Tsivilis et al. [8] investigated the fineness of both the clinker and Limestone constituents after intergrinding for various times. Particle size distribution was determined by the coarser fraction and Limestone in the finer. As the Limestone content and grinding time increase, the particle size distribution becomes wider and finer. Longer grinding times significantly change the amount of coarsest particles for lower Limestone contents, but affect the entire size distribution at higher Limestone contents.

Hydration Reaction

Some researchers studied cement pastes made from various properties of Type I clinker, Limestone, and gypsum and hydrated in sealed vials for 1, 3, 7, 14, and 28 days. X-ray diffraction showed that for a combination of 2 % gypsum, 6 % Limestone, and 92 % clinker, the $CaCO_3$ reacts with the C_3A in the clinker. Care was taken to avoid exposing the pastes to atmospheric (CO_2). They believe that the reaction begins with a $C_3A.CaCO_3.12H_2O$ product which with continued hydration forms compounds containing a molar ratio of C_3A to $CaCO_3$ between 0.5 and 0.25. At later stages, the product appears to stabilize as $C_3A.X\ CaCO_3.11H_2O$, where X ranges from 0.5 to 0.25. During the course of this reaction ettringite formation proceeds normally.

Klemm and Adams [9] studied the reaction of 5 of 15 % calcium carbonate (either as reagent grade $CaCO_3$ or as ground Limestone) and Type II cement with hydration times up to one year. They found that crystalline monocarboaluminate hydrate is slower to form than ettringite, and after 129 days of hydration, 80–90 % of the Limestone of calcium carbonate remains unreacted. They concluded that with Type II cements Limestone acts primarily as an inert diluent. Based on the solubility products of the possible reaction phases, they predicted that the most stable reaction phase would be ettringite, followed by monocarboaluminate and the monosulfoaluminate forming in preference to monosulfate. They also found that for cement with higher C_3A contents the amount of monocarbonate increase at all ages with increasing levels of Limestone as compared with cements having lower C_3A contents.

A research shows that the hydration of 90 % pure C_3A is suppressed by $CaCO_3$ due to the formation of the low form of calcium carboaluminate ($C_3A.CaCO_3.11H_2O$) on the surface of the C_3A grains. They concluded that C_3A reacts with $CaCO_3$ by a direct mechanism.

Of gypsum, C_3A reacts with $CaCO_3$ in Limestone to form both "hexagonal prism" phase tricarbonate $C_3A.3CaCO_3.30H_2O$ and the "hexagonal plate" phase monocarbonate $C_3A.3CaCO_3.11H_2O$. In Portland cement, ettringite is formed during early hydration, with monosulfate becoming significant only after the first 16 h of hydration, when the concentration of sulfate ions is not sufficient for the formation of ettringite. Tricarbonate is much less stable than ettringite at ambient temperatures; thus tricarbonate does not form in cement hydration. Bested examined the possibility of substituting Limestone for some or all of the gypsum used to control the early hydration of C_3A and concluded that because of differences in their stereochemistry, sulfate ions enter more readily than carbonate ions into solid solution. Thus sulfate ions are more effective in controlling setting carbonate ions. However, he concluded that in a given system it is possible to substitute Limestone for 25 % or even 50 % of the gypsum without deleterious effect. The exact amount of substitution depends on the cement. A research shows that Limestone at 500 m^2/kg Blaine has some effect on the setting of cement, but is less effective than gypsum in controlling flash set. For high-sulfate clinker,

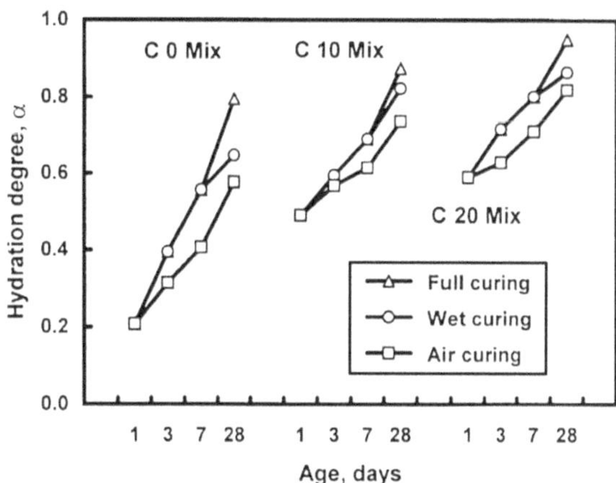

Fig. 7.5 Degree of hydration for limestone concretes under various curing conditions (Bonavetti et al. 2000)

however, times of set were not markedly different from those for cement in which the gypsum content was optimized.

As limestone acts as nucleation sites for hydration products [10], it is not surprising that the inclusion of limestone increases the rate of hydration. Bonavetti et al. [11] investigated hydration of cement pastes at various w/c ratios (0.25–0.50) with approximately 10 and 20 % limestone replacements. It was found that the degree of hydration was markedly more rapid during the first 7 days in the higher w/c pastes containing limestone. At lower w/c (~0.30), the differences were not as noticeable. However, it must be noted these pastes were designed to have similar strength. The fineness increased with additional limestone, so it is difficult to separate the effects of the fineness from the influence of the limestone content. Increases in hydration with increasing limestone contents (up to 35 %) were determined by non-evaporable water contents of 0.30 w/c cement pastes [12]. Similar results were observed in concrete mixtures at 0.34 and 0.50 w/c containing approximately 10 and 20 % limestone [13, 14]. For both water to cement ratios investigated, the addition of limestone increased the degree of hydration at all ages as can be seen in Fig. 7.5. Again, cement was interground with limestone at a cement plant to obtain equivalent strength class; the fineness of the mixture increased as the limestone content increased.

Effects Limestone on the Properties of Fresh Concrete

The literature contains contradictory conclusions in regards to the effect of limestone on workability of concrete mixes. While some researchers have found that as limestone addition is increased, more water is required to maintain the same level of consistency for cement pastes [15, 16], others have found no significant change in the consistency of mixes when limestone is added [17] (see Fig. 7.6).

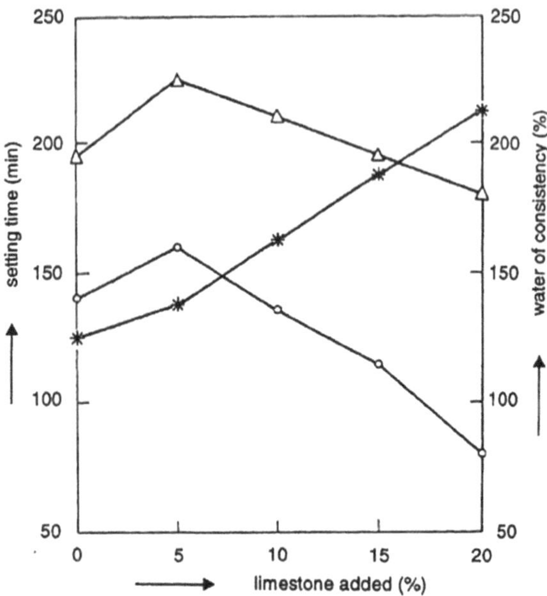

Fig. 7.6 Water demand and setting of limestone cement pastes (El-Didamony et al. 1995)

Ramezanianpour et al. [18] found that the workability in PLC concretes were greater than that for PC concretes on similar mixture designs.

Guemmadi et al. (2005) found that the consistency of low w/c cement pastes (w/c 0.24–0.26) varied with both the fineness and replacement level of limestone, but no clear trend was observed.

One fact that could somehow explain this discrepancy is the relationship between particle size distribution and water demand. According to Detwiler [19], narrow particle size distributions generally result in high water demand, while wide particle size distributions lead to reduced water demand. Limestone is easily ground, and as such, it usually has a wide particle size distribution which allows the limestone particles to fill the gaps between the clinker particles, reducing the water demand [4]. Hence, the particle sizes of the two components, i.e. cement and limestone, influence workability. Also, whether the limestone is interground with clinker or it is ground separately and added to the mix plays a role to some extent. This influence has not been clearly discussed in the studies.

Bleeding

Portland limestone cements are observed to result in less bleeding of mortars or concrete samples. Schmidt [20] compared the bleeding in mortars made with Portland cement and Portland-limestone cement after 2 h standing time. He found that the amount of bleed water from the Portland-limestone cement mortar was less than half that from the Portland cement mortar. Albeck and Sutej [21] made Portland-limestone cement designated as PKZ 35 F by intergrinding 18 % limestone ($CaCO_3$ content of 90.8 %) with the clinker of ordinary Portland cement, PK 35 F, both of which met the requirements of a strength class 35. Then, they measured the bleeding for a series of concretes made from these two types of cement and found that those made with PKZ 35 F always resulted in less bleeding and stopped bleeding sooner than those made with PZ 35 F. Kanazawa et al. [22] found that adding finely ground limestone reduced bleeding in mass concrete.

Setting Time

Several studies have shown that although setting time of cements containing limestone in the amounts less than 5 % increases the setting time, adding limestone to cement beyond 5 % would decrease the setting time of cement paste. Alunno-Rossetti and Curcio [23] compared the setting time of four different cements, two of which had 20 % limestone, and observed that the setting time of cements containing limestone were greater than that of Portland cement. According to El-Didamony et al. (1995), the set time of cement decreases with increasing the limestone additions greater than 5 % even though the amount of water needed to keep the consistency of the paste constant has to be increased.

Similar results have been reported by other researchers who compared the setting time of cement pastes made from Portland cement and Portland-limestone cements with limestone contents of 15, 25, 35, and 45 %. Nevertheless, some studies have shown that the other parameters have influence on the setting time of Portland-limestone cement. Hawkins (1986) conducted two series of test. First, he ground clinker and gypsum with 0, 3, 5.5, and 8 % limestone to a more or less constant Blaine fineness, and observed that the use of limestone had no significant effect on the setting time. Then, he repeated the procedure with 0, 2, 5, and 8 % limestone, except that he kept the <325 mesh value constant. This series indicated a reduction in setting time with the use of limestone. Also, Vuk et al. [7] noted a decrease in setting time with the use of 5 % limestone in cements made from two different clinkers. With a clinker C3S content of 35 %, initial setting was reduced by 50 min while it was decreased by 25 min with a clinker C3S content of 46 %. The authors have stated that the effects of clinker composition were less significant when the cement fineness was higher. Therefore, it seems that the fineness of the clinker and the limestone play an important role in the setting behavior of Portland-limestone cement.

Heat of Hydration

According to Péra et al. [24], limestone accelerates the hydration of C3S, resulting in a higher heat of hydration. However, similar to the case of setting time, other parameters are involved in the heat of hydration of Portland-limestone cements. Albeck and Sutej [21] found that the heat of hydration of Portland and Portland-limestone cements having the same strength grade was almost identical. They concluded that the reduced clinker content was compensated by finer grinding of the Portland-limestone cement, which resulted in equal heat evolution. Studies of Barker and Matthews [25] indicated that in general, the use of 5 % limestone reduces the peak rate of heat evolution; however, the timing of the peak is dependent upon the method of preparing the Portland-limestone cement. In case the cements were prepared by blending, the peak time remained unchanged or a bit retarded while those prepared by intergrinding, which usually have higher finenesses than the Portland cements, showed an acceleration of peak heat evolution. Using isothermal conduction calorimetry, Livesey [26] determined the heat release of concretes made from cements having various limestone contents, and found that increasing limestone content reduced both the rate and total amount of heat released.

Moreover, Portland-limestone cements having fineness similar to that of the Portland cement showed a retarded maximum heat evolution, but the heat evolution was accelerated where the Portland-limestone cements had greater fineness. According to Vuk et al. [7], the heat of hydration of cements with and without 5 % limestone were similar after 3 days, but when finer cements were used, a slight drop in heat of hydration of cements with 5 % limestone relative to the control was noted.

Effects of Limestone on Mechanical Properties of Hardened Concrete

Compressive Strength

Based on the investigations of Schmidt [20], if limestone is interground with cement, the cement and concrete strengths are not normally reduced where the level of addition is between 5 and 10 %. Yet, he found a reduction in strength at larger proportions of limestone addition, and attributed it to a strength-reducing "dilution effect". To overcome this issue, he suggested that cement be ground finer so that the reactivity of the Portland cement clinker would be increased, and blended cements would generally reach the same 28-day standard strengths as comparable Portland cements. It is noteworthy that limestone is softer than clinker, and therefore, the clinker is much coarser than Portland cement at equal Blaine values. Sprung and Siebel [27] realized that at early ages, the use of inert material such as limestone as a very fine filler can lead to an increase in strength due to improved packing of the particles. However, they perceived a reduction in strength when limestone was included in quantities of 15–25 %, where the dilution effect was considerable. They suggested finer grinding to offset the loss of strength due to dilution, just like Schmidt. Conducting two series of tests, Hawkins (1986) realized that although increasing the limestone content would result in reduction of strength, Portland-limestone cements would give comparable strengths to that of Portland cement when the fineness is increased. Similar results in regards to the effect of fineness of Portland cement and limestone on the strength of Portland-limestone cement has also been stated by other researchers [7, 8, 28].

Hawkins et al. [4] summarized the available data on strength of Portland-limestone cements from various sources. They plotted the ratio of compressive strength for samples of concrete or mortar made with and without limestone in Portland as a function of the age of the sample at testing, as shown in Fig. 7.7. They included the limestone contents of up to 6 %. Based on the results, they found out that the mean value of this ratio for the data was between 97 and 105 % at every age where more than two data points are available. Overall, based on the literature, it is concluded that for concrete compared at equal water–cement ratio, strength is relatively unaffected by limestone contents up to 5–10 % when limestone has been interground to achieve higher fineness. However, the strength is reduced as the limestone content is increased above 5–10 %. This reduction in strength is generally explained as the result of "diluting" the cement.

Ramezanianpour et al. [29] investigated the effect of various limestone replacements with cement and up to 20 percent on mechanical properties of limestone cement concretes. As expected, the compressive strength of all concrete specimens increases with the period of curing. In addition, generally, when limestone is increased, compressive strength decreases. However, up to 10 % limestone replacement is not significant in this reduction (Fig. 7.8).

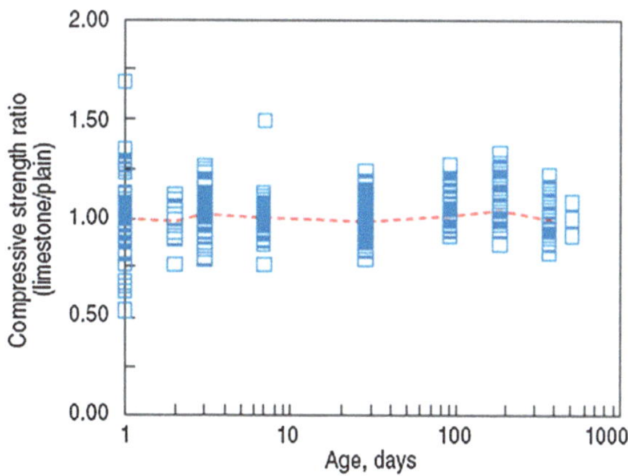

Fig. 7.7 The ratio of compressive strength for samples of concrete or mortar made with and without limestone in Portland cement [4]

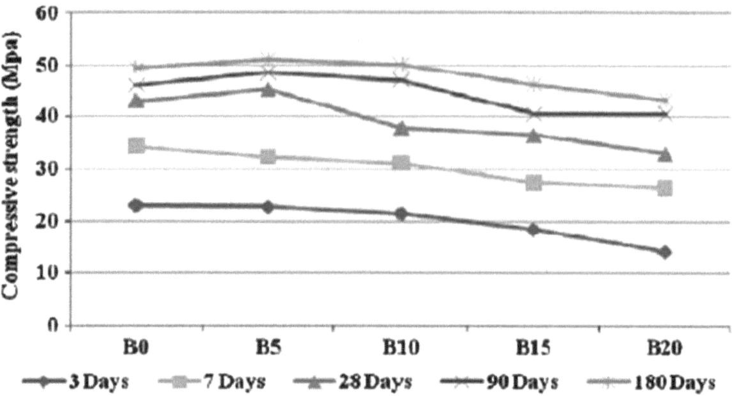

Fig. 7.8 The effect of limestone on the compressive strength at various ages [21]

Tensile and Flexural Strengths, Modulus of Elasticity

Studies of tensile (cylinder splitting) and flexural strength, and modulus of elasticity have been made by a number of authors [13, 23, 30, 31]. Generally the trend in behaviour is the same as that observed for compressive strength and predictive equations used to estimate these properties from the compressive strength (e.g. relationships in Eurocode 2) are valid for concrete produced using PLC.

Shrinkage and Creep

A study by Adams and Race [9] on the effect of blended limestone on the drying shrinkage of Type I and Type II cements found slight increases in drying shrinkage as limestone was added. Also, they found the increases shrinkage could be offset by optimization of the sulfate content and/or the use of functional additions. Shrinkage of concrete specimens made with cements from two different plants with and without 20 % limestone was measured by Alunno-Rossetti and Curcio [23]. Their data revealed that even though the shrinkage was increased as the cement content was increased, there existed no differences between the cements from the same plant in either evolution of shrinkage or final values, i.e. limestone had no effect on shrinkage. Nonetheless, Dhir et al. [30] compared the shrinkage of moist-cured prisms made from cements with different limestone contents, and observed that the shrinkage of the specimens decreased as the limestone content increased.

Dhir et al. (2007) also reported reduced shrinkage and similar creep for concretes produced with cements containing up to 45 % ground limestone

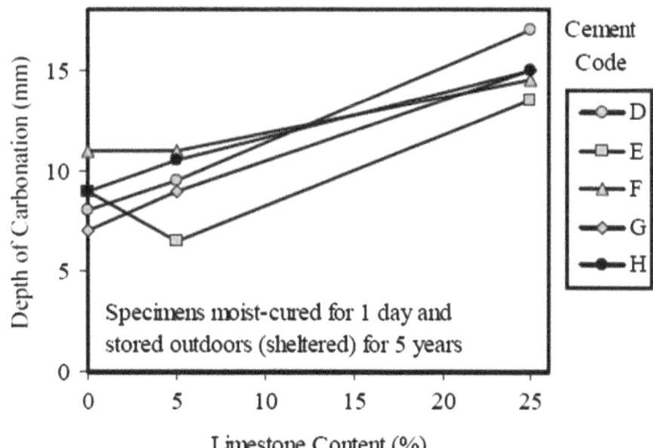

Fig. 7.9 Effect of limestone addition on the carbonation of concrete (W/CM ~ 0.60) moist-cured for 1 day and then stored outdoors-sheltered for 5 years (modified from Matthews [23])

(blended not interground). The data are shown in Fig. 7.9. The concretes were produced with 310 kg/m^3 and W/CM = 0.60. Creep tests were performed by loading specimens to 40 % of the cube strength at 28 days and drying shrinkage tests were performed by storing specimens at 55 % RH 24 h after casting.

Effect of Limestone on Durability of Concrete

Permeability

Permeability, the key to durability of a porous material like concrete, is highly related to the connectivity of the pore structure. The more the pores in concrete are disconnected, the less the permeability of concrete would be. As far as Portland-limestone cement is concerned, due to the nucleation effect of fine particles of $CaCO_3$, the pore structure of pastes with such cements is refined, the connectivity of the pore structure is reduced, and the pore structure is improved [4]. Several studies have reported results which confirm this hypothesis. Moir and Kelham [32] found that the presence of limestone slightly reduced the permeability to oxygen for a series of concretes made with cements with or without 5 or 25 % limestone. Also, they found the porosity and water sorptivity of the control sample very similar to that of 5 % limestone cement. Matthews [33] tested reinforced concrete samples with 0, 5, and 25 % limestone contents for oxygen permeability. He found that the addition of limestone decreased the permeability of samples; however, the amount of limestone had no significant effect on the permeability. Besides the improvement in the pore structure due to addition of limestone, the grinding and fineness of particles seem to influence the permeability of concrete. Schmidt [20] tested the permeability of air-entrained concretes made with Portland cement and Portland-limestone cement, and realized that in all cases, the permeability coefficients for the Portland-limestone concretes were slightly lower than for the comparable concretes made with Portland cement. Due to the limited number of tests, he was not certain whether this was due to finer grinding and/or more efficient particle packing. Nevertheless, the results of a study by Tsivilis et al. [8] showed that while having lower water permeability and sorptivity, Portland-limestone cements showed higher gas permeability and porosity compared to Portland cement. They explained this finding by the theory that in the case of Portland-limestone cement, which has a higher fineness and a lower clinker content than Portland cement, the pore system of concrete is affected by the filler effect of limestone and the (clinker) dilution effect. These two factors influence the total volume and size distribution of pores and finally affect the concrete permeability. They believed that gas permeability is closely related with the porosity, whereas water permeability and sorptivity is affected by the size and kind of pores. Nonetheless, they have stated that further investigation is required on this issue.

Carbonation

Baron [34] found the depth of carbonation for standard mortars made with 15 % limestone cements similar to that of the sample without limestone. Results from Sprung and Siebel [27] suggest that in general, concretes made with Portland-limestone cement with limestone contents of 6–20 % have increased rates of carbonation compared to those made with Portland cements, even when they have the same strength. In his study, Matthews [33] prepared concrete prisms having 0, 5 and 25 % limestone contents with different water-cement ratios. He moist-cured the prisms for 1, 3 or 28 days, exposed them either indoors or outdoors sheltered from precipitation, and measured the depth of carbonation after 5 years. Based on the results, he found that the depth of carbonation increased as more limestone was added to the specimens (see Fig. 7.9).

Also, further moist curing helped to reduce the depth of carbonation in all cases (see Fig. 7.10).

In general, he found that the depth of carbonation had a linear relationship with the water-cement ratio and an inverse linear relationship with the compressive strength of the specimens, regardless of the limestone content. Moir and Kelham [32] came to the same conclusion as they measured the carbonation depth of samples with containing 5 or 25 % limestone, as well as fly ash or slag. Testing two series of samples with limestone contents of 0–24 %, one with constant cement content and water cement ratio and one with equal strength, Barker and Matthews [25] also found an inverse correlation between the strength of the specimen and the depth of carbonation, in spite of the composition of the cement.

Schmidt [1] observed that carbonation depths for concretes made with Portland limestone cement with 13–17 % limestone contents were higher than that for concretes made with Portland cement, but lower than concretes made with slag cement.

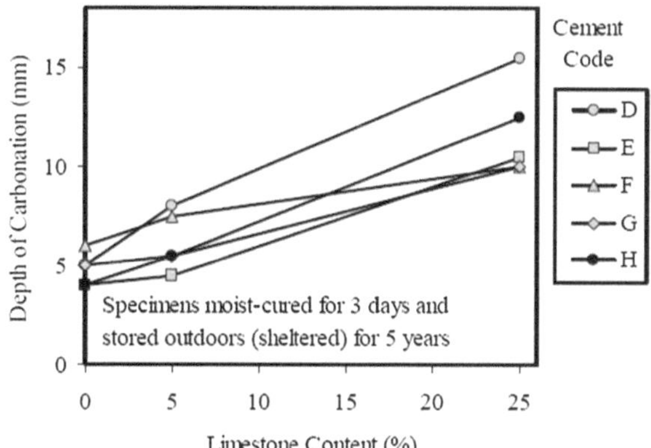

Fig. 7.10 Effect of limestone addition on the carbonation of concrete (W/CM ~ 0.60) moist-cured for 3 days and then stored outdoors-sheltered for 5 years (modified from Matthews [23])

Also, he found the increases in carbonation depths measured after 3 years of exposure to be minimal. Results reported by Tezuka et al. [35] show that the depth of carbonation for mortars containing various quantities of limestone were comparable to those for Portland cement mortars as long as the limestone contents were less than 10 %, but increased beyond this point. Presenting data on carbonation depth in concretes made with cements with and without 20 % limestone, Alunno-Rosetti and Curcio [23] found that use of limestone in cement did not impact depth of carbonation.

Similar findings were recently reported by Dhir et al. (2007) using blended PLC containing up to 45 % limestone as shown in Fig. 7.11. Even concrete produced with a PLC containing 45 % limestone showed similar resistance to carbonation when compared with PC concrete of the same strength grade. Concrete produced

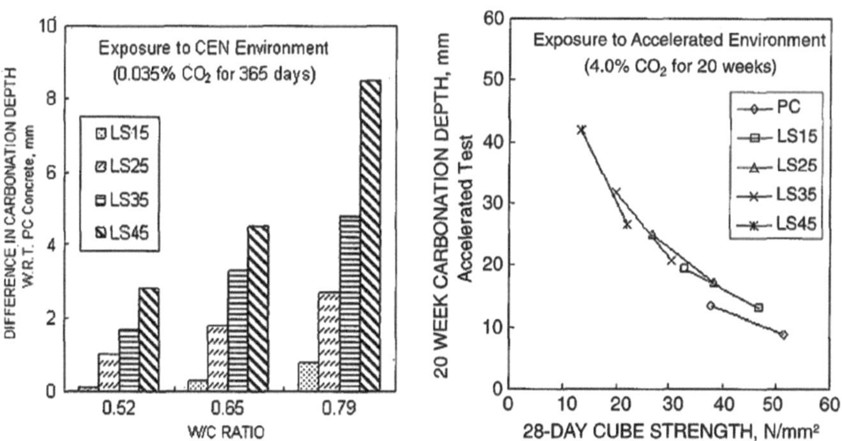

Fig. 7.11 Effect of W/CM ratio and strength on the carbonation of concrete with varying levels of limestone exposed to normal (*Left*) and elevated (*Right*) levels of carbon dioxide (Dhir et al. 2007)

with a PLC with 15 % showed little increase in carbonation over the control, especially at the lower W/CM (0.52) used in the carbonation tests.

Freeze/Thaw

Sprung and Siebel [27] tested the frost resistance of concrete samples containing 15 % limestone. They wet cured the concretes, which had a cement content of 300 kg/m^3 and a water-cement ratio of 0.60, for 6 days, then moist cured for an additional 28 or 56 days, and subjected them to 100 freeze/thaw cycles. Moreover, they tested concretes having different water-cement ratios for frost resistance using three commercial Portland limestone cements having 11, 26, and 12 % limestone. Concrete samples that experienced a mass loss of less than 10 % were considered to be frost resistant. They found that concretes made with Portland-limestone cement showed reduced resistance to frost damage as compared with those made with Portland cements, even when the strengths were the same. Also, they found that concretes having a water–cement ratio greater than 0.60 were not frost resistant, while those with water-cement ratios less than or equal to 0.60 were adequately frost resistant. Therefore, they concluded that concretes made from Portland-limestone could be frost resistant provided that the limestone content does not exceed 20 % by mass of cement, the limestone meets the criteria for composition limits specified by EN 197-1, and sufficient concrete strength is reached prior to exposure to freeze/thaw cycles. Schmidt [20] tested the frost resistance of specimens made from Portland-limestone cement with limestone contents of 13–17 %, and found that they performed as well as or slightly better than those made from Portland cement. Having found the tests of frost resistance performed at different laboratories conflicting results, Baron [34] followed a new approach that involved determining the required air void spacing factor for mortars made with different cements individually, rather than prescribing the same spacing factor for all cements. Data from his study suggested that in order to provide good frost resistance, Portland-limestone cements with 15 % limestone require smaller spacing factors than Portland cements. Dhir et al. (2007) studied the resistance of concrete samples made with cements containing 0, 15, 25, 35, and 45 % limestone with and without air entrainment against freeze–thaw damage. The results of their tests indicated that the inclusion of limestone did not affect the freeze–thaw resistance of air-entrained concrete. However, in the case of non-air entrained concrete, the mass loss due to exposure to freeze–thaw cycles increased with the limestone content, becoming between 3 and 4 times the amount for Portland cement for 45 % limestone.

Overall, the literature suggests that the freeze–thaw resistance of air-entrained concrete is not influenced by limestone content; however, if the concrete is not air-entrained, concretes made with Portland-limestone cements perform similar to Portland cement only at low amounts of limestone, and the frost resistance decreases as the limestone content is increased [1, 8, 33], Dhir et al. (2007).

Chloride Resistance

Results on the chloride resistance of Portland-limestone cements vary significantly between studies. Ramachandran et al. [36] studied mortars containing 0, 2.5, 5, and 15 % precipitated $CaCO_3$ (particle size 1–5 Mm) or ground limestone (particle size 1–40 Mm) at water-cement ratios of 0.42 and 0.60. They hydrated the specimens in limewater or in laboratory prepared "seawater" for up to one year and monitored the length change periodically. They found that specimens with water-cement ratio of 0.60 or the ones that contained precipitated $CaCO_3$ exhibited much higher expansions than the control samples when exposed to seawater, whereas the specimens with a water-cement ratio of 0.42 containing ground limestone showed similar expansions to the controls.

Matthews [33] designed concrete samples with equal water-cement ratios containing limestone in the amounts of 0, 5, and 25 %. He stored the samples in the tidal zone of a marine exposure site for 5 years and plotted the chloride concentration at different depths (Fig. 7.12).

He found that in terms of chloride resistance, the 5 % limestone samples performed better than the control samples, while the 25 % limestone samples performed the worst. Mortar specimens containing 5 % ground limestone were subjected to a low pressure steam treatment followed by immersion for up to one year in a mixed salt solution loosely based on the composition of seawater by Deja et al. [37]. They tested three types of specimens for strength tests, passivation of reinforcement, and mass loss. Their data suggested that exposure is deleterious to the strength of mortars regardless of the presence of limestone. Also, limestone was effective in protecting the steel from corrosion. Tezuka et al. [35] determined the steady-state diffusion coefficient for chloride ions for a series of paste samples with water-cement ratio of 0.4 containing different quantities of limestone from 0 to 35 %. The diffusion coefficients for the 10 % limestone samples were found comparable to that of the control sample; the 5 % limestone had the lowest diffusion coefficient; and increase in the limestone content beyond 10 % increased

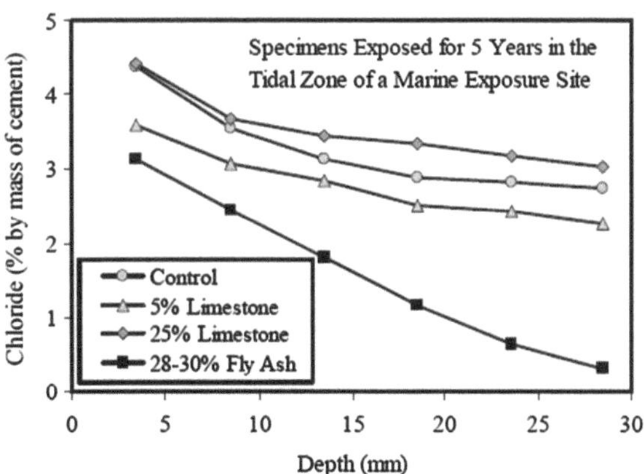

Fig. 7.12 Effect of limestone additions on chloride penetration into concrete [23]

the diffusion coefficient. Bonavetti et al. [14] prepared concrete prisms with 0, 9.3, and 18.1 % limestone cements with water-cement ratio of 0.5. Next, they wet-cured some of the samples for 7 days in normal lab environmental while the rest were air-cured in the lab environment after 1 day. They immersed the samples in 3 % NaCl solution and measured the depth of chloride in the samples. They found the chloride penetration less sensitive to the lack of curing in concretes containing limestone cements. Moreover, 7 days moist curing considerably enhanced the resistance to chloride penetration, and the depth of penetration increased with the increase of limestone content, which the authors attributed it to the predominance of the dilution effect over the hydration acceleration. Comparing two sets of 20 % Portland-limestone cements from two different sources, Alunno-Rosetti and Curcio [23] found that the resistance against chloride penetration was better than that of Portland cement in the first set while worse in the second, showing larger differences between cements from different plants.

Dhir et al. [30] produced 5 series of concretes with W/CM ranging from 0.45 to 0.80 and, within each series, concretes were produced with a PC and the same PC blended with 15, 25, 35 or 45 % ground limestone. These concretes were subjected to tests to determine, among other properties, water absorption (using the initial surface absorption test or ISAT) and chloride diffusion (using an electrical migration test). The results are shown in Figs. 7.13 and 7.14. At a given W/CM there was little difference in the ISA or chloride diffusion coefficient between concrete produced with PC or the PLC with 15 % limestone. At higher levels of limestone there was an increase in both the ISA and chloride diffusion. However, if the concretes are compared on the basis of compressive strength it can be seen that there is no difference in the performance of PC or PLC concretes of the same 28-day strength.

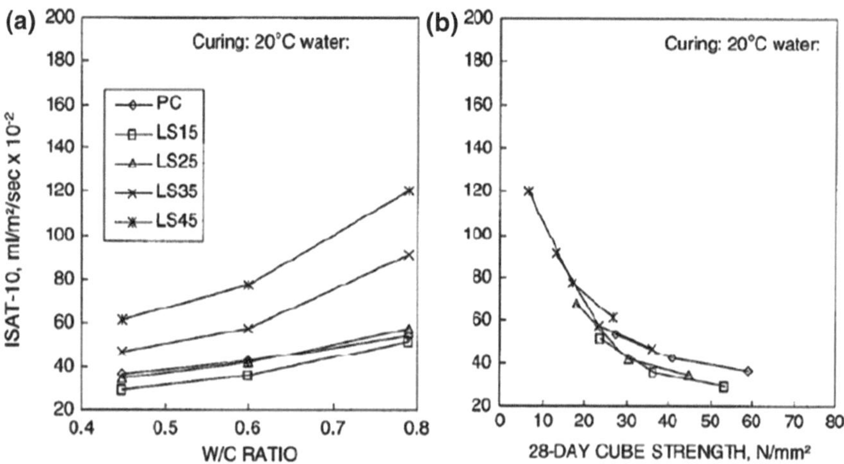

Fig. 7.13 Effect of limestone addition on the initial surface absorption of water into concrete (Dhir et al. 2007)

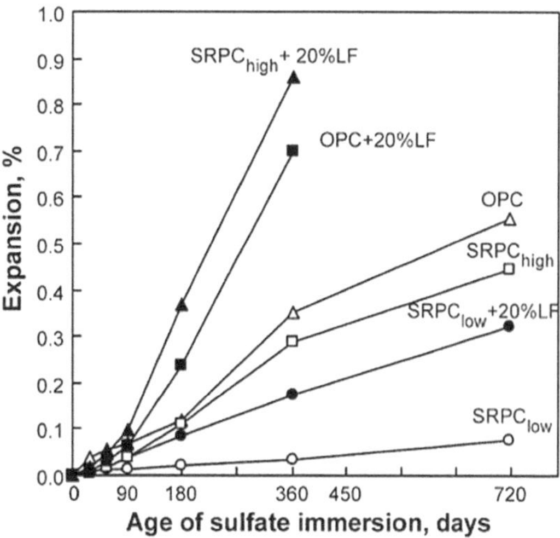

Fig. 7.14 ASTM C 1012 expansions of cements with and without 20 % limestone [38]

In general, considering the variability of the test used by different researchers, it could be concluded that there is not much difference between Portland cement and Portland-limestone cement in terms of chloride resistance. Nevertheless, further investigation is definitely required, especially in the case of the combination of Portland-limestone cements with supplementary cementitious materials (SCMs) as they have proven to be extremely beneficial in enhancing the resistance of concrete against chloride penetration.

Corrosion

Matthews [33] studied the effect of limestone on corrosion of reinforced concrete. He prepared reinforced specimens from Portland-limestone cements with limestone content of 5 and 25 %. One set of specimens were placed in the tidal zone of a marine exposure site while another set were subject to sheltered outdoor exposure, and the weight loss of the rebars were measured after 2 years of exposure. He concluded that chloride induced corrosion of reinforcing steel in concretes subject to marine exposure correlated with the chloride concentration at the rebars and showed no clear dependence upon cement type. Also, carbonation-induced corrosion of reinforcing steel in externally-stored specimens correlated well with the depth of carbonation, which in turn, was correlated to the compressive strength of concrete. Overall, it seemed that although using Portland limestone cements may affect the onset of corrosion by influencing the rate of chloride ingress or carbonation, the rate of corrosion appeared to be unaffected once the process is initiated. Tsivilis et al. [8] prepared concrete prisms made with 10, 15, 20, and 35 % Portland-limestone cements and embedded four cylindrical steel rebars inside them. The placed the specimens in a 3 % NaCl solution and measured the mass loss of the rebars after 9 and 12 months of exposure. The found that there is an explicit decrease of corrosion in specimens made with Portland-limestone cements compared to those made from Portland cement. Also, the mass loss of rebars was decreased as the limestone content of cements was increased up to 20 %, to the extent that that the corrosion rate of 20 % Portland-limestone cement was a third of that of the one made with Portland cement. Considering that the Portland-limestone cement specimens showed no carbonation, as opposed to the Portland cement ones, the authors attributed the corrosion behavior of the Portland-limestone cement concrete to the lower total porosity and the negligible carbonation depth of the samples, which had lead to a significant reduction of the corrosion potential and reduced mass loss of the steel rebars.

The corrosion-resisting performance of concrete is influenced by its electrical resistivity, which refers to the resistance that an electrical charge experiences while passing through the concrete. The increased electrical resistivity of concrete hinders the movement of electrons from the anodic to the cathode regions, and thereby retards the propagation of the corrosion process. Ramezanianpour et al. [18] investigated the electrical resistivity of concretes containing various amounts of limestone at three different W/B ratios. Results are shown in Table 7.2. It can be seen that electrical resistivity decreases with increasing the w/b ratio. The electrical resistivity decreased in the B and C series with increasing limestone replacement. However in the A series, PLC concretes containing 10 and 20 % limestone replacement show maximum and minimum resistivity values, respectively. The highest value of electrical resistivity is 31.5 kΩ cm for the A-10 mixture after 180 days and the minimum is 15.0 kΩ cm for the C-20 mixture.

Table 7.2 The effect of limestone on the electrical resistivity (kΩ cm) at various ages [10]

Mix	28 days	90 days	180 days
A-0	16.7	26.3	30.5
A-5	18.0	24.0	30.0
A-10	19.3	28.3	31.5
A-15	20.5	25.3	29.5
A-20	16.0	23.3	28.3
B-0	16.5	20.0	21.7
B-5	15.0	20.3	20.7
B-10	14.7	19.7	20.3
B-15	13.2	20.0	20.0
B-20	13.9	19.0	20.2
C-0	12.9	16.8	16.8
C-5	11.5	16.3	16.3
C-10	11.0	14.7	15.3
C-15	11.7	13.7	15.3
C-20	11.7	14.7	15.0

Sulfate Resistance

The effect of limestone on the sulfate resistance of Portland cement has been studied by several researchers. Soroka and Stern [10] prepared mortars from a cement to sand mix, and studied their sulfate resistance when $CaCO_3$ and was added as a filler by 10, 20, 30, and 40 percent of the cement weight, replacing a corresponding absolute value of the sand. To determine the sulfate resistance, they immersed the mortars at the age of 28 days in a 5 % Na_2SO_4 solution and measured their length at 14-day intervals until the prisms showed visible signs of cracking, i.e. the onset of cracking. According to their findings, the addition of $CaCO_3$ has a beneficial effect in terms of the sulfate resistance of the mortars. Following up the previous study, Soroka and Setter [10] examined the expansion and deterioration of cement mortars containing various amounts of ground limestone immersed in 5 % Na_2SO_4 solution for up to 11 months. In their experiment, they obtained the fillers by grinding limestone, dolomite, and basalt gravel to the fineness of 1150–1300, 3700–3980, 6600–7100, and 9600–11200 cm^2/g (Blaine). Then they added them to the mortars by 10, 20, 30, and 40 weight percent of the cement. Based on their findings, the authors concluded that using the length of time to the onset of cracking as a criterion, the addition of calcareous fillers, obtained by grinding limestone, improved the sulfate resistance of Portland cement mortars. However, they stated that this addition did not produce completely sulfate-resistant mortars since the intensity of cracking of the limestone-containing mortars was essentially the same as that of those containing the other fillers after long exposure periods. Moreover, they realized that the fineness of the additive was also significant as increasing the fineness of the limestone filler improved sulfate resistance of the mortars. However, they did not observe any improvement with the use of coarser grindings in the range of 1150–6600 cm^2/g, or similarly with the filler content that varied from 10 to 40 weight percent of the cement, making them to believe in the effect to be inconsistent. Sawicz and Heng (1996) investigated the influence of powdered limestone and water cement ratio on the sulfate resistance of concrete. In their mixes, they used limestone powder as a mineral addition to mortars with water cement ratio of 0.5–0.7. Then, they immersed the samples in a 5 % Na_2SO_4 solution and measured their swelling in intervals of 14 days. Simultaneously, the investigated the phase composition of cement pastes with water cement ratio of 0.5, with and without addition of 20 % limestone powder, by means of X-ray diffraction (XRD).

According to their findings, for each water cement ratio, a progressive increase of limestone powder to concrete would first improve the sulfate resistance of concrete (decrease the amount of swelling to a minimum value) and then impair it. Also, a higher content of limestone powder was required to achieve the optimum sulfate resistance as the water cement ratio was increased. Taking advantage of the results of the XRD, the authors attributed the beneficial effects of limestone on sulphate resistance to the reversion of the initially formed ettringite (during curing) to carboaluminate phases (monocarbonate and hemicarbonate), which are more

Table 7.3 Sulphate resistance of limestone mixtures [28] as Presented in Hawkins et al. [4]

Cement	Type V			Type V			Type II		
C_3A content, % mass	0			1			6		
C_3S content, % by mass	40			74			51		
Limestone replacement	0	10	20	0	10	20	0	10	20
Time to 0.10 % expansion, days	1260	857	208	148	164	92	165	209	108
Reduction in compressive strength (1 year in sulfate solution, %)	3	4	5	29	17	50	8	25	40

resistant to sulfate attack, rather than sulfoaluminates (monsulfate, hemisulphate, and solid solutions). Gonzáles and Irassar [39] investigate the effect of limestone filler on the sulfate resistance of low C_3A Portland cement. They added limestone filler containing 85 % of $CaCO_3$ with a Blaine fineness of 710 kg/m^2 as 0, 10, and 20 % of replacement by cement weight. Their test method was based on the ASTM C 1012 procedure, in which they immersed the samples in a 5 % sodium sulfate solution and determined the expansion, flexural and compressive strengths, solution consumption, and X-ray diffraction analysis at different exposure times up to 1 year. The results of their study showed a 10 % replacement of limestone filler caused no significant changes in the sulfate performance of cements while sulfate performance of mortars containing 20 % of limestone filler was worsened in all cases, especially when high C_3S or moderate C_3A contents were in the Portland cement (Table 7.3).

They concluded that the addition of limestone filler may increase or decrease the sulfate performance of blended cements depending on the mineralogical composition of the Portland clinker, the amount of filler replacement, and the equilibrium between the increase of hydration degree before the exposure and the increase in water cement ratio by addition of filler. Also, they stated that sulfate attack on Portland cements containing limestone filler occurs without noticeable changes in the mineralogical nature of sulfate attack products, gypsum and ettringite being the main.

Irassar et al. [38] evaluated sulphate resistance of a Type II and two Type V cements, each with and without 20 % limestone replacement, using ASTM C 1012 mortar bars. All three Portland cements passed ASTM C 452, using the CSA expansion limits. Expansions are shown in Fig. 7.14. The data looks like it might be part of the same data presented in Table 7.4, above. In this paper, with the exception of the SRPC (Type V) with low C3S content, they observed thaumasite, which had formed after ettringite.

With all 3 cements, addition of 20 % limestone increased expansions, but it should be noted that only the SRPC with 0 % C_3A and only 40 % C3S passed the ASTM C 1157 expansion limit of 0.1 % at one year.

Alkali-Silica Reaction

Hobbs [40] studied the effect of limestone on the expansion due to alkali-silica reaction. He prepared mortar bars made from Thames Valley sand and a Beltane opal rock with and without 5 % limestone, and measured the time to cracking and the expansion of the specimens after 200 days. He found that the expansion at 200 days averaged 0.009 % for the Portland cement specimens as opposed to 0.021 % for the specimens with 5 % limestone. Based on the findings, he concluded that although the average expansion is affected, the use of 5 % limestone neither reduces the time to cracking below the minimum observed for Portland cement mortars nor increases the expansion at 200 days above the maximum observed for Portland cement mortars. As such, he stated that the use of Portland-limestone cement does not increase the likelihood of deleterious expansions due to alkali-silica reaction.

Thaumasite Sulfate Attack

Thaumasite is a calcium-silicate-sulfate-carbonate mineral ($CaSiO_3.CaCO_3.CaSO_4.15H_2O$), and the formation of which is considered to be a potential deterioration mechanism in concrete. Thaumasite forms in concrete as a result of a reaction between calcium carbonate, calcium silicates, and sulfates, usually from an external source such as surrounding ground or sea water. The sulfates react with the calcium silicates and carbonates and form thaumasite. This reaction happens slowly, and eventually results in a soft, white, pulpy mass which causes total disintegration of the concrete. It should be noted that cold weather, especially below 5 °C, are more favorable for the formation of thaumasite. Since presence of soluble carbonate is essential for thaumasite sulfate attack (TSA) to happen, and limestone is basically calcium carbonate, the use of Portland-limestone cements might be of concern in regards to thaumasite sulfate attack. According to the literature, the performance of Portland-limestone cement in conditions where thaumasite sulfate attack is a concern seems to be highly dependent on the limestone content of the cement. Several studies [41–44] have reported that the use of Portland-limestone cements with a limestone content of less than 5 % seems to have no significant adverse effect on the performance of concrete in thaumasite sulfate attack. However, in cases where the limestone content was greater than 5 %, the results appear to be inconsistent. There is little data on the performance of 10 % Portland-limestone cement, and data for 15 % limestone content is mixed. For instance, a study by Barker and Hobbs [45] indicated that there was little difference in the performance of concrete produced with high-C_3A (around 10 %) Portland cements with and without 15 % limestone. Nevertheless, other studies [12, 46] have reported damage due to thaumasite sulfate attack in mortars made from Portland-limestone cements with 15 % limestone content. Also, there seems

Cement	Aggregate	Free w/c ratio	Mix number
High C_3A PLC	Dolomitic limestone	0.45	A00/246
Year	1	2	3
Wear rating	2	4	7
Photograph			
Comment	Very minor exposure of aggregate on struck face and sides	Aggregate becoming exposed around top of side faces, layer of soft white paste on base	Aggregate exposed around top of cube with blisters developing elsewhere

Fig. 7.15 Photos of concrete cubes: 12 % limestone cement at 0.45 w/c (Hooton 2003)

to be greater concern when the limestone content of Portland-limestone cement is 20 % or greater.

On contract to the British Cement Association, the Building Research Establishment [47] evaluated the resistance of concretes to TSA over a 3 year period at 5 °C in Class 2 sulphate exposures 1400 mg/L sulphate as calcium sulphate and Class 3 Exposures 3000 mg/L sulphate (mixed calcium and magnesium). Evidence of thaumasite damage was found after 3 years in both types of sulphate exposures. Concretes had w/c from 0.40 to 0.50 and cement contents ranging from 330 to 460 kg/m^3 were seal cured at 20 °C for 28 days prior to exposure. Concretes were evaluated by visual and depth of wear ratings. The resistance to TSA of Portland-limestone cement concretes (CEM II/A-L class 42.5 with 12 % limestone and a clinker C_3A = 12 %) with w/c ratios in the range 0.4–0.45 (Fig. 7.15) and minimum cement contents of 380 kg/m^3, was comparable to that of high C_3A PC concretes of 0.5 w/c ratio and 330 kg/m^3 minimum cement content (Fig. 7.16), with a range of high and low carbonate content aggregates. However, TSA was not prevented in these concretes. The C_3A content of the Portland-limestone cement had minimal effect on sulphate resistance. To quote Holton et al. [48], "In terms of UK practice the results lend support to extending the use of CEM II/A-L cements to concrete in class DS-2 conditions (0.4–1.4 g/l SO_4) where magnesium sulfate is frequently absent". Holton et al. [48] also suggested, "that a reduced C_3A level in

Cement	Aggregate	Free w/c ratio	Mix number
High C_3A PC	Dolomitic limestone	0.5	A00/229
Year	1	2	3
Wear rating	1	2	4
Photograph			
Comment	Some exposed aggregate on sides	Thinly exposed aggregate on side faces	Aggregate thinly exposed on side faces but top and base still intact

Fig. 7.16 Photos of concrete cubes: High C_3A Portland cement at 0.50 w/c (Hootton 2003)

a Portland limestone cement may not be advantageous in preventing TSA. This is in agreement with the results of thermodynamic studies carried out by Juel et al. [49], who have shown that, at a given level of sulfate, thaumasite formation is favoured in cement systems that contain less aluminate".

Summarizing the studies on the performance of Portland-limestone cements against thaumasite sulfate attack, Hooton and Thomas [50] concluded that:

Field experience with up to 5 % limestone in Portland cement in Canada and Europe for over 20 years has not produced any known cases where this has contributed to thaumasite sulfate attack. Based on the literature reviewed, there does not appear to be any significantly increased susceptibility to sulfate attack with respect to use of up to 5 % limestone in Portland cements. Research does exist concerning much higher levels of limestone (15–35 %, or where the carbonate fines originate from the aggregate), where used in cold temperatures combined with wet and aggressive sulfate environments, that indicates more susceptibility to the thaumasite form of sulfate attack. Overall, the data available support the proposal to permit use of up to 5 % limestone in Portland cements.

References

1. M. Schmidt, Cement with interground additives–capabilities and environmental relief, part 1. Zement-Kalk-Gips **45**(2), 64–69 (1992)
2. G.K. Moir, Limestone cements, gaining acceptance. Int. Cement Rev. (2003)
3. CEMBUREAU, *Cement Standards of the World 1991* (CEMBUREAU, Brussels, 1991)
4. P. Hawkins, Personal Communication to R. E. Gebhardt, 10 October 1986. P. Hawkins, P. Tennis, R. Detwiler, *The Use of Limestone in Portland Cement: A State-of-the-Art Review* (Portland Cement Association, EB227, 2005)
5. L. Opoczky, Progress of particle size distribution during the intergrinding of a clinkerlimestone mixture. Zement Kalk Gips **45**(12), 648 (1992)
6. N. Voglis, G. Kakali, E. Chaniotakis, S. Tsivilis, Portland-limestone cements. Their properties and hydration compared to those of other composite cements. Cement Concr. Compos. **27**, 191–196 (2005)
7. T. Vuk, V. Tinta, R. Gabrovšek, V. Kaučič, V, The effects of limestone addition, clinker type, and fineness on properties of Portland cement. Cement Concr. Res. **31**(1), 481–489 (2001)
8. S. Tsivilis, E. Chaniotakis, E. Badogiannis, G. Pahoulas, A. Ilias, A study on the parameters affecting the properties of Portland limestone cements. Cement Concr. Compos. **21**(2), 107–116 (1999)
9. L.D. Adams, R.M. Race, in *Effect of Limestone Additions Upon Drying Shrinkage of Portland Cement Mortar*, ed. by P. Klieger, R.D. Hooton. Carbonate Additions to Cement, ASTM STP 1064 (American Society for Testing and Materials, Philadelphia, 1990), pp. 41–50
10. I. Soroka, N. Stern, Effect of calcareous fillers on sulfate resistance of Portland cement. Am. Ceram. Soc. Bull. **55**(6), 594–595 (1976)
11. V. Bonavetti, H. Donza, G. Menéndez, O. Cabrera, E.F.Irassar, Limestone filler cement in low w/c concrete: A rational use of energy. Cement Concr. Res. **33**, 865–871 (2003)
12. G. Kakali, S. Tsivilis, A. Skaropoulou, J.H. Sharp, R.N. Swamy, *Parameters affecting thaumasite formation in limestone cement mortar.* Proceedings of the 1st International

Conference on Thaumasite in Cementitious Materials (Building Research Establishment, U.K, 2002)
13. V. Bonavetti, H. Donza, V. Rahhal, E.F. Irassar, *High-strength concrete with limestone filler cements*. Proceedings of the 2nd CANMET/ACI International Conference on High-Performance Concrete and Performance and Quality of Concrete Structures, ACI SP-186, American Concrete Institute, Detroit, 567–580 (1999)
14. V. Bonavetti, H. Donza, V. Rahhal, E. Irassar, Influence of initial curing on the properties of concrete containing limestone blended cement. Cement Concr. Res. **30**(5), 703–708 (2000)
15. H. El-Didamony, T. Salem, N. Gabr, T. Mohamed, Limestone as a retarder and filler in limestone blended cement. Ceramics—Silikaty **39**(1), 15–19 (1995)
16. M. Heikal, I. Helmy, H. El-Didamony, F. Abd El-Raoof, Electrical properties, phsysio-chemical and mechanical characteristics of fly ash-limestone-filled pozzolanic cement. Ceramics—Silikáty **48**(2), 49–58 (2004)
17. Z. Guemmadi, G. Escadeillas, H. Houari, P. Clastres, B. Toumi, Influence of limestone fillers on the mechanical performances of cement pastes. Proceedings International Conference on Cement Combinations for Durable Concrete. 339–350 (2005)
18. A.A. Ramezanianpour et al., Influence of various amounts of limestone powder on performance of Portland limestone cement concrete. Cement Concr. Compos. **31**, 715–720, ISI (2009)
19. R.J. Detwiler, *Effects on Cement of High Efficiency Separators* (Research and Development Bulletin RD110T, Portland Cement Association, Skokie, 1995)
20. M. Schmidt, Cement with interground additives–capabilities and environmental relief, part 2. Zement-Kalk-Gips **45**(6), 296–301 (1992)
21. J. Albeck, B. Sutej, *Characteristics of Concretes Made of Portland Limestone Cement*, Beton, vol. 41, no. 5, pp. 240–244. (In German, English translation by Susan U. Lauer available from PCA Library) (1991)
22. K. Kanazawa, K. Yamada, S. Sogo, *Properties of low-heat generating concrete containing large volumes of blast-furnace slag and fly ash*, ed. by V.M. Malhotra. *Fly Ash, Silica Fume, Slag, and Natural Pozzolans in Concrete, Proceedings, Fourth International Conference, Istanbul, Turkey*, ACI SP-132 May 1992, pp. 97–117
23. V. Alunno-Rosetti, F. Curcio, *A Contribution to the Knowledge of the Properties of Portland-Limestone Cement Concretes, with Respect to the Requirements of European and Italian Design Code*. Ed. by H. Justnes. Proceedings of the 10th International Congress on the Chemistry of Cement, Gothenburg, Sweden, 2–6 June 1997, p. 3v026, 6 p
24. J. Péra, S. Husson, B. Guilhot, Influence of finely ground limestone on cement hydration. Cement Concr. Compos. **21**(2) 99–105 (1999)
25. A.P. Barker, J.D. Matthews, Heat release characteristics of limestone-filled cements. Performance of limestone-filled cements: Report of joint BRE/BCA/Cement Industry Working Party, 28 November 1989. (Building Research Establishment, Garston, Watford, England, 1993)
26. P. Livesey, in *Performance of Limestone-Filled Cements*, ed. by R.N. Swamy. (Blended Cements in Construction, Elsevier, 1991) pp. 1–15
27. S. Sprung, E. Siebel, Assessment of the suitability of limestone for producing Portland limestone cement (PKZ). Zement-Kalk-Gips **44**(1), 1–11 (1991)
28. P. Livesey, Strength characteristics of Portland-Limestone cements. Constr. Build. Mater. **5**(3), 147–150 (1991)
29. A.A. Ramezanianpour, E. Ghiasvand, I. Nikseresht, F. Moodi, M.E. Kamel, *Engineering Properties and Durability of Concretes Containing Limestone Cements*. Second International Conference on Sustainable Construction Materials and Technologies, Ancona, Italy 28–30 June 2010
30. R.K. Dhir, M.C. Limbachiya, M.J. McCarthy, A. Chaipanich, Evaluation of portland limestone cements for use in concrete construction. Mater. Struct. **40**, 459–473 (2007)
31. E.F. Irassar, V.L. Bonavetti, G. Menhdez, H. Donza, O. Cabrera, Mechanical properties and durability of concrete made with portland limestone cement. Proceedings of the 3rd

CANMET/ACI International Symposium on Sustainable Development of Concrete, ACI SP-202, American Concrete Institute, Detroit, 431–450 (2001)
32. G.K. Moir, S. Kelham, *"Durability 1", Performance of Limestone-Filled Cements: Report of Joint BRE/BCA/Cement Industry Working Party, 28 November 1989* (Building Research Establishment, Garston, 1993)
33. J.D. Matthews, in *Performance of limestone filler cement concrete*, ed. by R.K. Dhir, M.R. Jones. In Euro-Cements–Impact of ENV 197 on Concrete Construction (E&FN Spon, London, 1994), pp. 113–147
34. J. Baron, in *The Durability of Limestone Composite Cements in the Context of the French Specifications*, ed. by L.-O. Nilsson. Durability of Concrete: Aspects of Admixtures and Industrial Byproducts, International Seminar, April 1986 Swedish Council for Building Research (1988), pp. 115–122
35. Y. Tezuka, D. Gomes, J.M. Martins, J.G. Djanikian, in *Durability Aspects of Cements with High Limestone Filler Content.* 9th International Congress of the Chemistry of Cement, New Delhi, India, vol. V, pp. 53–59 (1992)
36. V.S. Ramachandran, R.F. Feldman, J.J Beaudoin, Influence of sea water solution on mortar containing calcium carbonate. Mater. Struct. **3**(138), 412–417 (1990)
37. J. Deja, J. Malolepszy, G. Jaskiewicz, in *Influence of Chloride Corrosion of the Durability of Reinforcement in the Concrete*, ed. by V.M. Malhotra. Durability of Concrete, Second International Conference, Montréal, Canada 1991, ACI SP-126, pp. 511–525
38. E.F. Irasser, V.L. Bonavetti, M.A. Trezza, M.A. González, Thaumasite formation in limestone filler cements exposed to sodium sulphate solution at 20°C. Cement Concr. Compos. **27**, 77–84 (2005)
39. M.A. Gonzáles, E.F. Irassar, Effect of limestone filler on the sulfate resistance of low C_3A Portland cement. Cement Concr. Res. **28**(11), 1655–1667 (1998)
40. D.W. Hobbs, *Possible influence of small additions of PFA, GBFs and limestone flour upon expansion caused by the alkali-silica reaction.* Mag. Concr. Res. **35**(122), 55–58 (1983)
41. M.A. Halliwell, N.J. Crammond, A.P. Barker, *The thaumasite form of sulfate attack in limestone-filled cement mortars* (BRE Laboratory Report - BR307, Construction Research Communications Ltd (CRC), UK, 1996)
42. R.D. Hooton, in *Effects of Carbonate Additions on Heat of Hydration and Sulfate Resistance of Portland Cement*, ed. by P. Klieger, R.D. Hooton. Carbonate Additions to Cement, ASTM STP 1064 (American Society for Testing and Materials, Philadelphia, 1990), pp. 73–81
43. D. Stark, Performance of concrete in sulfate environments. PCA R&D Bull. **RD129**, 23 p (2002)
44. P.C. Taylor, Sulfate resistance tests on type V cements containing limestone. PCA R&D Serial No. **2182a**, 8 p (1998)
45. A.P. Barker, D.W. Hobbs, Performance of Portland limestone cements in mortar prisms immersed in sulfate solutions at 5°C. Cement Concr. Compos. **21**, 129–137 (1999)
46. S.A. Hartshorn, J.H. Sharp, R.N. Swamy, Engineering properties and structural implications of Portland limestone cement mortar exposed to magnesium sulfate attack. Adv. Cement Res. **13**, 31–46 (2001)
47. I. Holton, Final report on 3-year investigation to establish sulfate resistance of Portland limestone cement concretes. BRE Report 209–164 for the British Cement Association (2003)
48. I.R. Holton, N.J. Crammond, W.F. Price, Comparative performance of portland limestone cement concretes and portland cement concretes in cold calcium sulfate solutions. (personal communication of unpublished draft paper) (2007)
49. I. Juel, D. Herfort, R. Gollop, J. Konnerup-Madsen, H.J. Jakobsen, J. Skibsted, A thermodynamic model for predicting the stability of thaumasite. Cement Concr. Compos. **25**(8), 867–872 (2003)
50. R.D. Hooton, M.D.A. Thomas, *The Use of Limestone in Portland Cements: Effect on Thaumasite Form of Sulfate Attack.* PCA R&D Serial No. 2658, (Portland Cement Association, Skokie, 2002)

Chapter 8
The Role of Supplementary Cementing Materials on Sustainable Development

Introduction

It is impossible to walk through cities without seeing concrete in some form. Whether it is in the latest high rise being constructed, new side walks being cured, in roads connecting the city, in dams, bridges, marine structures, industrial plants, etc. concrete is inescapable. When a material becomes as integral to the structure as concrete, it is important to analyze its environmental impacts to conclude if the material is as sustainable as it is prevalent. If the material does not satisfy the credential of sustainability it should be further developed, especially in present society when environmentally detrimental processes are currently subject to scrutiny.

It is often debated whether concrete can or should be considered as a sustainable option due to its particular properties and characteristics. Concrete is made of several different elements. In its simplest form this includes cement, water, and aggregate. Cement requires substantial amounts of energy to produce and releases large amounts of carbon dioxide. However, it can also be replaced in part by supplementary cementing materials.

Concrete has a long service life; buildings made of concrete can usually be expected to last hundreds of years with proper maintenance. Sine the structural lifetime is so long, potential waste if another type of building material was used is reduced. As well, once a building requires demolition, its material can be used for subsequent buildings. Truly, when a structure reaches the end of its useful life, its concrete component can be completely recycled into the aggregates to be used in other concrete mixtures. In spite of the amount of initial energy required to produce concrete, the material has the potential to be efficient over its estimated life expectancy. A list of credentials that should addressed when examining the sustainability of a material is provided:

- Energy required to produce the material
- CO_2 emissions resulting from the material's manufacture
- Toxicity of the material
- Transportation of the material during its manufacturing and delivery

A. A. Ramezanianpour, *Cement Replacement Materials*,
Springer Geochemistry/Mineralogy, DOI: 10.1007/978-3-642-36721-2_8,

- Degree of pollution resulting from the material at the end of its useful life
- Maintenance required and the materials required for maintenance
- Lifetime of the material and its potential for reuse if the building is demolished.

It is important to address all of these factors in deciding if a material can be viewed wholly as a sustainable material. These factors can be separated into two main components, embodied energy is low, but the operational energy is high then a material cannot be deemed sustainable. An ideal solution would be a material with a low embodied energy which results in high operational energy savings. If these two factors can be satisfied then a material, when used with sustainability in mind, is well on its way to reducing its environmental impact.

Embodied Energy

Embodied energy is the amount of energy required to produce a material. This includes the energy required for the raw material extraction; the energy required to process and manufacture the material; and transportation for all stages of production. It is impossible to assign a value to the embodied energy of concrete on a whole because mix designs vary widely which subsequently changing the embodied energy. Table 8.1 provides a summary of different elements found in concrete and the approximated values of the relative embodied energy.

Although there is variation in the values a general idea about the embodied energy of cement can be established. From the table it can be estimated that 7 GJ of energy is required to produce one tonne of cement. The mix used to produce concrete can evidently have a large effect on the amount of embodied energy required. In general, however, the magnitude of this parameter can be understood as well as the importance of attempting to minimize it. The process required to manufacture cement is fairly consistent regardless of the plant type. Conversely, the transportation component and the individual plant part efficiencies can differ and thus have room for improvement. Therefore, the embodied energy can be assumed to be a realistic estimation of the actual embodied energy.

Accordingly, the result of a large embodied energy quantity is a significant production of carbon dioxide. Both of these components of concrete are of concern because they are unsustainable. It must be concluded then that the manufacturing process of cement must be developed further to increase the sustainability of

Table 8.1 Embodied energy for cement and concrete

Source	Material	Embodied energy (GJ/ton)
The new ecological home	Poured on site concrete	1.0–1.6
	Cement	7–8
Ecohouse 2: a design guide	Cement	4.3–7.8
	Natural aggregates	
Cement association of Canada	Clinker	3.0–6.0

cement and thereby increase the sustainability of the concrete itself. There are two obvious ways in which to reduce the impact that concrete has on the environment. The first is to increase the efficiency during production, and the second is to reduce the amount of cement in the concrete mix. In combination, these two ideas could become a powerful means to creating a versatile material that is practical, beneficial and sustainable in our society.

Greenhouse Gas Emissions and Global Warming

Carbon dioxide emissions are the main greenhouse gas emissions affecting the sustainable development of the world. Some other greenhouse gases such as methane and nitrous oxide also affect the sustainable development to some extent but not as high as carbon dioxide development of the world. Some other greenhouse gases such as methane and nitrous oxide also affect the sustainable development to some extent but not as high as carbon dioxide.

Global warming is the consequence of greenhouse gas emissions. According to the weather scientists a linear relationship exists between the earth's surface temperature and the atmospheric concentration of carbon dioxide. The estimated carbon dioxide concentration was about 380 ppm in 2005. The concentration of the CO_2 is expected to increase at an exponential rate resulting in rapid temperature rise of the earth.

In the reports of the United Nations Intergovernmental Panel on Climate Change, leading weather scientists of the world have stated that global warming is occurring, and that it has been triggered by human activities. They have warned about devastating consequences of global warming if immediate action is not taken by national and industry leaders to reduce the carbon dioxide emissions to the 1990 level or less. According to Kyoto Protocol, proposed in 1990 and signed in 2005 by 141 countries, world countries agreed to stabilize the greenhouse gas emissions by 2012 to 6 % below the 1990 level.

The published information in 2004 shows that the USA and Canada are the largest contributors of carbon dioxide emissions in the world with CO_2 emissions estimated at approximately 20 tonnes per capita per year, followed by the EU countries at about 9 tonnes, China about 3 tonnes and India nearly 1 ton. Similar to other developing countries, China and India have low emissions at present, but their rapid growth of population and industry in the future will result in significant increase in the carbon dioxide emissions per capita.

Contribution of Cement on CO_2 Emissions

Normal Portland cements are usually consisted of 95 % clinker and 5 % gypsum. Depending on the carbon content of fossil fuels used for clinkering, nearly 0.9 to 1.0 tonnes of CO_2 is directly released from cement kilns during the manufacture of clinker. The direct release of CO_2 occurs from two sources, namely the decomposition of calcium carbonate as the principal raw material and the combustion of fossil fuels. Calcium carbonate releases nearly 0.6–0.65 kg CO_2 and the fossil fuels depending on their carbon contents release about 0.25–0.35 kg CO_2 for the production of 1 kg of clinker.

The U.S. Geological Survey records show that the world consumption of cement in 1990 was 1,044 million tonnes. From the fragmentary information available it is estimated that, globally, the average clinker factor of cement (units of clinker per unit of cement) in 1990 was 0.9, which means that 940 million tonnes of clinker was used. Assuming the average CO_2 emission rate as 1.0 tonne CO_2/tonne clinker, in 1990 the direct CO_2 emission from clinker production were 940 million tonne.

In 1995, the production of cement was about 1.4 billion tonnes, thus emitting about 1.4 billion tonnes of CO_2 to the atmosphere. According to world energy outlook 1995, issued by the International Energy Authority (IEA), the worldwide CO2 emissions from all sources were 21.6 billion tonnes. Thus, the worldwide cement production accounts for almost 7 % of the total world CO2 emissions.

In 2005, due to increase in the use of alternate, low-carbon, fuels for burning clinker, the average CO_2 emission rate dropped to 0.9 tonne per tonne of clinker. This means that, in 2005, 1,900 million tonnes of clinker was produced, with 1,700 million tonnes of direct CO_2 release to the environment. In conclusion, the global cement industry has almost doubled its annual rate of direct CO_2 emissions during the last 15 years.

Concrete Production

To thoroughly assess sustainability and the implications of embodied energy, it is important to have a sufficient comprehension of the processes involved. Possessing this understanding gives depth to the concept of embodied energy, which aids in finding opportunities for improvement, as well as a concept of the environmental impacts. Cement has been marked as a possible threat to the sustainability of concrete. By delving into the nature of cement, this threat can be realized and addressed. The missions resulting from the calcination process cannot be reduced, and thus the CO_2 emissions must be reduced by other means. Improvements to kiln efficiencies; alternate fuel methods; and reduction in the cement quality requirement though intelligent mix design, can all work to aid the reduction of embodied energy and CO_2 emissions.

Reducing Energy and Emissions

To reduce the energy-intensity and pollution levels, change must occur on all levels of the concrete process. In Germany, effort was made to decrease the negative environmental effects of the industry by improving the manufacturing process. This was achieved by installing new plants for the purpose of ensuring that the kiln production operation was smoother and energy requirements were decreased. Besides, the existing kilns were optimized in a way that required minimum fan power. A waste heat recovery process was devised to utilize remaining energy losses of the new cement kilns. Finally direct electrical power requirements were decreased by improving the efficiency of the grinding system. Once the manufacturing process was optimized a greater amount of waste fuels was substituted for fossil fuels and alternatives to the cement composition via inclusion of blended cements were derived [1]. Finally, it was admitted that greater research and development was required in the field of "process technologies, use of secondary materials, properties and application of blended cements".

Malhotra [2] has estimated the cement production, supplementary cementing materials consumption and CO_2 emissions in the year 2020. Table 8.2 shows the estimated global cement consumption and CO_2 emission in the year 2020 by 3 options.

As stated before, global carbon emissions direct from Portland clinker production have already doubled in the past 15 years. Option 2 in Table 8.2 reveals the fact that if no serious measures are taken by the world's construction industry, the rate of direct carbon emissions from cement kilns will almost triple by the year 2020. Table 8.2 also includes data on two other options, the challenging and formidable options.

Comparing option 2 with option 1 it is seen that if the global concrete construction industry is able to reduce the concrete consumption by 20 % (compared to Option 1) and at the same time increase the CCM utilization to 30 % of the total

Table 8.2 Estimates of global cement consumption and CO_2 emission attributable to clinker production in the year 2020, million tonnes

Options	No. 1 Business as usual	No. 2 Challenging option	No. 3 Formidable option
Cement consumption/production	3,500	2,800	2,100
Complementary cementing materials (CCM)	700	840	1,050
Portland clinker requirement	2,800	1,960	1,050
CO_2 emissions (0.9 T/T clinker)	2,520	1,760	945
% CO_2 increased from the 1990 level	270	190	0

* Cement consumption goes up by about 50 % of the 2005 level and the use of CCM increased to 20 % of the total cementing material

** Cement consumption goes down by 20 % of the BAU level, and CCM increased to 30 % of the total cementing material

cement, these steps will have the effect of reducing the direct CO_2 emissions from cement kilns to 1,760 million tonnes.

According to Option 3, in 2020, the total cementing material (2,100 tonnes) would comprise 1050 million tonnes of Portland clinker and the same amount of complementary cementing materials. This shows the vital role of complementary cementing materials on sustainability in the cement and concrete industry.

Supplementary Cementing Materials and Sustainability

When concrete is the material of choice one can look forward to having endless options and opportunities in its composition. There are many types of cement, admixtures, aggregate, and supplementary cementing materials that can be incorporated in different quantities. By incorporating a higher quantity of supplementary cementing materials the amount of cement can be reduced, lowering the emissions and energy with a mix.

Supplementary cementing materials have proven to be economical environmental alternatives to typical concrete mixes. Fly ash and silica fume in particular have allowed for significant reductions in CO_2 emissions in cement production. This was possible because of the decreased amount of clinker used to produce the cement and concretes as well as decreasing the requirement of fuel for clinker burning. CO_2 emissions could also be decreased with the increase in the efficiency of clinker in concrete strength development. This can be done by including mineral admixtures to allow for a decrease in the water requirement with fly ash use and strength development by using silica fume. By incorporating 7.5 % of silica fume to cement during the grinding process, an increase in efficiency of 25 % resulted in the clinker [3].

The production of silica fume is estimated to be about two million tonnes. Based on the information available, the current annual production of fly ash is of the order of 900 million tonnes worldwide. In addition to this, there are millions of tonnes of fly ash that have been stockpiled over the years.

Another factor that can be manipulated using mix design is the durability. Increased durability decreases maintenance as well as increases lifetime. Longevity can be increased with a reduction in the heat of hydration; reduced porosity [4]. Most widely perhaps, fly ash has been accepted as a great supplement to 100 % cement use. It improves the most critical cement characteristics such as workability, impermeability and durability. However, regardless of fly ash's positive affect on cement production there have been barriers associated with its use. However, new technologies have been developed as well as newer cement formulations which overcome these previous barriers. Other alternatives have been used such as superplasticizers for water reduction, and silica fume or rice husk ash for porosity minimization and lengthening.

Rice husk ash (RHA) is a byproduct of burning rice husks. In 1982 about 406.6 million tonnes of paddy was produced on a global basis, which represents about 81 million tonnes of husk or 16 million tonnes of ash if an ash content of 20 % is assumed [5]. The available ash for cement is approximately 4.5 million tonnes. The husks are incinerated and an ash that is predominantly silica is residual. RHA is highly pozzolanic due to its extremely high surface area (50,000–100,000 m^2/kg). When 5–15 % RHA is incorporated by mass higher compressive strengths, decreased permeability, resistance to sulfate and acid attack, and resistance to chloride penetration can be expected. By adding in RHA the durability and recycled content on the mix can be increased [6].

Natural pozzolans are also viewed as effective substitutes for cement. However, it has been observed that the substitution rate, a less workable mortar was produced. Thus, a decrease in compressive strength was witnessed. Yet, as the level of fineness is increased, the comparable strength increases as well. This is most apparent when natural pozzolan substitutions are high [7]. Therefore, using natural pozzolans proves to be an optimal substitute for cement as it provides equivalent strength and reduces the per unit emissions of greenhouse gases. The production of natural pozzolans especially in developing countries is expected to reach 25 million tonnes by the year 2020. The consumption of natural pozzolans particularly in dams construction is growing rapidly.

Another supplementary cementing materials which is available in large quantities and can be used to replace portland cement in concrete is granulated blast-furnace slag (ggbfs). Granulated blast-furnace slag is derived as a by-product from the blast-furnaces manufacturing iron ant therefore its use has environmental benefits. The use of ggbfs in concrete significantly reduces the risk of damages caused by alkali-silica reaction, provides higher resistance to chloride ingress, reduces the risk of reinforcement corrosion, and provides high resistance to attacks by sulfate and other chemicals.

The worldwide production of slag is about 100 million tonnes but granulated blast-furnace slag is only produced approximately 25 million tonnes per year. The use of granulated blast-furnace slag in concrete has increased considerably in recent years and it is is expected to reach to 75 million tonnes by the year 2020.

Metakaolin as a manufactured product has been used to enhance the durability of concrete. Thus its contribution to CO_2 emission reduction is indirect because durable structures require less repair and maintenance. The performance of metakaolin in concrete is similar to that of silica fume, but unlike silica fume which is a by-product, its use in concrete contributes very little to reduction of greenhouse gas emissions. The production of metakaolin is estimated to be about 2 million tonnes annually.

The most readily available mineral additive for cement is limestone. In Europe, more limestone is used in Portland-based cements than all other mineral additions combined, notably in the European CEM II L class (24.6 % of all European cement manufactured in 2003) and to a lesser extent in the M class, and as minor addition of up to 5 % in almost all other Portland cements. It has been shown that much of the alumina from calcium carbo-aluminate hydrates, which can result in a significant decrease in porosity. Both the European cement standard, EN 197-1, and ASTM C150 allow up to 5 % limestone. Limestone added in excess of this amount, although constituting essentially a "filler", can also act as an accelerator for alite hydration, so that, with suitable grinding techniques, Cement strength up to 28 days are often not much reduced even at limestone contents as high as 20 %. In addition to this, Limestone additions can improve concrete consistency by reducing cement water demand, and provided that a low w/c concrete mix design is used, high limestone replacement as some pure Portland cements.

Durability Enhancement by Application of Supplementary Cementing Materials

The durability of a product can have one of the most significant influences on the environment. The less durable a product, the shorter the service life is. This results in new purchases or repairs which is a wasteful practice especially when the industry is as energy intense as cement production. Therefore the longer the service life is the better earth's natural resources are conserved. The challenge then is to produce concrete that is highly durable, and a high-performance building material for future structure.

The major causes of reinforced concrete deterioration in structures are "corrosion of the reinforcing steel, exposure to cycles of freezing and thawing, alkali-silica reaction and sulfate attack". The mechanism of concrete expansion and cracking is highly dependent on a high degree of water saturation. Therefore it is evident that water-tightness of concrete is a crucial step to ensure minimal damage. Thus this suggests that the soundness of concrete is closely related to its durability.

Early cracking in concrete is usually a function of either thermal contraction and/or drying shrinkage. The construction industry is driven by economics and thus the mentality of the faster it is done the more profit earned encourages cements with high-early strength, or rather, concrete with high levels of Portland cement to be used. These concretes have low cracking resistances as a result of an increase in "the shrinkage, and elastic modulus on one hand, and a reduction in the creep coefficient on the other hand". As such, this explains the high level of vulnerability to cracking of high-early-strength concretes to that of moderate or low-strength concrete mixtures.

Early cracking is commonly minimized by incorporation excessive steel reinforcement. However this solution simply replaces a small number of wide cracks with a numerous number of invisible and often unmeasurable micro-cracks. This does not in any sense constitute as a sufficient solution for durability.

If concrete is properly consolidated and cured it will remain watertight unless the pores and cracks within it form interconnected pathways to the surface surrendering the concrete to further deterioration. Evidently, however, the drive of economics does not allow for this. As such, when thermal cracking and durability are of primary concern, it has been shown that supplementary cementing materials (SCM) should be incorporated into the mix design as it will prove to be most cost-effective. The reasoning for this is the property of concrete mixes that contain SCM to have stronger transition zones and are thus less prone to microcracking. Ultimately, by incorporating byproducts such as fly ash or slag into mix design, the durability of concrete is augmented through prolonged water-tightness [8–10].

Operational Energy

As important as it is to reduce the embodied energy and emissions, it is just as important that when implementing a material that the energy requirements during its useful life are not increased as a byproduct of material selection. Concrete offers solutions to reduce the operational energy of structures such as buildings, dams, and roads.

Dams

Pozzolans are known scientifically to be both an environmental beneficial substitute in cement production as well as an economically feasible solution. Since the 1980's, roller-compacted concrete dams (RCC) have been known to be one of the most rapid and economical method for construction of medium-height dams. By 1992, ninety-six RCC dams were built in over seventeen countries. 85 % of these dams included pozzolans in the mix design. In fact, high paste content RCC most commonly uses 250 kg/m^3 of concrete with 70–80 % pozzolan addition. Of the 85 % of dams which incorporated pozzolans in the mix design, fly ash was incorporated in 90 %. This quantity hints at the capability pozzolans have in cement/concrete production and the effect it can have on the environment when waste products become commodities.

Conclusion

Concrete is taking leaps and bounds when it comes to sustainable development. The management of CO_2 emissions along with voiced concern regarding the negative environmental impact of cement production proves that the minds of the industry are in right place. Research involving supplementary cementing materials has continuously proven the benefits of incorporating what is often a waste product from industries into concrete mix design. This can be noted through the increase of durability and strength resulting particularly in greater sustainable practices but also economical ones. Developments in the cement production process suggest that the interest to make improvements is being realized. New innovative methods are also being created to reduce the quantity of cement in a mix which is proof of a new perspective on the role the concrete industry can play in sustainability.

In addition to the developments occurring directly with the production of cement and concrete the application of these materials is also being redefined. Many limitations once binding concrete from becoming sustainable are fading as its use is incorporated into newer areas. Buildings can be built to use concrete's thermal mass to help reducing energy requirements. The construction of dams is being optimized to use concrete to save energy during construction, and the lifetime of concrete is expanding with its reuse in aggregate form in roads. In combination, concrete has become multipurpose. As a result, although the initial energy production level is high, concrete can become more efficient.

The environmental and economic benefits of development in the direction of sustainability are inescapable. The continual search for opportunities to make this material, which has become such and integral part of our cities all over the world, a more sustainable option, proves that the minds of the masses are in the right place.

References

1. V. Hoeing, M. Schneider (2001) *German cement industry's voluntary efforts on the issue of climate change- A success story*, ed. by V.M. Malhotra. Third Canmet/ACI International. Sustainable development of cement and concrete (American concrete institute, Farmington Hills, Michigan), pp. 15
2. V.M. Malhotra (2010) Global warming, and role of supplementary cementing materials and superplasticisers inreducing greenhouse gas emissions from the manufacturing of portland cement. Int. J. Struct 1(2):116–130
3. K. Popovic (2001) *Reducing CO_2 emission into the atmosphere-achievements and experience of creation cement industry* ed. by V.M. Malhotra. Third Canmet/ACI international. Sustainable development of cement and concrete. (American concrete institute, Farmington Hills, Michigan)
4. R. Horton (2001). *Factor ten emission reductions: the key to sustainable development and economic prosperity for the cement and concrete industry* ed. by V.M. Malhotra. Third Canmet/ACI International. Sustainable development of cement and concrete. (American concrete institute, Farmington Hills, Michigan), pp. 1
5. Rice Husk ash cements: their development and applications. United nations industrial development organization. Vienna. 1985
6. A.D. Neuwald (2004) Supplementary cementitious materials, National precast concrete association
7. W. South, I. Hinczak (2001) *New Zealand Pozzolans-An ancient to a modern dilemma* ed. by V.M. Malhotra. Third Canmet/ACI International. Sustainable Development of Cement and Concrete. (American concrete institute, Farmington Hills, Michigan), pp. 97
8. 2006 Sustainability Report. Cement association of Canada, 2006, 10 Apr 2008
9. U. Meinhold, G. Mellmann, M. Maultzsch (2001) *Performance of high-grade concrete with full substitution of aggregates by recycled concrete* ed. by V.M. Malhotra. Third Canmet/ACI international. Sustainable development of cement and concrete. (American concrete institute, Farmington Hills, Michigan), pp. 85
10. H. Uchikawa (2000) *Sustainable development of the cement and concrete industry* ed. by E.G. Odd, S. Koji. Concrete Technology for a Sustainable Development in the 21st Century. (E & FN spon, New York), pp. 177